Automatic Autocorrelation and Spectral Analysis

Piet M.T. Broersen

Automatic Autocorrelation and Spectral Analysis

With 104 Figures

 Springer

Piet M.T. Broersen, PhD
Department of Multi Scale Physics
Delft University of Technology
Kramers Laboratory
Prins Bernhardlaan 6
2628 BW, Delft
The Netherlands

British Library Cataloguing in Publication Data
Broersen, Piet M. T.
  Automatic autocorrelation and spectral analysis
  1.Spectrum analysis - Statistical methods 2.Signal
  processing - Statistical methods 3.Autocorrelation
  (Statistics) 4.Time-series analysis
  I.Title
  543.5'0727

e-ISBN 1-84628-329-9
ISBN-13: 978-1-84996-581-1   e-ISBN-13: 978-1-84628-329-1

© Springer-Verlag London Limited 2010

MATLAB® is a registered trademark of The MathWorks, Inc., 3 Apple Hill Drive, Natick, MA 01760-2098, U.S.A. http://www.mathworks.com

Printed in Germany

9 8 7 6 5 4 3 2 1

Springer Science+Business Media
springer.com

To Rina

Preface

If different people estimate spectra from the same finite number of stationary stochastic observations, their results will generally not be the same. The reason is that several subjective decisions or choices have to be made during the current practice of spectral analysis, which influence the final spectral estimate. This applies also to the analysis of unique historical data about the atmosphere and the climate. That might be one of the reasons that the debate about possible climate changes becomes confused. The contribution of statistical signal processing can be that the same stationary statistical data will give the same spectral estimates for everybody who analyses those data. That unique solution will be acceptable only if it is close to the best attainable accuracy for most types of stationary data. The purpose of this book is to describe an automatic spectral analysis method that fulfills that requirement. It goes without saying that the best spectral description and the best autocorrelation description are strongly related because the Fourier transform connects them.

Three different target groups can be distinguished for this book.

Students in signal processing who learn how the power spectral density and the autocorrelation function of stochastic data can be estimated and interpreted with time series models. Several applications are shown. The level of mathematics is appropriate for students who want to apply methods of spectral analysis and not to develop them. They may be confident that more thorough mathematical derivations can be found in the referenced literature.

Researchers in applied fields and all practical time series analysts who can learn that the combination of increased computer power, robust algorithms, and the improved quality of order selection have created a new and automatic time series solution for autocorrelation and spectral estimation. The increased computer power gives the possibility of computing enough candidate models such that there will always be a suitable candidate for given data. The improved order-selection quality always guarantees that one of the best candidates will be selected automatically and often the very best. The data themselves decide which is their best representation, and if desired, they suggest possible alternatives. The automatic computer program ARMAsel provides their language.

Time series scientists who will observe that the methods and algorithms that are used to find a good spectral estimate are not always the methods that are preferred in asymptotic theory. The maximum likelihood theory especially has very good asymptotic theoretical properties, but the theory fails to indicate what sample sizes are required to benefit from those properties in practice. Maximum likelihood estimation often fails for moving average parameters. Furthermore, the most popular order-selection criterion of Akaike and the consistent criteria perform rather poorly in extensive Monte Carlo simulations. Asymptotic theory is concerned primarily with the optimal estimation of a single time series model with the true type and the true order, which are considered known. It should be a challenge to develop a sound mathematical background for finite-sample estimation and order selection. In finite-sample practice, models of different types and orders have to be computed because the truth is not yet known. This will always include models of too low orders, of too high orders, and of the wrong type. A good selection criterion has to pick the best model from all candidates. Good practical performance of simplified algorithms as a robust replacement for truly nonlinear estimation problems is not yet always understood.

The time series theory in this book is limited to that part of the theory that I consider relevant for the user of an automatic spectral analysis method. Those subjects are treated that have been especially important in developing the program required to perform estimation automatically. The theory of time series models presents estimated models as a description of the autocorrelation function and the power spectral density of stationary stochastic data. A selection is made from the numerous estimation algorithms for time series models. A motivation of the choice of the preferred algorithms is given, often supported by simulations. For the description of many other methods and algorithms, references to the literature are given.

The theory of windowed and tapered periodograms for spectra and lagged products for autocorrelation is considered critically. It is shown that those methods are not particularly suitable for stochastic processes and certainly not for automatic estimation. Their merit is primarily historical. They have been the only general, feasible, and practical solutions for spectral analysis for a long period until about 2002. In the last century, computers were not fast enough to compute many time series models, to select only one of them, and to forget the rest of the models. ARMAsel has become a useful time series solution for autocorrelation and spectral estimation by increased computer power, together with robust algorithms and improved order selection.

Piet M.T. Broersen

November 2005

Contents

1

Introduction

1.1 Time Series Problems

The subject of this book is the description of the main properties of univariate stationary stochastic signals. A univariate signal is a single observed variable that varies as a function of time or position. Stochastic (or random) loosely means that the measured signal looks different every time an experiment is repeated. However, the process that generates the signal is still the same. Stationary indicates that the statistical properties of the signal are constant in time. The properties of a stochastic signal are fully described by the joint probability density function of the observations. This density would give all information about the signal, if it could be estimated from the observations. Unfortunately, that is generally not possible without very much additional knowledge about the process that generated the observations. General characteristics that can always be estimated are the power spectral density that describes the frequency content of a signal and the auto-covariance function that indicates how fast a signal can change in time. Estimation of spectrum or autocovariance is the main purpose of time series identification. This knowledge is sufficient for an exact description of the joint probability density function of normally distributed observations. For observations with other densities, it is also useful information.

A time series is a stochastic signal with chronologically ordered observations at regular intervals. Time series appear in physical data, in economic or financial data, and in environmental, meteorologic and hydrologic data. Observations are made every second, every hour, day, week, month, or year. In paleoclimatic data obtained from an ice core in Antarctica, the interval between observations can even be a century or 1000 years (Petit *et al.*, 1999) for the study of long-term climate variations.

An example of monthly data is given in Figure 1.1. The observations are made to study the El Niño effect in the Pacific Ocean (Shumway and Stoffer, 2000). At first sight, this series can be considered a stationary stochastic signal. Can one be sure that this signal is a stationary stochastic process? The answer to this question is definitely NO. It is not certain, it is possible, but there are at least three different and valid ways to look at practical data like those in Figure 1.1.

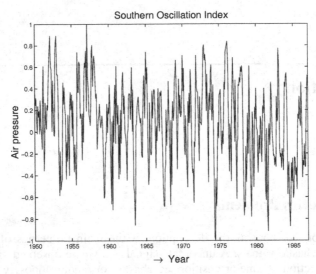

Figure 1.1. Monthly observations of the air pressure above the Pacific Ocean

- It is a historical record of deterministic numbers that describe the average air pressure was on certain days in the past at certain locations.

  Application: Archives loaded with data.

- The air pressure would perhaps have been slightly different at other locations, and perhaps some inaccuracy occurs in the measurement. Therefore, the data are considered deterministic or stochastic true pressure levels plus additive noise contributions.

  Application: Filtering out the noise.

- This whole data record is considered an example of how high and low pressures follow each other. The measurements are exact but they would be else if they had been made at other moments. Measuring from 1900 until 1950 would have given a different signal, possibly with the same statistical characteristics. The signal is treated as one realisation of a stationary stochastic process during 40 years.

  Application: Measure the power spectral density or the autocorrelation function and use that for a compact description of the statistically significant characteristics of the data. That can be used for prediction and for understanding the mechanisms that generate or cause such data.

All three ways can be relevant for the data in Figure 1.1. The correct practical question to be posed is which of the three ways will give the best answer for the problem that has to be solved with the data. Not the measured data but the intention of the experimenter decides the best way to look at the data. This causes a fundamental problem with the application of theoretical results of time series analysis to practical time series data. Most theoretical results for stationary stochastic signals are derived under asymptotic conditions for a sample size going to infinity; see Box and Jenkins (1976), Brockwell and Davis (1987), Hannan and Deistler (1988), Porat (1994), Priestley (1981), and Stoica and Moses (1997). The applicability of the theoretical results to finite samples is generally not part of the asymptotic theory. Nobody would believe that the data in Figure 1.1 are similar to data that would have been found millions of years ago. Neither it is probable that the data at hand will be representative of the air pressure in the future over millions of years. Broersen (2000) described some practical implications of spectral estimation in finite samples. This book will treat useful, automatic, finite-sample procedures.

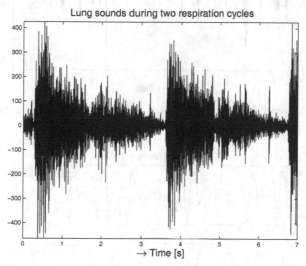

Figure 1.2. The microphone signal of the sound of healthy lungs during two cycles of the inspiration and the expiration phases. The amplitude during inspiration is much greater.

A second example shows the sound observed with a microphone on the chest of a male subject (Broersen and de Waele, 2000). This signal has been measured in a project that investigates the possibility of the automatic detection of lung diseases in the future. The sampling frequency in Figure 1.2 is 5 kHz. It is clear that this signal is not stationary. The first inspiration cycle starts at about 0.2 s and lasts until about 1.7 s. Selecting only the signal during the part of the expiration period between 2.2 and 3.0 s gives the possibility of considering that signal as stationary. Its properties can be compared to the properties at similar parts of other respiration cycles.

Speech coding is an important application of time series models. The purpose in mobile communication is to exchange speech of good quality with a minimal bit rate. Figure 1.3 shows 8 s of a speech signal. It is filtered to prevent aliasing and afterward sampled with 8 kHz, giving 4 kHz as the highest audible frequency in the digital signal. A lower sampling frequency would damage the useful frequency content of the signal and cause serious audible distortions. Therefore, crude quantisation of the speech signal with only a couple of bits per sample requires

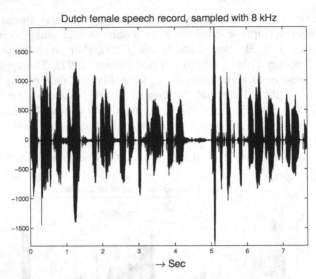

Figure 1.3. Speech fragment sampled with 8000 Hz; this signal is not stationary

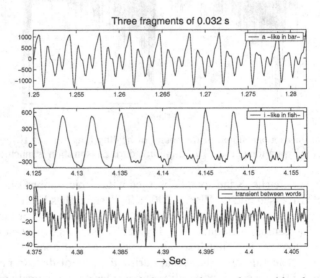

Figure 1.4. Three fragments of the speech fragment that can be considered stationary. Each fragment gets its own time series model in speech coding for mobile communications.

more than 20,000 bps (bits per second). It is obvious that the speech signal is far from stationary. Nevertheless, time series analysis developed for stationary stochastic signals is used in low-bit speech coding. The speech signal is divided into segments of about 0.03 s. Figure 1.4 shows that it is not unreasonable to call the speech signal stationary over such a small interval. For each interval, a time series model with some additional parameters can be estimated. Vowels give more or less periodic signals and consonants have a noisy character with a characteristic spectral shape for each consonant. In this way, it is possible to code speech with a bit rate of 4000 bps or even lower. This comes down to only one half bit per observation. This reduced bit rate gives the possibility of sending many different speech signals simultaneously over a single communication line. It is not necessary for efficient coding to recognize the speech. In fact, speech coding and speech recognition are different scientific disciplines.

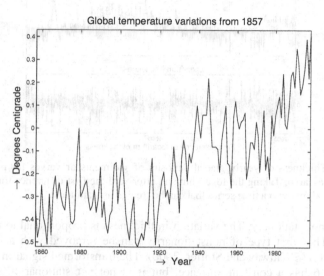

Figure 1.5. Global temperature time series indicating that the temperature on the earth increases. Comparing this record with other measurements shows that almost similar temperature variations over a period of a few centuries have been noticed in the last 400,000 years. However, if it continues to rise in the near future, the changes seem to be significantly different from what has been seen in the past.

Figure 1.5 shows some measurements of variations in global temperature. An important question for this type of climatic data is whether the temperature on the earth will continue to rise in the near future, as in most recent years. Considering the data as a historical record of deterministic numbers is safe but not informative. Extrapolating the trend with a straight line through the data obtained after 1975 would suggest dangerous global warming. However, extrapolating data without a verified model is almost never useful and always very inaccurate. This can be seen as treating the data as deterministic plus noise. The proposed third way to look at the data, as a stationary stochastic process, does not seem logical at first sight because there is a definite trend in this relatively short measurement interval. However, one should realise that there is a possibility that variations with a period of one or two centuries are more often found in paleoclimatic temperature data with a length of more than 100,000 years. In that case, the data in Figure 1.5 are just too short for any conclusions.

Figure 1.6 gives data about the thickness of varves of a glacier that has been melted down completely already long ago. The thickness of the sedimentary deposits can be used as a rough indicator of the average temperature in a year because the receding glacier will leave more sand in a warm year. This varve signal

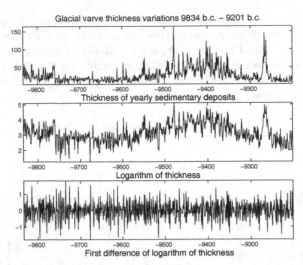

Figure 1.6. Thickness of yearly glacial deposits of sediments or varves for paleoclimatic temperature research. Taking the logarithm and afterward the first difference transforms the nonstationary signal into a time series that can be treated as stationary.

is typically not stationary. The variation in thickness is proportional to the amount deposited. That first type of nonstationarity can be removed with a logarithmic transformation (Shumway and Stoffer, 2000). The transformed signal in the middle of Figure 1.6 has a constant variance, but it is not yet stationary. Therefore, a method often applied to economic data is used (Box and Jenkins, 1976), taking the first difference of the signal, where the new signal is $x_n - x_{n-1}$. The final differenced signal at the bottom is stationary but misses most interesting details that are still present in the two preceding figures. The first two plots in Figure 1.6 show that there has been a period with gradually increasing temperature between 9575 b.c. and 9400 b.c. That period is longer than the measurements given in Figure 1.5. Hence, there has been a time when the global temperature increased for more than a century about 11,400 years ago. However, how much the temperature increased then cannot be derived from the given data because the calibration between varve thickness and degrees Centigrade is missing. Furthermore, a sharp shorter rise started about 9275 b.c. That large peak is still visible in the logarithm but is no longer seen in the differenced lower curve.

Hernandez (1999) warned of the "deleterious effects that the apparently innocent and commonly used processes of filtering, detrending, and tapering of data" have on spectral analysis. Transformations that improve stationarity should be used with care, otherwise a comparison with the results of raw data becomes difficult or impossible. Also the low-pass filtering operation that is often used to prevent aliasing in downsampling the signal should be treated with caution; Broersen and de Waele (2000b) showed that such filters destroy the original frequency content of the signal if the passband of the filter ends within half the resampling frequency. A higher cutoff frequency will allow some aliasing, but nevertheless it will often give the best spectral estimate over the reduced frequency range until half the resampling frequency.

This study of single univariate signals is not really decisive on the issue of global warming. An approach to explain global long-term atmospheric development with physical or chemical modeling uses input-output modeling (Crutzen and Ramanathan, 2000). A problem with the explanatory force of all approaches is that an independent verification of the ideas is virtually impossible with long-term climate data. Most research started after the first signs of global warming were detected and lack statistical independence: the supposition of global warming in the last 50 or 80 years was the reason to start the investigation. Unfortunately, the statistical significance or justification of *a posteriori* explanations is rather weak.

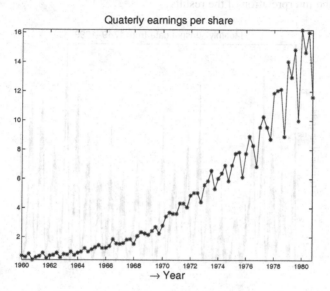

Quaterly earnings per share

→ Year

Figure 1.7. Economic time series with four observations per year. The series has a strong seasonal component and a trend.

Figure 1.7 shows the earnings of shareholders of the U.S. company, Johnson and Johnson (Shumway and Stoffer, 2000). Those data show a strong trend. Furthermore, the observations have been made each quarter, four times per year. That pattern is strongly present in the data. Modeling such data requires special care. Brockwell and Davis (1987) advised estimating a model for those data as the sum of three separate components:

$$X_t = m_t + s_t + Y_t$$

- X_t the measured observations
- m_t a slowly changing function, often called the trend component, which can be estimated as a polynomial in time; this class of functions includes the mean value as a constant
- s_t a seasonal component; see also Shumway and Stoffer (2000)
- Y_t a stationary stochastic process.

Many data from economics and business show daily, weekly, monthly, or yearly patterns. Therefore, the data are not stationary because the properties vary exactly with the period of the patterns. It is mostly advisable to use a seasonal model for those purely periodic patterns and a time series model for the residuals remaining after removing the seasonal component.

This book treats the automatic analysis of stationary stochastic signals. It is tacitly assumed that transformations and preprocessing of the data have been applied according to the rules that are specific for certain special areas of application. However, it should be realised that all preprocessing operations can deteriorate the interpretation of the results.

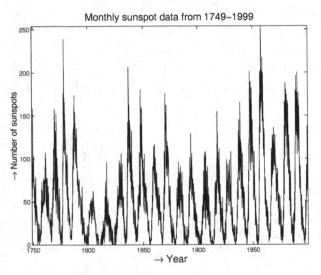

Figure 1.8. Sunspot numbers that indicate the activity in the outer spheres of the sun

The sunspot data in Figure 1.8 show a strong quasi-periodicity with a period of about 11 years. A narrow peak in the power spectral density rather than one exact frequency characterizes those data. The seasonal treatment that is useful in economic data would fail here. The period is not exact, the measurements are not synchronized, and they cannot be modeled accurately as a spectral peak with a finite bandwidth. Therefore, modeling as a stationary stochastic process is the preferred signal processing in this case. At first sight, it is clear that the probability density of the sunspot data is not normal or Gaussian. That would, among others, require symmetry around a mean value. For normally distributed data, the best prediction is completely determined by the autocorrelation function, which is a second-order moment. For other distributions, higher order moments contain additional information. Using only the autocorrelation gives already reasonable predictions in the sunspot series, but better predictions should be possible by using a better approximation of the conditional density of the data.

Signal processing is the intermediate in relating measured data and theoretical concepts. Theoretical physics gives a theoretical background and explanation for observed phenomena. For stochastic observations, it is always important that signal

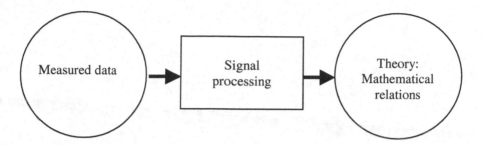

Figure 1.9. Signal processing as an intermediate between real-life data acquisition and theoretical explanations of the world

processing is objective without (too much) influence of the experimenter. Representing measured stochastic data by power spectral densities or autocorrelation functions is a good way to reduce the amount of data. When the data are a realisation of a normally distributed stationary stochastic process, the accuracy of the time series solution presented in this book will be close to the accuracy achievable for the spectrum and autocorrelation function of that type of data. If the assumptions about the normal distribution and the strict stationarity are not fulfilled, the time series solution is still a good starting point for further investigation.

It is considered a great advantage for different people to estimate the same spectral density and the same autocorrelation function from the same stochastic observations. That would mean that they draw the same theoretical conclusions from the same measured data. This book is an attempt to present the existing signal processing methods for stationary stochastic data from the point of view that a unique estimated spectrum or autocorrelation is the best contribution that signal processing can give to scientific developments in other fields. Of course, the accuracy of the spectral density and of the autocorrelation function must also be known, as well as which details are statistically significant and which are not.

2

Basic Concepts

2.1 Random Variables

The assumption is made that the reader has a basic notion about random or stochastic variables. A precise axiomatic definition is outside the scope of this book. Priestley (1981) gives an excellent introduction to statistical theory for those users of random theory who want to understand the principles without a deep interest in all detailed mathematical aspects.

Given a random variable X, the *distribution function* of X, $F(x)$, is defined by

$$F(x) = p[x \leq X] \tag{2.1}$$

which is the probability that the random variable X is less than or equal to the number x. From this, it follows that

$$p[a < X \leq b] = p[X \leq b] - p[X \leq a] = F(b) - F(a) \tag{2.2}$$

The *probability density function* $f(x)$ is defined as the derivative of $F(x)$, if that exists. Hence, for a continuous variable X

$$F(x) = \int_{-\infty}^{x} f(y)\, dy \tag{2.3}$$

and also

$$f(x)\, dx = p[x < X \leq x + dx] \tag{2.4}$$

By taking $dx = 0$, it follows that $p[X = x] = 0$, if X is a continuous random variable. The probability that X takes any precise given numerical value is zero. This can easily be understood because each small interval still contains infinitely many real numbers. Some properties of the probability density function are

$$f(x) \geq 0, \quad \forall x$$

$$\int_{-\infty}^{\infty} f(x)\, dx = 1 \tag{2.5}$$

The expectation operator defines the expected value of a function $g(X)$ as

$$E[g(X)] = \int_{-\infty}^{\infty} g(x) f(x)\, dx \tag{2.6}$$

This defines the mean μ_X as

$$\mu_x = \int_{-\infty}^{\infty} x f(x)\, dx \tag{2.7}$$

the variance σ_X^2 as

$$\sigma_X^2 = \mathrm{var}[X] = E\left[(X - \mu_X)^2\right] = \int_{-\infty}^{\infty} (x - \mu_X)^2 f(x)\, dx \tag{2.8}$$

and the rth central moment of X as

$$E\left[(X - \mu_X)^r\right] = \int_{-\infty}^{\infty} (x - \mu_X)^r f(x)\, dx \tag{2.9}$$

This gives the moments of a random variable. Also noncentral moments can be defined by leaving out μ_X in (2.9).

A bivariate probability density function of two random variables X and Y is defined by

$$f(x, y)\, dxdy = p\left[x < X \leq x + dx,\ y < Y \leq y + dy\right] \tag{2.10}$$

The definition of a multivariate or joint probability density function is straightforward, and it will not be given explicitly here. The covariance between two random variables X and Y is defined as

$$\mathrm{cov}(X, Y) = E\left[(X - \mu_X)(Y - \mu_Y)\right]$$

$$= \int_{-\infty}^{\infty} \int_{-\infty}^{\infty} (x - \mu_X)(y - \mu_Y) f(x, y)\, dxdy \tag{2.11}$$

The correlation coefficient is the normalized covariance, given by

$$\rho_{X,Y} = \frac{\mathrm{cov}(X, Y)}{\sigma_X \sigma_Y} \tag{2.12}$$

The correlation coefficient has the important property

$$\left|\rho_{X,Y}\right| \le 1 \tag{2.13}$$

A negative correlation coefficient has a tendency for the signs of X and Y to be opposite; more often, positive correlation gives a pair with the same sign. Figure 2.1 gives clouds of realisations of correlated pairs (X,Y), for various values of the correlation coefficient.

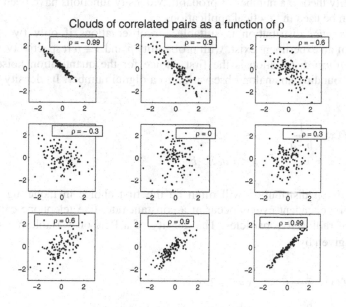

Clouds of correlated pairs as a function of ρ

Figure 2.1. Pairs of correlated variables X,Y, each with mean zero and variance one, for various values of the correlation coefficient

Two random variables are independent if the bivariate density function can be written as the product of the two individual density functions,

$$f(x, y) = f_X(x)f_Y(y) \tag{2.14}$$

In this formula, the univariate probability density functions have an index to indicate that they are different functions. Whenever possible without confusion, indexes are left out.

The covariance of two independent variables follows from (2.11) and (2.14) as

$$\begin{aligned}
\text{cov}(X,Y) &= E\left[(X - \mu_X)(Y - \mu_Y)\right] \\
&= \int_{-\infty}^{\infty}(x - \mu_X)f_X(x)\,dx \int_{-\infty}^{\infty}(y - \mu_Y)f_Y(y)\,dy \\
&= 0
\end{aligned} \tag{2.15}$$

Independence implies that the correlation coefficient equals zero. The converse result, however, is not necessarily true. Uncorrelated variables are not necessarily independent. This result and much more can be found in Priestley (1981) and in Mood *et al.* (1974).

2.2 Normal Distribution

In probability theory, a number of probability density functions have been introduced that can be used in practical applications.

The *binomial* distribution is suitable for observations if only two possible outcomes of an experiment exist, with probability p and $1 - p$ respectively.

The *uniform* distribution is the first choice for the quantisation noise that is caused by rounding an analog observation to a digital number. Its density function is given by

$$f(x) = \frac{1}{b-a} \qquad a \leq x \leq b$$
$$= 0 \qquad x < a \ or \ x > b \tag{2.16}$$

The *Poisson* distribution will often be the first choice in modeling the time instants if independent events occur at a constant rate, like telephone calls or the emission of radioactive particles. The density of a Poisson variable X with parameter λ is given by

$$f(x) = \frac{e^{-\lambda}\lambda^x}{x!} \qquad x = 0, 1, 2, \cdots$$
$$= 0 \qquad \text{otherwise} \tag{2.17}$$

This distribution is characterized by $E[X] = \lambda$ and var $[X] = \lambda$.

The *Gaussian* or *normal* distribution is the most important distribution in statistical theory as well as in physics. The probability density function of a normally distributed variable X is completely specified by its mean μ and its variance σ^2

$$f(x) = \frac{1}{\sigma\sqrt{2\pi}} \exp\left[-\frac{(x-\mu)^2}{2\sigma^2} \right] \tag{2.18}$$

The probability that a normal variable will be in the interval $\mu - 1.96\sigma < x < \mu + 1.96\sigma$ is 95%. The normal distribution is important because it is completely determined by its first- and second-order moments. Also a practical reason can be given why many measured variables have a distribution that at least resembles the normal. Physical phenomena can often be considered a consequence of many independent causes, *e.g.*, the weather, temperature, pressure, or flow. The central limit theorem from statistics states roughly that any variable generated by a large

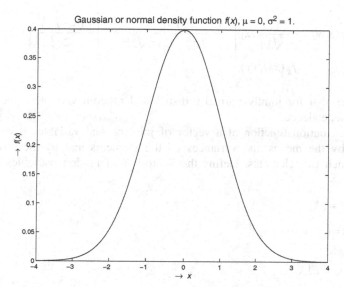

Figure 2.2. The Gaussian or normal probability density function with zero mean and unit variance

number of independent random variables of arbitrary probability density functions will tend to have a normal distribution.

Apply this result to dynamic processes, which may be considered the convolution of an impulse response with an input signal. Suppose that the input is a random signal with an arbitrary probability density function. The output signal, the weighted convolution sum of random inputs, is closer to a normal distribution than the input was. If the input were already normally distributed, it remains normal, and if it were not normal, the output result would tend to normal or Gaussian.

The bivariate normal density for two joint normal random variables X and Y is given by

$$f(x, y) = \frac{1}{2\pi\sigma_X\sigma_Y\sqrt{1-\rho_{XY}^2}}$$

$$\times \exp\left\{-\frac{1}{2(1-\rho_{XY}^2)}\left[\frac{(x-\mu_X)^2}{\sigma_X^2} - \frac{2\rho_{XY}(x-\mu_X)(y-\mu_Y)}{\sigma_X\sigma_Y} + \frac{(y-\mu_Y)^2}{\sigma_Y^2}\right]\right\} \quad (2.19)$$

The marginal density function of X can be retrieved by integrating out Y

$$f_X(x) = \int_{-\infty}^{\infty} f(x, y)dy = \frac{1}{\sigma_X\sqrt{2\pi}}\exp\left[-\frac{(x-\mu_X)^2}{2\sigma_X^2}\right] \quad (2.20)$$

This is already given in (2.18), without index. For uncorrelated normally distributed X and Y, it follows by substituting zero for ρ_{XY} in (2.19) that

$$f(x, y) = \frac{1}{\sigma_X \sqrt{2\pi}} \exp\left[-\frac{(x - \mu_X)^2}{2\sigma_X^2}\right] \frac{1}{\sigma_Y \sqrt{2\pi}} \exp\left[-\frac{(y - \mu_Y)^2}{2\sigma_Y^2}\right]$$

$$= f_X(x) f_Y(y) \tag{2.21}$$

This proves that for jointly normally distributed random variables, uncorrelated implies independence.

The distribution function of a vector of joint normal variables is completely specified by the means and variances of the elements and by the covariances between each two elements. Define the vectors X of random variables and x of numbers as

$$X = \left(X_1 \ X_2 \cdots\cdots X_m\right)^T$$

$$x = \left(x_1 \ x_2 \cdots\cdots x_m\right)^T$$

with

$$\mu_X = E[X]$$

$$R_{XX} = E\left[(X - \mu_X)(X - \mu_X)^T\right]$$

The joint normal density is given by

$$f(x_1, \cdots, x_m) = \frac{1}{(2\pi)^{m/2} |R_{XX}|^{1/2}} \exp\left\{-\frac{1}{2}\left[(x - \mu_X)^T R_{XX}^{-1}(x - \mu_X)\right]\right\} \tag{2.22}$$

This reflects the property that normally distributed variables are completely specified by the first- and second-order moments that are the mean values, the variances, and the covariances. An important consequence is that all higher dimensional moments can be derived from the first- and second-order moments.

A useful and simple practical result for a fourth-order moment of four jointly normally distributed zero mean random variables A, B, C, and D is

$$E[ABCD] = E[AB]E[CD] + E[AC]E[BD] + E[AD]E[BC] \tag{2.23}$$

A fourth-order moment can be written as a sum of the products of second-order moments. Likewise, all even higher order moments can be written as sums of second-order moments. All odd moments of that zero mean normally distributed variables are zero; see Papoulis (1965). With (2.23), it is easily derived that

$$E\left[X^2 Y^2\right] = \sigma_X^2 \sigma_Y^2 + 2\text{cov}(X, Y)$$

The normal distribution has important properties for use in practice. The popular least-squares estimates for parameters are maximum likelihood estimates with very favourable properties if the distribution of measurement errors is Gaussian.

The chi-square, or χ^2 distribution, is derived from the normal distribution for the sum of K independent, normalized, squared, normally distributed variables, each with mean zero and variance one. The main property of this χ^2 for increasing K is that for $K \to \infty$, the χ^2 density function becomes approximately normal with mean K and variance $2K$. This approximation is already reasonably accurate for K greater than 10.

The Gumbel distribution is derived from the normal distribution to describe the occurrence of extreme values like the probability of the highest water levels of seas and rivers.

The true distribution function is often unknown if measurement noise or physical fluctuations cause the stochastic character of the observations. For that reason, stochastic measurements are often characterized by some simple characteristics of the probability density function of the observations. The three most important are

- mean
- variance
- covariance matrix

Those characteristics are all there is to know for normally distributed variables. Furthermore, they are also the most important simply obtainable characteristics for unknown distributions.

2.3 Conditional Densities

The conditional density $f_{X|Y}(x|y)$ is the probability density function of X, given that the variable Y takes the specific value y. With the general definition of the conditional density function, the joint probability density function of N arbitrarily distributed variables $X = (X_1 \; X_2 \cdots\cdots X_m)^T$ can be written as (Mood et al., 1974)

$$f_X(x) = f_{X_{k+1},\cdots,X_N|X_1,\cdots,X_k}(x_{k+1},\cdots,x_N \mid x_1,\cdots,x_k)$$
$$\times f_{X_1,\cdots,X_k}(x_1,\cdots,x_k) \tag{2.24}$$

In particular, it follows that

$$f_{X_1,X_2,\cdots,X_k}(x_1,x_2,\cdots,x_k)$$
$$= f_{X_k|X_1,\cdots,X_{k-1}}(x_k \mid x_1,\cdots,x_{k-1}) f_{X_1,\cdots,X_{k-1}}(x_1,\cdots,x_{k-1}) \tag{2.25}$$

With those results for the intermediate index k, the joint density $f_X(x)$ for arbitrary distributions can be written as a product of the probability density function of the first observation with conditional density functions

$$f_X(x) = \left[\prod_{k=2}^{N} f_{X_k|X_1,\cdots,X_{k-1}}(x_k \mid x_1,\cdots,x_{k-1}) \right] f_{X_1}(x_1) \tag{2.26}$$

2.4 Functions of Random Variables

It is sometimes possible to derive the probability density of a function of stochastic variables theoretically, like for the sum of variables (Mood *et al.*, 1974). Sometimes also the distribution of a nonlinear function of a stochastic variable can be determined exactly, but computationally this solution is often not very attractive, albeit the most accurate. The expectation of the mean and of the variance of a nonlinear function of a stochastic variable can be approximated much more easily with a Taylor expansion. The Taylor approximations are accurate only if the variations around the mean are small in comparison to the mean itself. For a single stochastic variable, the expansion of a function $g(X)$ becomes

$$g(X) \approx g(\mu_X) + \left[\left. \frac{\partial g}{\partial X} \right|_{(X=\mu_X)} (X - \mu_X) \right] + \left[\left. \frac{1}{2} \frac{\partial^2 g}{\partial X^2} \right|_{(X=\mu_X)} (X - \mu_X)^2 \right] \quad (2.27)$$

This can be used to find an asymptotic approximation for the mean,

$$E[g(X)] \approx g(\mu_X) + \left. \frac{\partial^2 g}{\partial X^2} \right|_{(X=\mu_X)} \frac{\sigma_X^2}{2} \quad (2.28)$$

Sometimes, only the first term is used as an approximation, but the second term can significantly improve the accuracy. The variance can be approximated with

$$\operatorname{var}[g(X)] = E\left[g(X) - E[g(X)] \right]^2$$

$$\approx \left(\left. \frac{\partial g}{\partial X} \right|_{(X=\mu_X)} \right)^2 \sigma_X^2 \quad (2.29)$$

Hence, the variance of the linear relation $g(X) = aX$ becomes $a^2 \sigma_x^2$.

The Taylor expansion formula for two stochastic variables becomes

$$g(X,Y) = g(\mu_X, \mu_Y)$$

$$+ \left. \frac{\partial g}{\partial X} \right|_{(X=\mu_X)(Y=\mu_Y)} (X - \mu_X) + \left. \frac{\partial g}{\partial Y} \right|_{(X=\mu_X)(Y=\mu_Y)} (Y - \mu_Y)$$

$$+ \left. \frac{1}{2} \frac{\partial^2 g}{\partial X^2} \right|_{(X=\mu_X)(Y=\mu_Y)} (X - \mu_X)^2 + \left. \frac{1}{2} \frac{\partial^2 g}{\partial Y^2} \right|_{(X=\mu_X)(Y=\mu_Y)} (Y - \mu_Y)^2$$

$$+ \left. \frac{\partial^2 g}{\partial X \partial Y} \right|_{X=\mu_X, Y=\mu_Y} (X - \mu_X)(Y - \mu_Y) . \quad (2.30)$$

An asymptotic approximation for the expectation of the mean follows as

$$E[g(X,Y)] \approx g(\mu_x,\mu_y) + \frac{1}{2}\frac{\partial^2 g}{\partial X^2}\bigg|\sigma_X^2 + \frac{1}{2}\frac{\partial^2 g}{\partial Y^2}\bigg|\sigma_Y^2 + \frac{\partial^2 g}{\partial X \partial Y}\bigg|\text{cov}(X,Y) \quad (2.31)$$

The variance can be approximated with the definition

$$\text{var}[g(X,Y)] = E\{g(X,Y) - E[g(X,Y)]\}^2.$$

Substitution of (2.31) in (2.30) gives

$$\text{var}[g(X,Y)] \approx E\left[\frac{\partial g}{\partial X}\bigg|_{(X=\mu_X)(Y=\mu_Y)}(X-\mu_x) + \frac{\partial g}{\partial Y}\bigg|_{(X=\mu_X)(Y=\mu_Y)}(Y-\mu_y)\right]^2$$

$$\approx \left(\frac{\partial g}{\partial X}\bigg|_{(X=\mu_X)(Y=\mu_Y)}\right)^2 \text{var}\{X\} + \left(\frac{\partial g}{\partial Y}\bigg|_{(X=\mu_X)(Y=\mu_Y)}\right)^2 \text{var}\{Y\} +$$

$$2\left(\frac{\partial g}{\partial X}\bigg|_{(X=\mu_X)(Y=\mu_Y)}\right)\left(\frac{\partial g}{\partial Y}\bigg|_{(X=\mu_X)(Y=\mu_Y)}\right)\text{cov}\{X,Y\}$$

$$\approx \sigma_X^2 \frac{\partial g}{\partial X}\bigg|^2 + \sigma_Y^2 \frac{\partial g}{\partial Y}\bigg|^2 + 2\text{cov}(X,Y)\frac{\partial g}{\partial X}\bigg|\frac{\partial g}{\partial Y}\bigg| \quad (2.32)$$

The use of those formulas is illustrated with an example: $g(X,Y) = X/Y$. The first and second derivatives are given by

$$\frac{\partial g}{\partial X} = \frac{1}{Y}$$

$$\frac{\partial g}{\partial Y} = -\frac{X}{Y^2}$$

$$\frac{\partial^2 g}{\partial X^2} = 0$$

$$\frac{\partial^2 g}{\partial Y^2} = \frac{2XY}{Y^4}$$

$$\frac{\partial^2 g}{\partial X \partial Y} = -\frac{1}{Y^2}$$

Substituting these derivatives in the approximate formulas for the expectation yields

$$E\left[\frac{X}{Y}\right] \approx \frac{\mu_X}{\mu_Y} + \frac{\mu_X}{\mu_Y^3}\sigma_Y^2 - \frac{1}{\mu_Y^2}\text{cov}(X,Y) \quad (2.33)$$

and the variance becomes

$$\text{var}\left[\frac{X}{Y}\right] \approx \frac{1}{\mu_Y^2}\sigma_X^2 + \frac{\mu_X^2}{\mu_Y^4}\sigma_Y^2 - 2\frac{\mu_X}{\mu_Y^3}\text{cov}(X,Y) \quad (2.34)$$

2.5 Linear Regression

A simple example of linear regression is the estimation of the slope of a straight line through a number of measured points. A mathematical description uses the couples of variables (x_i, y_i). The *regressor* or independent variable x_i is a *deterministic* variable that has been adjusted to a specified value for which the *stochastic* response variable y_i is measured. The response is also denoted the dependent variable. The true relation for a straight line is given by

$$y_i = \beta_0 + \beta_1 x_i + \varepsilon_i \tag{2.35}$$

where ε_i is a stochastic measurement error. Suppose that N pairs of the dependent and the independent variable have been observed. The parameters of a straight line can be estimated by minimizing the sum of squares of the residuals RSS defined as

$$\text{RSS} = \sum_{i=1}^{N} \left(y_i - \hat{b}_0 - \hat{b}_1 x_i \right)^2 \tag{2.36}$$

The regressor variable for the parameter $\hat{b}_0$ is the constant one, the same value for every index i. Hats are often used to denote estimates of the unknown parameter value. It is obvious that the sequence of the indexes of the variables (x_i, y_i) has no influence on the minimum of the sum of squares. Also the estimated parameters are independent of the sequence of the variables in linear regression. The least-squares solution in (2.36) is a computationally and attractive method for estimating the parameters $\hat{b}_0$ and $\hat{b}_1$ if the ε_i are statistically independent. It is also the best possible solution if the measurement errors ε_i are normally distributed.

Priestley (1981) gives a survey of estimation methods. In general, the most powerful method is the maximum likelihood method. That method uses the joint probability density function of the measurement errors ε_i to determine the most plausible values of the parameters, given the observations and the model (2.35). It can be proved that the maximum likelihood solution is obtained precisely with least squares if the errors ε_i are normally distributed. This is another reason to assume a normal distribution for errors. The simple least-squares estimator has some desirable properties then.

It is important that the observed input independent variables are considered to be known exactly, without observational errors and that all deviations from the relation between x and y are considered independent measurement errors in y. Linear regression analysis treats the theory of observation errors that are linear in the parameters, as in the example of the straight line. Extending that example to a polynomial,

$$y_i = \beta_0 + \beta_1 x_i + \beta_2 x_i^2 + \cdots + \beta_p x_i^p + \varepsilon_i \tag{2.37}$$

conserves the property that the error is linear in the parameters. Now, the polynomial regressors are nonlinear functions of the independent variable x_i, but the error is still a linear function of the parameters. The solution for the parameters is found by minimising

$$\text{RSS} = \sum_{i=1}^{N} \left(y_i - \hat{b}_0 - \hat{b}_1 x_i - \hat{b}_2 x_i^2 \cdots - \hat{b}_p x_i^p \right)^2 \qquad (2.38)$$

Minimization of the RSS is the optimal estimation method if the errors are normally distributed. However, often the distribution function of the errors is not known. Then, it is not possible to derive the optimal estimation method for the parameters in (2.35) or (2.37). Nevertheless, an important property of the least-squares solution which minimises (2.38) remains that minimising the RSS gives a fairly good solution for the parameters in most practical cases, $e.g.$, if the errors are not normally distributed but still independent.

With a slight change of notation, general regression equations are formulated in matrix notation. In this part, the index of the observations is given between brackets. The following vectors and matrices are defined:

- $N \times K$ matrix X of deterministic regressors or independent variables $x_1(i), \ldots, x_K(i)$, with $i = 1, \ldots, N$.
- $N \times 1$ vector y which contains the observed dependent variables $y(i)$, $i = 1, \ldots, N$.
- $N \times 1$ error vector ε which are i.i.d. (independent identically distributed) random variables, zero mean, and variance σ^2.
- $K \times 1$ vector β of the true regression coefficients with the $K \times 1$ vector $\hat{b}$ as an estimate.

With those definitions, the individual observed variables can be written as

$$y(i) = x_1(i)\beta_1 + x_2(i)\beta_2 + \cdots + x_K(i)\beta_K + \varepsilon(i) \qquad (2.39)$$

In matrix notation, this can be written concisely as

$$y = X\beta + \varepsilon$$

Estimates $\hat{b}$ are found by minimizing the sum of squared errors

$$\text{RSS} = \left[y - X\hat{b} \right]^T \left[y - X\hat{b} \right] \qquad (2.40)$$

where X^T is the transpose of X. The solution of the parameters can be written explicitly as

$$\hat{b} = (X^T X)^{-1} X^T y \qquad (2.41)$$

if $(X^T X)^{-1}$ exists. It has to be stressed that (2.41) is an explicit notation for the solution, not an indication of how the parameters are calculated. No numerically efficient computation method involves inversion of the $(X^T X)$ matrix. Efficient solutions of linear equations can be found in many texts (Marple, 1987).

The variance of the estimated parameters is for jointly normally distributed independent errors ε with variance σ^2 given by the $K \times K$ covariance matrix:

$$\text{cov}(\hat{b}, \hat{b}) = E\left[(\hat{b} - \beta)(\hat{b} - \beta)^T \right] = \sigma^2 (X^T X)^{-1} \tag{2.42}$$

The diagonal elements are the variances of the estimated parameters and the off-diagonal elements represent the covariance between two parameters. The regression equations have been derived under the assumption that the residuals are uncorrelated.

Otherwise, with correlated errors ε, the best linear unbiased estimate is computed with *weighted least squares* (WLS). If the errors ε are correlated with the $N \times N$ covariance matrix V, with the elements $v_{ij} = E\{\varepsilon_i \varepsilon_j\}$, the WLS equations become

$$\mathbf{RSS} = \left[y - X\hat{b} \right]^T V^{-1} \left[y - X\hat{b} \right]$$

with solution

$$\hat{b} = (X^T V^{-1} X)^{-1} X^T V^{-1} y \tag{2.43}$$

The residual covariance matrix now influences the accuracy of the parameters in

$$\text{cov}(\hat{b}, \hat{b}) = (X^T V^{-1} X)^{-1} \tag{2.44}$$

The variance of the parameters is made smaller by using the weighting matrix with the covariance of the ε. Equations (2.41) and (2.42) with independent identically distributed residuals can be considered as weighted with the unit diagonal covariance matrix for the ε, multiplied by the variance of the residuals. This variance is incorporated in the covariance matrix V in the weighted least-squares expression.

In nonlinear regression, the parameters belong to nonlinear functions of the regressors

$$y(i) = g\left[x_1(i), x_2(i), \cdots, x_K(i), \beta_1, \beta_2, \cdots \beta_L \right] + \varepsilon(i) \tag{2.45}$$

For this type of equations, no simple explicit analytical expression for the solution is possible. The least-squares solution is found by minimising

$$\mathbf{RSS} = \sum_{i=1}^{N} \left\{ y(i) - g\left[x_1(i), x_2(i), \cdots, x_K(i), \hat{b}_1, \hat{b}_2, \cdots \hat{b}_L \right] \right\}^2 \tag{2.46}$$

Numerical optimisation algorithms can be used to find a solution, but that takes generally much longer computation time, convergence is not guaranteed, and starting values are necessary.

2.6 General Estimation Theory

Priestley (1981) gives a good introduction to the theory of estimation. Some main definitions and concepts will be given here briefly.

Observed random data may contain information that can be used to estimate unknown quantities, such as the mean, the variance, and the correlation between two variables. We will call the quantity that we want to know θ. For convenience in notation, θ is only a single unknown parameter, but the estimation of more parameters follows the same principle. Suppose that N observations are given. They are just a series of observed numbers, as in Figure 1.1. Call the numbers x_1, $x_2, x_3, \ldots, x_{N-1}, x_N$. They are considered a realisation of N stochastic variables $X_1, X_2, X_3, \ldots, X_{N-1}, X_N$. The mathematical form of the joint probability distribution of the variables and the parameter is supposed to be known. In practice, often the normal distribution is assumed or even taken without notice. The joint probability distribution is written as

$$f(x_1, x_2, \cdots, x_{N-1}, x_N, \theta) \tag{2.47}$$

where θ is unknown and where the x_i are the given observations. The question what the measured data can tell about θ, is known as statistical inference.

Statistical inference can be seen as the inverse of probability theory. There, the parameters, say the mean and the variance are assumed to be known. Those values are used to determine which values x_i can be found as probable realisations for the stochastic variables X_i. In inference, we are given the values of $X_1, X_2, X_3, \ldots, X_{N-1}, X_N$ which actually occurred, and we use the function (2.47) to tell us something about the possible value of θ. There is some duality between statistical inference and probability theory. The data are considered random variables; the parameter is not random but unknown. The data can give us some idea about the values the parameter could have.

In estimation, no *a priori* information about the value of the parameter θ is given. The measured data are used to find either the most plausible value for the parameter as a point estimate or a plausible range of values, which is called interval estimation.

Hypothesis testing is a related problem. A hypothesis specifies a value for the parameter θ. Then, (2.47) is used to find out whether the given realised data agree with the specified value of θ.

An estimator is a prescription for using the data to find a value for the parameter. An estimator for the mean is defined as

$$\bar{X} = \frac{1}{N}\sum_{i=1}^{N} X_i \qquad (2.48)$$

The *estimator* is defined with random variables and is itself a random variable. Substitution of the realisation gives the *estimate*

$$\bar{x} = \frac{1}{N}\sum_{i=1}^{N} x_i \qquad (2.49)$$

which is simply a number. This number $\bar{x}$ may be close to the real true value, say μ, or further away. The estimator $\bar{X}$ as a random variable has a distribution function that describes the probability that certain values of $\bar{x}$ will be the result in a realisation. The distribution function of $\bar{X}$ is called the sampling distribution.

More general with less mathematical precision, an estimator for a parameter $\hat{\theta}$ is defined as

$$\hat{\theta} = \hat{\theta}(X_1, X_2, \cdots, X_N) \qquad (2.50)$$

A particular estimate is found by substituting the realisation of the data in (2.50). It is also usual to call the estimate $\hat{\theta}$, as long as confusion is avoided. Suppose that the true value of $\hat{\theta}$ is θ. It would be nice if the estimator would converge to the true value for increasing N, if more and more observations are available.

An estimator $\hat{\theta}$ is called unbiased if the average value of $\hat{\theta}$ over all possible realisations is equal to the true value θ. The bias is defined as

$$\text{bias}(\hat{\theta}) = E(\hat{\theta}) - \theta \qquad (2.51)$$

The bias will sometimes depend on N. If the bias disappears only for $N \to \infty$, the estimator $\hat{\theta}$ is called asymptotically unbiased. In most cases, being unbiased is a desirable property of an estimator. If two different unbiased estimators exist, preference is given to the unbiased estimator with the smallest estimation variance.

If biased estimators are compared, it is necessary to combine the bias and the variance of the estimators in a single accuracy measure. That is the mean square error $\text{MSE}(\hat{\theta})$, defined as

$$\text{MSE}(\hat{\theta}) = E\left[\hat{\theta} - \theta\right]^2$$
$$= \text{bias}^2(\hat{\theta}) + \text{var}(\hat{\theta}) . \qquad (2.52)$$

Another desirable property of an estimator is that it gets better if the number of data grows. This property is called consistency, which can be defined in many subtle mathematical ways. Here, an estimator is called consistent if

$$E\left[\hat{\theta}-\theta\right]^{2} \to 0, \quad \text{as } N \to \infty \tag{2.53}$$

Together with (2.51), it follows that both the bias and the variance of consistent estimators vanish for N going to infinity.

It was very easy to guess a good estimator for the mean in (2.48). Another unbiased estimator for the mean value would have been to average the largest and the smallest of all observations. However, the variance of (2.48) will be smaller than the variance of this two-point estimator for almost all distributions. Therefore, (2.48) is a better estimator. The question is how a good estimator for $\hat{\theta}$ can be found in general.

For many quantities θ, a simple estimator can be formulated. That is the maximum likelihood estimator, which is the most general and powerful method of estimation. It requires knowledge of the joint probability distribution function of the data as a function of θ, as given in (2.47). For unknown distributions, it is quite common to use or to assume the normal distribution and still to call the result a maximum likelihood estimator, although that is not mathematically sound. For a given value of θ, $f(x_1, x_2, \cdots, x_{N-1}, x_N, \theta)$ describes the probability that a certain realisation of the data will appear for that specific value of θ. If the resulting value of the joint distribution is higher for a different realisation, that realisation is more plausible for the given value of θ. However, in estimation problems we observe the data and want to say something about θ. That means that $f(x_1, x_2, \cdots, x_{N-1}, x_N, \theta)$ is considered a function of θ. Then, it is called the likelihood function of θ. If

$$f(x_1, x_2, \cdots, x_{N-1}, x_N, \theta_1) > f(x_1, x_2, \cdots, x_{N-1}, x_N, \theta_2), \tag{2.54}$$

we may say that θ_1 is a more plausible value than θ_2. The method of maximum likelihood is based on the principle that the best estimator for θ is the value that maximises the plausibility $f(x_1, x_2, \cdots, x_{N-1}, x_N, \theta)$ of θ. Generally, the natural logarithm of the likelihood function is used that is defined as

$$L = L(x_1, x_2, \cdots, x_{N-1}, x_N, \theta) = \ln\left[f(x_1, x_2, \cdots, x_{N-1}, x_N, \theta)\right] \tag{2.55}$$

and is called the log-likelihood function.

With the definition of the log-likelihood, a very important result can be formulated for unbiased estimators. If $\hat{\theta}$ is an unbiased estimator of θ, the Cramér-Rao inequality says that

$$\text{var}(\hat{\theta}) \geq \frac{1}{E\left[\partial L(X_1, X_2, \cdots, X_{N-1}, X_N, \theta) / \partial \theta\right]^{2}} \tag{2.56}$$

This can be interpreted as follows. Any unbiased estimator that tries to estimate θ has a variance. The minimum bound for that variance is given by the Cramér-Rao lower bound, which is the right-hand side of (2.56). An estimator whose variance is equal to the right-hand side is called efficient.

In many texts, use is made of a general property of the log-likelihood function

$$E\left[\frac{\partial L(X_1, X_2, \cdots, X_{N-1}, X_N, \theta)}{\partial \theta}\right]^2 = -E\left[\frac{\partial^2 L(X_1, X_2, \cdots, X_{N-1}, X_N, \theta)}{\partial \theta^2}\right] \qquad (2.57)$$

This is equal to the inverse of the minimum of the variance in (2.56). The smaller the variance, the better the estimator. If the quantity in (2.57) is high, it may be possible to obtain estimates with a small variance. The quantity is often called the Fisher information on θ, given by the observations.

Maximum likelihood estimation looks for the parameter that maximises the likelihood of (2.55). It has been proved that the maximum likelihood (ML) estimators have the following properties:

- ML is asymptotically unbiased
- ML is asymptotically efficient

Furthermore, it has been shown by Zacks (1971) that under rather mild mathematical conditions the invariance property of maximum likelihood can be applied

> the maximum likelihood estimator of a function of a parameter
> is equal to
> the function of the maximum likelihood estimator of the parameter

This invariance property will play a key role in the estimation of spectra and autocorrelation functions. By expressing them as functions of a small number of parameters, efficient estimates for the functions can be determined as functions of efficient parameter estimates.

2.7 Exercises

2.1 A random variable X has a uniform distribution between the boundaries a and b. Find an expression for the expectation and for the variance of X.

2.2 A random variable X has a normal distribution with mean μ and variance σ^2. Find an expression for the expectation of X^2, X^3 and X^4.

2.3 Give an example of two stochastic variables that are completely dependent and have zero correlation at the same time.

2.4 A random variable X has a normal distribution with mean zero and variance σ^2. Find an approximate expression for the expectation and for the variance of $\ln(X^2)$. Use a Taylor expansion of $\ln(X^2)$ around $\ln(\sigma^2)$.

2.5 The acceleration of gravity is estimated in a pendulum experiment. The pendulum formula is $T = 2\pi\sqrt{L/g}$. The length of the pendulum is measured repeatedly with an average result of 1.274 m with a standard deviation of 3 mm. The measured oscillation time averaged 2.247 s with a standard deviation of 0.008 s. Calculate the expectation and the variance of g from those experimental results.

2.6 Is it possible that the standard deviation of the sum of two stochastic variables is the sum of their individual standard deviations. What is the condition?

2.7 Is it possible that the variance of the sum of two stochastic variables is the sum of their individual variances. What is the condition?

2.8 N independent random variables $X_1, X_2, ..., X_N$ all have a normal distribution with the same mean zero and variance σ^2. Derive the maximum likelihood estimator for the mean of the variables. Derive the maximum likelihood estimator for the variance of the variables.

2.9 How many independent observations of a random variable with a uniform distribution between one and two are required to determine the mean of those observations with a standard deviation that is less than 1% of the true mean.

2.10 A careless physicist repeats a measurement of a random variable 15 times. Unfortunately, he loses five results. He determines the average and the standard deviation of the remaining 10 measurements and throws them away. Afterward, he finds the other five results. Can he still determine the average and the standard deviation of all 15 measurements? Did he lose some accuracy for the mean and the standard deviation with his carelessness?

2.11 A star has an unknown temperature X. Experiments in the past have yielded α as average for the temperature, with an estimation variance β. New experiments with a satellite give N unbiased observations

$$Y_i = X + W_i , \quad i = 1, \cdots, N$$

The measurement errors W_i are independent stochastic variables with variance σ^2. Determine the optimal unbiased estimate for X and the variance of that unbiased estimator.

3

Periodogram and Lagged Product Autocorrelation

3.1 Stochastic Process

A distinction is made between the stochastic or random *variables* in Chapter 2 and stochastic *processes*. A *process* or *signal* can be viewed as a carrier of information. A random or stochastic process $X(n)$ is a chronologically ordered family of random variables indexed by n, with $n = 0, \pm 1, \pm 2, \ldots$. Signals may be ordered in time or in place. Priestley (1981) gives a good introduction for users of random processes. Suppose that $X(n)$ arises from an experiment which may be repeated under identical conditions. The first time, the experiment produces a record of the observed variable $X(n)$ as a function of n. Due to the random character of $X(n)$, the next time the experiment will produce a different record of observed values. An

Stochastic process, ensemble

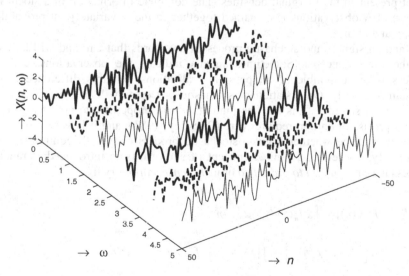

Figure 3.1. Six possible realisations of a stochastic process, which is an ensemble of all possible realisations with different ω that could have been observed. In practice, the argument ω is suppressed, whenever possible, because a single realisation is all that is available.

observed record of a random or stochastic process is merely one of a whole collection of records that could have been observed. The collection of all possible records is called the ensemble and an individual record is a realisation of the process. One experiment gives a single realisation that can be indexed with ω. Various realisations are $X(n,\omega_1)$, $X(n,\omega_2)$, ..., but the fact that generally only a single realisation is available gives the possibility of dropping the argument ω. Figure 3.1 shows six records of an ensemble with $\omega = 0, 1, 2, 3, 4, 5$.

According to the definition, the stochastic process for every n could be characterized by a different type of stochastic variable with a different probability density function $f_n(x)$. The mean at index n is given by

$$\mu(n) = E[X(n)] = \int_{-\infty}^{\infty} x f_n(x) \, dx \tag{3.1}$$

and the variance is given by

$$\sigma^2(n) = E\left\{[X(n)-\mu(n)]^2\right\} = \int_{-\infty}^{\infty} [x-\mu(n)]^2 f_n(x) \, dx \tag{3.2}$$

In principle, the mean and the variance can vary with n. Here, only random processes are considered where the marginal probability density function $f_n(x)$ is the same for all values of n. They are called stationary.

The joint probability distribution at two arbitrary time indexes n_1 and n_2 cannot be derived from the marginal distributions at n_1 and n_2; see the bivariate normal (2.19) where the two-dimensional distribution requires a correlation coefficient that is not present in the marginal densities. The complete information in a stochastic process of N observations is contained together in the N-variate joint probability density at all times.

Random signals and stochastic processes are words that can and will be used for the same concepts. Sometimes signals indicate the observations, and the process is the ensemble of all possible realisations, but this difference is not maintained strictly. Only stationary stochastic processes will be treated. A loose definition of stationarity is that the joint statistical properties do not change over time; a precise definition requires care (Priestley, 1981), and it is very difficult to verify in practice whether a given stochastic process obeys all requirements for stationarity. Therefore, a limited concept of stationarity is introduced: a random process is *stationary up to order two* or *wide sense stationary* if

- $E[X(n)] = \int_{-\infty}^{\infty} x f_n(x) dx = \mu$, $\forall n$

- $E\left\{[X(n)-\mu(n)]^2\right\} = \int_{-\infty}^{\infty} [x-\mu(n)]^2 f_n(x) d = \sigma^2$, $\forall n$

- $E[X(n)X(m)] =$ function only of $(n-m)$, $\forall n,m$

In words, a process is said to be stationary up to order two or wide sense stationary if

- the mean is constant over all time indexes n
- the variance is constant over all time indexes n
- the covariance between two arbitrary time indexes n and m depends only on the difference $n - m$ and not on the values of n and m themselves.

All signals in this book are defined only for discrete equidistant values of the time index, unless specified otherwise. A new notation x_n is introduced for this class of processes that are stationary up to order two. Jointly normally distributed variables, however, are completely stationary if they are stationary up to order two. Unless stated otherwise, the mean value of all variables is taken to be zero. In practice, this is reached by subtracting the average of signals before further processing.

If the properties of a process do not depend on time, it implies that the duration of a stationary stochastic process cannot be limited. Each possible realisation in Figure 3.1 has to be infinitely long. Otherwise, the first observations would have a statistical relation to their neighbours different from the observations in the middle. If a measured time series is considered as a stationary stochastic process, it means that the observations are supposed to be a finite part of a single infinitely long realisation of a stationary stochastic process.

3.2 Autocorrelation Function

The covariance between two observations x_n and x_{n+k} of a stationary stochastic process is defined in (2.11) as

$$r(k) = \text{cov}(x_n, x_{n+k}) = E\left[(x_n - \mu)(x_{n+k} - \mu)\right] \tag{3.3}$$

Only covariances in stationary processes are considered. The quantity $r(k)$ is defined for all integral values of k, and together it is called the *autocovariance function* of x_n. It measures the covariance between pairs at a distance or lag k, for all different values of k. This makes it a function of lag k. A long autocovariance function indicates that the data vary slowly. A short autocovariance function indicates that the data at short distances are not related or correlated.

The autocovariance function represents all there is to know about a normally distributed stochastic process because together with the mean, it completely specifies the joint probability distribution function of the data. Other properties may be interesting, but they are limited to the single realisation of the stochastic signal or process at hand. If the process is approximately normally distributed, the autocovariance function will describe most of the information that can be gathered about the process. Only if the distribution is far from normal, might it become interesting to study higher order moments or other characteristics of the process. That is outside the scope of this book.

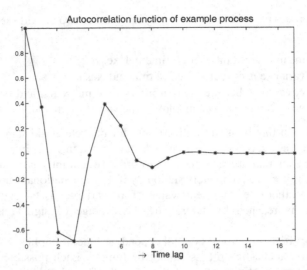

Figure 3.2. Autocorrelation function of example process. The dots represent the auto-correlation function, and the connecting lines are drawn only to create a nicer picture. The autocorrelation is symmetrical around zero, but generally only the part with positive lags is shown.

From (3.3), it follows that

$$r(0) = E\left[(x_n - \mu)^2 \right] = \sigma_x^2 \qquad (3.4)$$

Like the covariance between two variables, the autocovariance function $r(k)$ also can be normalized to give the autocorrelation function $\rho(k)$

$$\rho(k) = \frac{r(k)}{r(0)} = \frac{r(k)}{\sigma_x^2} \qquad (3.5)$$

The value for the autocorrelation at lag 0 is 1. It follows from (2.13) that $|\rho(k)| \leq 1$, and it can be seen in (3.3) that $\rho(k) = \rho(-k)$. This property also follows from the definition of stationarity where the correlation should be only a function of the time lag between two observations; the lags $-k$ and k are equal in that respect. Thus, the autocorrelation function is symmetrical about the origin where it attains its maximum value of one.

Figure 3.2 gives an example of an autocorrelation function. Usually, only the part with positive lags is represented in plots, because the symmetrical negative part gives no additional information. This example autocorrelation has a finite length: it is zero for all lags greater than 13. Most physical processes have an autocorrelation function that damps out for greater lags. This means that the relationship at a short distance in time is greater than the relation over longer distances. A damping power series is a common autocorrelation function that decreases gradually and has an infinite length theoretically. If the autocorrelation function equals precisely one at lag K, it will also be one at all lags nK that are multiples of K and the signal has to be periodic, with period K.

The $p \times p$ dimensional autocovariance matrix R_p is defined as

$$R_p = E\left\{ \begin{bmatrix} x_n \\ x_{n-1} \\ \vdots \\ \vdots \\ x_{n-p+1} \end{bmatrix} \begin{bmatrix} x_n & x_{n-1} & \cdots & \cdots & x_{n-p+1} \end{bmatrix} \right\} \tag{3.6}$$

which can be expressed in the individual autocovariances:

$$R_p = \begin{bmatrix} r(0) & r(1) & \cdots & r(p-1) \\ r(1) & r(0) & \ddots & \vdots \\ \vdots & \ddots & \ddots & r(1) \\ r(p-1) & \cdots & r(1) & r(0) \end{bmatrix} \tag{3.7}$$

The symmetry property of the autocovariances and of stationarity has been used. This matrix is positive-semidefinite (Priestley, 1981), which is a prerequisite for a valid autocovariance matrix. This result seems like an innocent property. However, most symmetrical functions do not fulfill this requirement. It is very demanding for a function to be a possible autocovariance function.

3.3 Spectral Density Function

The Fourier transform of the autocovariance function is the power spectral density function $h(\omega)$, also called the spectrum. The Wiener-Khintchine theorem (Wiener, 1930; Khintchine, 1934) defines conditions for valid autocovariances to have a transform that is nonnegative everywhere; see also Priestley (1981).

$$h(\omega) = \frac{1}{2\pi} \sum_{k=-\infty}^{\infty} r(k) e^{-j\omega k} \ , \ -\pi \leq \omega \leq \pi$$

$$r(k) = \int_{-\pi}^{\pi} h(\omega) e^{j\omega k} d\omega \ , \ k = 0, \pm 1, \pm 2, \cdots \tag{3.8}$$

The name power spectral density function is made clear by looking at the integral for the value $k = 0$:

$$\int_{-\pi}^{\pi} h(\omega) d\omega = r(0) = \sigma_x^2 . \tag{3.9}$$

The variance is the total power in the signal. The power spectral density gives the distribution of the total power over the frequency range. Moreover, $h(\omega)\,d\omega$ is the infinitesimal power in the frequency band from ω until $\omega + d\omega$. The discrete time Fourier transform in (3.8) is an infinite summation. Therefore, finite fast Fourier transform (FFT) algorithms cannot exactly represent those equalities; they transform N input numbers to N other numbers. Generally, the finite FFT can only compute approximations for infinite summations. Some properties of the spectral density are

$$h(\omega) \geq 0 \ , \ \forall \omega \tag{3.10}$$

and

$$h(\omega) = h(-\omega) \tag{3.11}$$

The fact that $h(\omega)$ is nonnegative everywhere is a consequence of the positive-semidefinite requirement of the autocovariance function. The symmetry property simply follows with elementary Fourier theory, because $h(\omega)$ is the Fourier transform of the real symmetrical autocorrelation function $r(k)$.

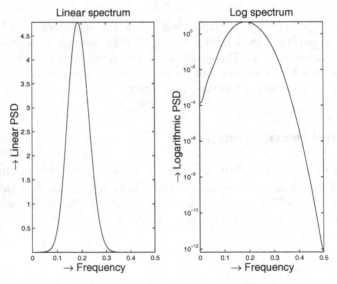

Figure 3.3. Power spectral density function of the process of Figure 3.2, represented in a linear and a logarithmic plot. The linear representation does not show the details with less power. This example has a large spectral variation and logarithmic representations are important for the details. The spectrum is symmetrical around zero frequency. Generally, only the part for positive frequencies is shown without loss of information.

Figure 3.3 gives an example of a power spectral density function. It is shown in two representations, linear and logarithmic. If the spectral density is rather flat over the frequency range, say the highest and lowest value differ by less than a factor of 100, it is not very important which representation is chosen; both show the important details. If the dynamic range in the spectrum is greater, logarithmic plots show more details. Generally, logarithmic plots are preferable. In the example of Figure 3.3, linear plots are preferable if only the frequency range between 0.15 and 0.25 is of interest. Details can be seen more easily in that range because it is elongated in the linear plot. For all other cases, the logarithmic plot gives more visible information, especially in the large part of the frequency range where the linear plot touches the bottom axis. There, no visible information is available about possible slopes in the spectrum.

The frequency range from 0 - 0.5 Hz is the range between zero and half the sampling frequency, which standard is chosen as one. If the sampling distance is T, the frequency range is from zero to $1/2T$. If the frequency ω is expressed in rad/s, the range is from $0 < \omega < \pi/T$. In evaluating the information in the autocorrelation function of Figure 3.2 and in the spectrum of Figure 3.3, it should be realised that they contain the same amount of information about the process because they are a Fourier transform pair. In Figure 3.2, a period of around five lags is visible. This agrees with the frequency of about 0.2 for which the spectrum has a maximum.

The normalized autocorrelation function $\rho(k)$ and the normalized power spectral density $\varphi(\omega)$ are found by dividing (3.8) by the variance of the signal

$$\varphi(\omega) = \frac{1}{2\pi} \sum_{k=-\infty}^{\infty} \rho(k) e^{-j\omega k} , \qquad -\pi \le \omega \le \pi$$

$$\rho(k) = \int_{-\pi}^{\pi} \varphi(\omega) e^{j\omega k} d\omega , \qquad k = 0, \pm 1, \pm 2, \cdots \qquad (3.12)$$

The formulas describe relations between **true** theoretical autocorrelations and **true** spectral densities. They certainly do not always describe the relations between arbitrary estimates. Estimated autocorrelations should be positive-semi-definite.

Often, a normalized frequency f is used, with $-0.5 \le f \le 0.5$. This frequency is equal to $\omega/2\pi T$, where T is the sampling frequency, by default taken as one. The integral

$$\int_{-0.5}^{0.5} h(f) d f = r(0) = \sigma_x^2 \qquad (3.13)$$

remains the same with the substitution

$$h(f) = 2\pi h(\omega) , \qquad \omega = 2\pi f \qquad (3.14)$$

This explains the absence of the often appearing factor 2π in many computational programs, and it does not create confusion in visual representations if the spectrum is given for f between -0.5 and $+0.5$, or between 0 and 0.5 for real data with the implicit convention that the mirrored image at negative frequencies gives no extra information and is left in the display. This type of transform, leaving the integral a constant, is also used in frequency scaling where $T \ne 1$. If the sampling distance is T and the spectrum $h^*(\omega)$ has been computed without this information, using the default value 1 for T, the spectrum with the proper frequency scale is then given by

$$h(\omega) = T h^*(\omega T), \quad |\omega| < \pi / T \tag{3.15}$$

Another valid definition for a spectrum can be given that is equivalent to the previous definition with mild mathematical restrictions:

$$h(\omega) = \lim_{N \to \infty} E\left\{ \frac{1}{2\pi N} \left| \sum_{n=1}^{N} x_n e^{-j\omega n} \right|^2 \right\}, \quad -\pi \le \omega \le \pi \tag{3.16}$$

The sequence of taking the Fourier transform, the square of the absolute value, the expectation, and the limit is essential in this definition. Only in this specific way is strict equivalence with the definition of the spectral density in (3.8) possible. A necessary condition for this equivalence is

$$\lim_{N \to \infty} \frac{1}{2\pi N} \sum_{k=-N}^{N} |k| |r(k)| = 0 \tag{3.17}$$

With this disappearing limit it can be derived that

$$h(\omega) = \lim_{N \to \infty} E\left\{ \frac{1}{2\pi N} \left| \sum_{n=1}^{N} x_n e^{-j\omega n} \right|^2 \right\}$$

$$= \lim_{N \to \infty} \frac{1}{2\pi N} \sum_{n=1}^{N} \sum_{m=1}^{N} E\{x_n x_m\} e^{-j\omega n} x_n e^{j\omega m}$$

$$= \lim_{N \to \infty} \frac{1}{2\pi N} \sum_{k=-(N-1)}^{N-1} (N - |k|) r(k) e^{-j\omega k}$$

$$= \frac{1}{2\pi} \sum_{k=-\infty}^{\infty} r(k) e^{-j\omega k} \tag{3.18}$$

The first line proves that the spectral density $h(\omega)$ is nonnegative-definite because the square cannot be less than zero. The last line of this derivation uses the limit for $N \to \infty$ in (3.17). That term represents a bias that disappears only asymptotically. Therefore, it is interesting to see the influence of the triangular bias on finite N in the third line of (3.18). Figure 3.4 gives the result for the example process of this section. Instead of transforming the true autocovariance function $r(k)$ to determine the spectrum, the biased $(1 - |k|/N) r(k)$ has been transformed to show the influence of the bias on the expectation of the spectrum.

The large variations in the spectral density of this example cause considerable bias for a finite number of observations. That would be much less if an example were chosen where the difference between highest and lowest spectral values is less than a factor of 100. Figure 3.4 shows once more the advantage of a logarithmic scale in spectral representations. The linear plot indicates that the triangular bias is negligible. That would also seem to be the case in the autocorrelation function of Figure 3.2, which is the Fourier transform of this spectral density. The bias would be less than the line thickness.

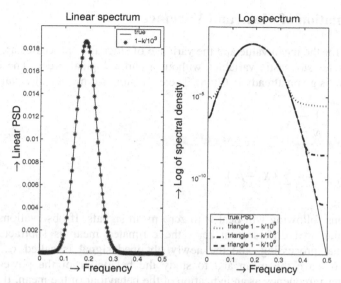

Figure 3.4. The influence of triangular bias in the autocorrelation function on the power spectral density function of the example process of Figure 3.2, represented in linear and logarithmic plots. The triangular bias is demonstrated by transforming $(1-k/N)\, r\,(k)$ for $N = 10^3$, 10^6, and 10^9. The linear representation of the power spectral density (PSD) does not show the bias that is very much present in the logarithmic plot of the spectrum. The bias would hardly be visible in the autocorrelation function of Figure 3.2. Small deviations in the autocorrelation can give large deviations in the spectrum.

The disappearance of the visual effect of the triangular bias occurs in both the linear spectral representations and in autocovariance or autocorrelation functions. That follows from Parseval's theorem (Priestley, 1981). This states roughly that the sums of squares are the same in both the time and the frequency domains. This can be applied to the difference signal between the true and the biased autocorrelation function and the difference between true and biased spectra. The fact that some differences are obvious in one type of plot and seem negligible in another representation is a strong reason to be careful with accuracy measures for autocorrelation and spectral estimates. It is obvious in Figure 3.4 that large differences in the logarithm of the spectrum can have an invisible influence on the linear spectrum.

Asymptotic theory may be of limited use in finite-sample practice. It is clear from Figure 3.4 that the property of the triangular bias to disappear asymptotically in (3.17) and (3.18) has not given a negligible bias in this finite- sample example. The bias, however, would disappear almost completely in rather flat spectra, with variations less than a factor of 100 over the frequency range. This example is an incentive to avoid relying on asymptotic properties in practice. Only after it has been verified that the sample size is large enough, may asymptotic theory be applicable. Of course, most relations have to be without asymptotic bias as a necessary condition. It really would be wrong if properties remain biased for increasing sample sizes. However, asymptotic properties are sometimes not sufficient in practice, if they are not applicable with any accuracy to finite sample sizes.

3.4 Estimation of Mean and Variance

Estimators for the mean value and the variance of stochastic processes are straight-forward, as for stochastic variables without a chronological order. The estimator for the mean is given already in (2.48). The estimators for mean and variance are

$$\hat{\mu}_x = \frac{1}{N} \sum_{n=1}^{N} X_n \tag{3.19}$$

$$\hat{\sigma}_x^2 = \frac{1}{N-1} \sum_{n=1}^{N} (X_n - \hat{\mu}_x)^2 \tag{3.20}$$

All equations following will deal with zero mean signals. If observations have an average value that differs from zero, the estimated mean is subtracted before further signal processing. and the newly obtained signal is called x_n. It is not advisable to leave the mean and to study the behaviour of the power spectral density at zero frequency as an indication of the behaviour of the mean. If the mean is interesting, it should be studied separately and not as a periodic signal that happens to have the spectral peak at zero frequency. Afterward, the mean can be added again to the signal if that is desirable. More general, higher-order trends and other deviations from a stationary appearance can be removed before the signal may be treated as stationary stochastic. In econometric data, it is often necessary to apply some seasonal adjustment and trend correction before the measured data can reasonably be considered a realisation of a stationary stochastic process. However, the differencing operation that is common practice in the treatment of economic data cannot be reversed afterward.

The best estimator for the mean value of different estimates of the same stochastic variable is not always the simple average defined in (3.19). Suppose that every estimator of the stochastic variable Y_i has as expectation the same value μ with some unknown errors ε_i, which are possibly correlated and may have different variances. It is much better to formulate this problem as the estimation of an unknown constant with the regression equation (2.35) where the linear term is omitted, as

$$Y_i = \mu + \varepsilon_i. \tag{3.21}$$

The general solution for this estimation problem is obtained with (2.43) if the measurement errors ε_i are correlated and/or have different variances. The covariance matrix of the errors ε_i is used as a weighting matrix. Taking just the average of the measured y_i as in (3.19) is a good choice only if the errors ε_i are known to be uncorrelated and all errors have the same variance. In other circumstances, the weighted least-squares solution of (2.43) might be a better choice.

It is an instructive exercise to derive the variance of the estimator of the mean of a correlated stochastic signal. It is given here, for its own sake as well as an example of how the variance of autocorrelation and periodogram estimators can be derived with simple principles and a lot of index manipulations. As in (2.50), the properties of an estimator are derived from the properties of stochastic variables and not from the given observations in the realisation that provides the estimate as a number. The expectation of the estimated mean is given by

$$E[\hat{\mu}_x] = E\left[\frac{1}{N}\sum_{n=1}^{N} X_n\right] = \frac{1}{N}\sum_{n=1}^{N} E[X_n]$$

$$= \frac{1}{N}\sum_{n=1}^{N} \mu = \mu \qquad\qquad (3.22)$$

The derivation of (3.22) shows that the estimator for the mean (3.19) is unbiased. With the level of mathematical rigor used here, summations, expectations, limits, and integrals may be interchanged in sequence "as long as all intermediate expressions remain properly defined." The definition of (3.16) is an example where the sequence of operations may not be changed because the Fourier transform of a stationary stochastic signal does not exist. That is because the infinite sum of absolute values of a stochastic signal is not finite for N increasing to infinity. The limit of the sum itself does not exist. It exists only after the expectation has been taken.

The derivation of the variance of the estimator of the mean requires exact knowledge of the true autocovariance function $r(k)$ of the signal. The properties of an estimator for the first-order moment of a signal require knowledge of the second-order moments. The variance of the mean is derived from the definition of the variance:

$$\mathrm{var}(\hat{\mu}_x) = E[\hat{\mu}_x - \mu]^2 \qquad\qquad (3.23)$$

where (3.22) has been used for the expectation of the estimator. Substitution of (3.19) gives

$$\mathrm{var}(\hat{\mu}_x) = E\left[\frac{1}{N}\sum_{n=1}^{N} X_n - \mu\right]^2$$

$$= \frac{1}{N^2} E\left[\sum_{n=1}^{N}(X_n - \mu)\sum_{j=1}^{N}(X_j - \mu)\right] \qquad\qquad (3.24)$$

This expression can be simplified by writing the products as autocovariances $r(j-n)$ and noting how many times the constant values $j-n = 0, 1,\dots$ are found. Limits for $j-n$ are $N-1$ and $-(N-1)$ which appear only once. The zero difference between the indexes appears precisely N times. Hence, it follows with some manipulation that

$$\operatorname{var}\left(\hat{\mu}_x\right) = \frac{1}{N^2}\left[\sum_{n=1}^{N}\sum_{j=1}^{N}r(j-n)\right]$$

$$= \frac{1}{N^2}\sum_{k=-(N-1)}^{N-1}\left(N-|k|\right)r(k)$$

$$= \frac{1}{N}\sum_{k=-(N-1)}^{N-1}\left(1-\left|\frac{k}{N}\right|\right)r(k) \tag{3.25}$$

This formula is often given in an asymptotic approximation, as a limit for $N \to \infty$. The triangle in the summations is left out then. As in (3.17), it is assumed that either k/N or $r(k)$ is very small. Therefore, their product is always small. For large lags, the autocovariance $r(k)$ is negligible and the triangle has almost no influence for small lags. The limiting expression for the variance of the estimated mean value becomes

$$\operatorname{var}\left(\hat{\mu}_x\right) = \frac{1}{N}\sum_{k=-\infty}^{\infty}r(k)$$

$$= \frac{2\pi}{N}h(0) \tag{3.26}$$

The last step follows from the definition of (3.8). The variance of the estimated mean is the sum of all **true** autocovariances, divided by the number of observations. Broersen (1998b) showed that it is generally not allowed or very inaccurate to use estimated lagged product autocovariances to approximate summations like that in (3.26).

The bias of the mean is zero for all N and the variance disappears for $N \to \infty$. Therefore, (3.19) is a consistent estimator for the mean and also for correlated observations. It will also be efficient for stationary error processes because then the variances of all ε_i in (3.21) are the same. However, if the error process has a time-dependent variance for ε_i, or if some of the ε_i are correlated, the weighted least-squares estimator for the mean would have a smaller variance. The estimator (3.19) would still be consistent, but not efficient because another estimator with a smaller variance would exist. This extensive and critical treatment of the estimation of the average is a preparation for the study of the estimation of an autocorrelation function.

3.5 Autocorrelation Estimation

The estimation of the autocovariance function requires some care because a true autocovariance *function* should be positive-semidefinite (Priestley, 1981). That is a prerequisite for a positive Fourier transform at all frequencies. Because that Fourier transform should represent the power spectral density, it is necessary that it is not negative for any frequency. If a function is not positive-definite, it cannot be an autocovariance function. Just taking estimates at different lags and combining them into a function can be a problem for the positive-semidefinite property.

Nevertheless, the estimator that has been used mostly for computations is based on the definition (2.11) of the covariance between two stochastic variables, applied to each lag individually. By taking the average covariance, as in definition (3.3), between the two stochastic variables x_n and x_{n+k} for different values of index n, an estimate for $r(k)$ is found. This estimator is often called the "sample autocovariance" or the "lagged product autocovariance." Combining all individual estimates for different values of k gives the estimated autocovariance function. It is regrettable that the concept that the estimates together should be a positive-semidefinite *function* does not lead to an estimator for the autocovariance function as a whole.

In principle, a maximum likelihood estimator can be formulated if the joint probability density function can be given of the observations and the function that has to be estimated. For the autocovariance function of normally distributed variables, (2.22) gives the probability density function that contains just observations and their autocovariance matrix. Unfortunately, that autocorrelation matrix has the size $N \times N$ with N^2 elements. This can be reduced to N elements by using the knowledge that the signal is stationary. But it cannot be expected that an estimator that transforms N observations into N estimated autocorrelation lags can be very accurate. Therefore, this maximum likelihood principle can be applied only if a more efficient way to express the autocorrelation function of a signal in only a couple of parameters can be found. For example, if it were known that the true autocorrelation function is a power series with $r(k) = a^{|k|}$, estimation of the single parameter a would provide an estimation for the whole autocorrelation function for all lags $k = 0$ until ∞.

An estimator for the autocovariance function follows from the definition of stochastic processes as the estimator over different realisations as in Figure 3.5. The estimator for a single autocovariance lag k at time m would become

$$\hat{r}_{ensemble}(k) = \frac{1}{K} \sum_{\omega=1}^{K} x_m(\omega) x_{m+k}(\omega) \tag{3.27}$$

where $x_m(\omega)$ denotes x_m in the realisation ω. Each product in (3.27) has the expectation $r(k)$ and all contributions are independent. Therefore, the average in (3.27) is an efficient estimator for the autocovariance of lag k at time m from the ensemble. However, in practice, only one single, finite realisation of a stochastic process will be available for an estimator of the autocovariance function.

In practice, only one realisation ω_l of N observations x_1 until x_N is given, and the index ω_l can be dropped without possible confusion. The autocovariance function has to be estimated from that single time series. Each product $x_n x_{n+k}$ in that realisation has the true value $r(k)$ as expectation. With N observations, only $N-k$ products are available for the shifted product at lag k, starting with $x_1 x_{k+1}$ and ending with $x_{N-k} x_N$. Taking the unweighted mean value of the lagged products, as in (3.19), gives an unbiased estimator:

Stochastic process, ensemble

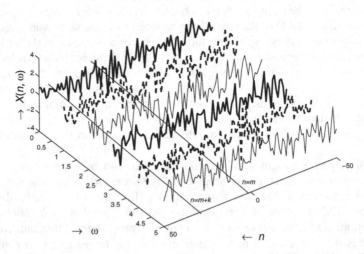

Figure 3.5. Estimation of the autocovariance $x_m x_{m+k}$ in an ensemble of possible realisations of the stochastic process

$$\hat{r}_{\text{unbiased}}(k) = \frac{1}{N-k} \sum_{n=1}^{N-k} x_n x_{n+k} \qquad (3.28)$$

Taking the average of a number of individual estimates $x_n x_{n+k}$ each of which has the desired expectation is not always a sound method from estimation theory. Such a naïve unbiased estimator is not necessarily efficient nor even useful. In this case, no statistical reasons have been given in the literature why this type of estimator is suitable. Therefore, care is required. No maximum likelihood expression leads to the estimator (3.28). To be more specific, the autocovariance estimator (3.28) has no general derivation from a sound statistical principle. The only justification is that any single contribution to the summations has the desired expectation $r(k)$, like the mean in (3.21). For a relation $V_j = i_j R + \varepsilon_j$, an estimator for R is found with the least-squares solution of (2.41) or (2.43), depending on the properties of ε_j. It is not found as the average of the individual estimates V_j / i_j, although each contribution would have the correct expectation for the autocovariance, as in (3.28). It is clear that observations with a small value of i_j would have too much influence on the average of V_j / i_j, for nonzero ε_j.

Suppose that $r(1)$ has to be estimated from a normally distributed signal. The first two contributions in the sum of (3.28) are the products $x_1 x_2$ and $x_2 x_3$. Then,

$$x_1 x_2 = r(1) + \varepsilon(1) \ , \ x_2 x_3 = r(1) + \varepsilon(2)$$

$$E[\varepsilon(1)\varepsilon(2)] = E\{[x_1 x_2 - r(1)][x_2 x_3 - r(1)]\} = E\left[x_1 x_2 x_2 x_3 - r(1)^2\right]$$

$$E[\varepsilon(1)\varepsilon(2)] = r(1)^2 + r(0)r(2) \qquad (3.29)$$

where (2.23) has been used to evaluate the joint fourth-order expectation. The same type of calculation can be made to find the complete covariance matrix of errors ε_i. This proves that the successive errors in the product in (3.28) are not independent and the statistically based estimator should be a weighted least-squares estimate of the unknown constant $r(k)$ based on the simple regression equation (3.21). The estimator (3.28) is unbiased but not necessarily efficient. Using the covariance matrix of errors ε_i, it might be possible to find a better estimator with a smaller variance of the estimated autocorrelation than with (3.28). Unfortunately, no better practical estimator can be based on this reasoning. One should know the true autocovariance function of the process in advance to construct the weighting matrix for the errors. The reasoning shows only that the estimator (3.28) is not necessarily statistically efficient.

The estimator (3.28) does not look for an autocovariance *function*, but only for estimates at individual lags. Despite the lack of statistical foundation, the lagged product estimator is considered more or less a definition of the autocovariance, which is a historical misconception. It will be shown later that for most signals and for most values of k, the efficiency of lagged product estimators is at best the same but generally less than what can be obtained with a completely different type of estimator. That type is defined in parametric time series analysis. Time series models represent parametric estimators for the autocorrelation function and for the power spectral density function, and those estimators can be derived from the maximum likelihood principle.

The estimator (3.28) is not positive-semidefinite, when combined with a function of the lag k. This can be demonstrated by a simple example. Suppose that three observations are available

$$x_1 = 1, \ x_2 = 0, \ x_3 = -1$$

With (3.28), this would give the estimates $\hat{r}_{\text{unbiased}}(k) = 2/3$, 0 and -1 for the lags k = 0, 1, and 2, respectively. The absolute value at lag 2 is larger than that at lag 0, which is not allowed in a positive-semidefinite function. The Fourier transform would become negative for some frequencies and hence the estimated autocovariance is not related to a possible spectral estimate. For that reason, the unbiased estimator (3.28) is not often used. Its performance as a function is not always that of an autocovariance function.

To guarantee the property of positive-definite that an autocovariance function must have, the improved estimator becomes

$$\hat{r}(k) = \frac{1}{N} \sum_{n=1}^{N-k} x_n x_{n+k} \tag{3.30}$$

Dividing the sum of the $N - k$ products by N is the same as multiplying the unbiased estimator (3.28) by a triangular window $1 - k/N$. The same bias also plays a role in (3.17) and in (3.18). The example of three observations 1, 0, and -1 would give the values $\hat{r}(k) = 2/3$, 0, and $-1/3$ for the first three lags k of this biased estimator. It can be proved that this estimator is positive-semidefinite (Priestley, 1981).

Instead of the autocovariance, the autocorrelation function also is often used, which is the normalized autocovariance. It is estimated as

$$\hat{\rho}(k) = \frac{1}{N} \sum_{n=1}^{N-k} x_n x_{n+k} \bigg/ \frac{1}{N} \sum_{n=1}^{N} x_n^2 = \frac{\hat{r}(k)}{\hat{r}(0)} \tag{3.31}$$

In the literature, this estimator is very popular. It is often called the lagged product estimator and also the nonparametric estimator for the autocovariance. Generally, (3.31) might be the only estimator that is considered for the autocorrelation. Its biased expectation is often approximated by (Priestley, 1981)

$$E\left[\hat{\rho}(k)\right] = \left(1 - \frac{k}{N}\right)\rho(k) \tag{3.32}$$

However, this expression omits the second-order contributions that appear in the expectation of the quotient of two stochastic variables in (2.31). Using (2.33), a better expression for the biased expectation is given by

$$E\left[\hat{\rho}(k)\right] = \left(1 - \frac{k}{N}\right)\rho(k) + \frac{r(k)}{r(0)^3}\operatorname{var}\left[\hat{r}(0)\right] - \frac{\operatorname{cov}\left[\hat{r}(k),\hat{r}(0)\right]}{r(0)^2} \tag{3.33}$$

The two additional contributions are of order $1/N$ and they are generally not small in comparison with k/N. The literature often refers to (3.30) as "the asymptotically unbiased estimator for the autocovariance function," without mentioning any of the problems or questions that arise in the derivation of this estimator in (3.29) or (3.33). The author considers those omissions historical misconceptions.

The question is whether the lagged product estimates for the autocovariance provide a good estimator for the autocovariance structure of stochastic data. That will be studied by looking at the variance of this estimator. Bartlett has given some asymptotic approximations for the variance and the covariance of the lagged product estimator of the autocovariance function from normally stochastic data; see Priestley (1981). Here, only the asymptotic results are presented for normally distributed signals:

$$\operatorname{var}\left[\hat{r}(k)\right] \approx \frac{\sigma_x^4}{N} \sum_{m=-\infty}^{\infty} \left[\rho^2(m) + \rho(m+k)\rho(m-k)\right] \tag{3.34}$$

The term after the summation is a constant for each given value of k. Hence, the estimator (3.30) gets a smaller variance for increasing N. The approximation of the true autocovariance function can become very accurate if the length N of the observed data is much greater than the length of the true autocovariance function. However, for the same number of observations, it will never become as accurate as the time series estimator for the autocovariance that will be introduced later.

The covariance of two estimates at different lags indicates how rough the estimated function will look. For a lag difference v, Priestley (1981) gives

$$\text{cov}\left[\hat{r}(s),\hat{r}(s+v)\right] \approx \frac{\sigma_x^4}{N} \sum_{m=-\infty}^{\infty} \left[\rho(m)\rho(m+v) + \rho(m+s+v)\rho(m-s)\right] \qquad (3.35)$$

Some results for specific values of the lags s and v are

$$\text{var}\left[\hat{r}(0)\right] \approx \frac{2\sigma_x^4}{N} \sum_{m=-\infty}^{\infty} \rho^2(m) \qquad (3.36)$$

$$\text{var}\left[\hat{r}(\infty)\right] \approx \frac{\sigma_x^4}{N} \sum_{m=-\infty}^{\infty} \rho^2(m) \approx \frac{1}{2}\text{var}\left[\hat{r}(0)\right] \qquad (3.37)$$

The reassuring property of the autocovariance is that its accuracy becomes better proportional with $1/N$. However, for all lags s for which the true autocorrelation is zero, the approximation (3.37) for infinite s applies. The accuracy is the same for all those lags, no matter how high. At very high lags where the true autocovariance is completely damped out, estimates also keep the variance of (3.37). Especially for a larger lag s, parametric estimators for the autocovariance function will be much more accurate.

Some other results for lagged product estimators are

$$\text{var}\left[\hat{r}(1)\right] \approx \frac{\sigma_x^4}{N} \sum_{m=-\infty}^{\infty} \left[\rho^2(m) + \rho(m+1)\rho(m-1)\right] \qquad (3.38)$$

$$\text{cov}\left[\hat{r}(0),\hat{r}(1)\right] \approx \frac{2\sigma_x^4}{N} \sum_{m=-\infty}^{\infty} \rho(m)\rho(m+1) \qquad (3.39)$$

Using the Taylor approximation (2.30) for nonlinear functions of stochastic variables, an approximate variance expression can also be derived for the autocorrelation function as the quotient of two random variables in (3.31):

$$\text{var}\{\hat{\rho}(k)\} \cong \frac{1}{N} \sum_{m=-\infty}^{\infty} \left\{ \begin{array}{l} \rho^2(m)\left[1 + 2\rho^2(k)\right] \\ + \rho(m+k)\,\rho(m-k) \\ - 4\,\rho(k)\,\rho(m)\,\rho(m-k) \end{array} \right\} \qquad (3.40)$$

These accuracy measures will be used later to compare the quality of lagged product estimators with parametric time series estimators for autocorrelations.

Figure 3.6 shows a true autocorrelation function and the estimates obtained from a realisation of 100 and one of 10,000 observations. It is clear that the accuracy increases for more observations. That is a consequence of the factor $1/N$ in (3.34). The covariance (3.35) between neighbouring points is the reason that estimates quite far away from their expectation still show a rather smooth function. As an example, the estimates for N equal to 100 in Figure 3.6 are all much below their expectation for lags between 25 and 36, and for lags between 38 and 48, they

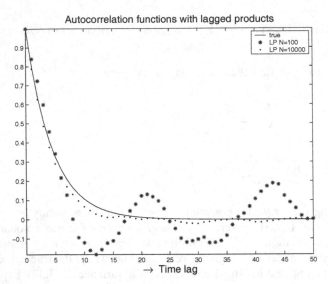

Figure 3.6. Estimation of the autocorrelation function with lagged products (LP), from 100 and from 10,000 observations. The errors in the estimates are strongly covariant and give estimated autocorrelation functions a smooth appearance.

are above. Although the individual estimates are rather poor, the complete function constructed with those estimates shows a deceptive smoothness. This often gives too much confidence in estimates, if no reliable statistical analysis is carried out. One important problem with such a statistical analysis of the accuracy is that substitution of the measured autocovariances of lagged product estimators in the asymptotic accuracy measures from (3.34) - (3.40) does not give a reliable answer. Even for large N, this substitution is not very useful. In other words, it is difficult to establish practical accuracy boundaries for estimated lagged product autocorrelation functions.

A rule of thumb for accuracy can be based on (3.37). The variance is equal for all estimates at lags where the true autocovariance is zero. Looking at Figure 3.6, this would indicate for $N = 100$ that the estimated autocorrelation wiggles between − 0.2 and + 0.2 at lags until half the sample size. Now suppose that the true autocorrelation is zero in the region with higher lags. As a consequence, all estimated values of the autocovariance between − 0.2 and + 0.2 are considered unreliable and probably only different from zero due the estimation variance. Only the estimates at the first lags below seven may be reliable according to this rule of thumb.

For $N = 10.000$, all estimates for lags above five are smaller than the true expectation. This illustrates the deceptive smoothness of inaccurate lagged product estimates. The rule of thumb replaces a more thorough investigation that is hampered because the estimated lagged product autocovariances may not be substituted in the asymptotic accuracy equations like (3.36). At least, it shows that taking more observations improves the accuracy of the estimated biased autocorrelation function. Furthermore, the accuracy can be as good as desired by increasing N.

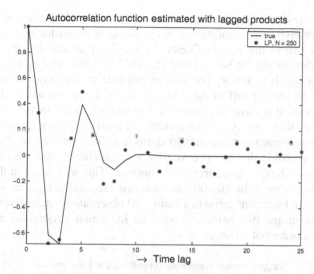

Figure 3.7. Estimate of a biased autocorrelation function with lagged products from 250 observations. The true autocorrelation is also shown in Figure 3.2.

Figure 3.7 gives the autocorrelation of an example for which the spectrum is sensitive to triangular bias, as shown in Figure 3.4. The theoretical triangular bias at lag 5 is only 0.008 for $N = 250$, much less than the statistical deviations in the estimate. No accuracy analysis can be given that can be applied convincingly to estimated autocorrelations without knowledge of the true autocorrelation function.

A general but inaccurate method to get a rough idea of the autocorrelation estimates with lagged product estimates for measured data is dividing the data into two halves and estimating the autocorrelation functions separately for the first and

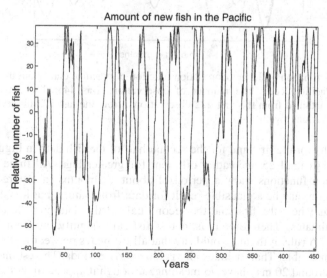

Figure 3.8. Monthly observations of estimated new fish in the Pacific Ocean

second half of the data. If differences in the estimated autocorrelation functions are small, this may give some confidence. An example is given for measurements of the amount of fish in the Pacific Ocean. Figure 3.8 presents the data, measured once per month between the years 1950 and 1987 and also analysed by Shumway and Stoffer (2000). It is always possible to estimate the autocorrelation from the first and from the second half of the data separately. If the process is stationary, it might be expected that those two estimated autocorrelation functions would look similar because they are two independent realisations of the same stochastic process. If the estimates are similar, no definite conclusion can be drawn, but it becomes very probable that the two similar autocorrelation functions are reliable representations of the true autocorrelation function. This will occur in the example of Figure 3.6 if more than 10,000 observations are available. The same figure shows that two independent estimates from 100 observations may diverge a lot. If the estimates from the two halves diverge, an indication is obtained as to which correlation details are not reliable.

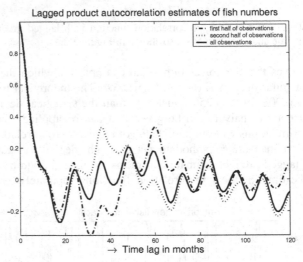

Figure 3.9. Estimation of the autocorrelation function of practical data about the amount of fish in the Pacific Ocean. The number of observations was 445. The lagged product estimates are obtained from the first 222 observations, from the last 223 observations, and from all observations together.

The results for the first and for the second half of the fish data in Figure 3.9 are similar for lags until 25 and rather different for greater lags. All three estimated autocorrelation functions show a period of about 12 months in Figure 3.9. That yearly variation may be a plausible result for data from nature, but it is only found for those lags where the first and the second half of the data produce completely different estimates. Therefore, it is not statistically significant from the given observations. A rule of thumb would say that all estimates between -0.3 and $+0.3$ are totally unreliable. That includes all visible yearly periods. The estimated valley for the lags around 20 may have no meaning, although it appears in both estimates: it can just incidentally be a part where both estimates coincide, like the lags between 80 and 100. The fact that estimates for neighbouring lags are strongly covariant because of (3.35) will often lead to rather regular but insignificant patterns in estimated lagged product correlation functions. It is not impossible that this effect causes a seeming periodicity as in Figure 3.9. The same example will be treated extensively in Section 10.1, where time series estimators will be applied to the same fish data.

It is evident that the autocorrelation function estimated from all observations together is not exactly the average of the two partial estimates in Figure 3.9. The difference is caused by the omission of lagged products around the middle of the observation interval. If one observation is in the first interval and the other in the second, the product contributes to the full length estimate and not to the two halves.

Dividing the observations into two halves and estimating two separate auto-correlation functions has been applied to many practical data as well as to simulated data where the true autocorrelation function is known. In many examples, the difference between the two halves is disappointingly great. That can be expected, given the inaccuracies in Figure 3.6. The differences (or the similarity) are for many reasons. It can be caused by a nonstationary signal, it can be the statistical inaccuracy given by (3.37), or there may be an outlier in the real-life data. Of course, taking more observations will improve the accuracy and also the similarity between estimates from two halves of the data.

3.6 Periodogram Estimation

It is attractive to base an estimator of the power spectral density function on the definition (3.18) of $h(\omega)$:

$$h(\omega) = \lim_{N\to\infty} E\left\{ \frac{1}{2\pi N} \left| \sum_{n=1}^{N} x_n e^{-j\omega n} \right|^2 \right\} = \frac{1}{2\pi} \sum_{k=-\infty}^{\infty} r(k) e^{-j\omega k} \tag{3.41}$$

The sequence of taking the Fourier transform, the square, the average over N, the expectation, and the limit was essential. A fundamental problem in estimating spectra of stationary stochastic signals, however, is that the Fourier transform of a stationary stochastic process does not exist, at least not in the domain of ordinary mathematical functions (Priestley, 1981). This means that

$$\lim_{N\to\infty} \sum_{k=-N}^{N} |x_k| \tag{3.42}$$

fails to converge to any finite value.

Priestley (1981) introduced the subject of the spectral analysis of random signals with the treatment of periodic signals. It is quite logical to use the discrete Fourier transform as a basis for spectral analysis for those periodic signals. It has excellent properties if the periodicity of the signal fits exactly in the observation interval.

Without a proper statistical derivation, the periodogram $S(\omega)$ is also defined as a basic estimator for the spectrum of stationary stochastic processes:

$$S(\omega) = \frac{1}{N}\left|\sum_{n=1}^{N} x_n e^{-j\omega n}\right|^2 \tag{3.43}$$

With a derivation similar to that in (3.18), it follows that exactly the same result $S(\omega)$ can be found by transforming the biased estimate of the lagged product autocovariance estimate of the process:

$$S(\omega) = \sum_{k=-(N-1)}^{N-1} \hat{r}(k)\, e^{-j\omega k} \tag{3.44}$$

Note that this transform uses $2N - 1$ estimated autocovariances, obtained from N observations. The relation of the periodogram with the lagged product estimator to the autocovariance is a reason to call the periodogram a nonparametric spectral estimate. The periodogram is a continuous function of ω, defined for all frequencies. With FFT algorithms on the computer, however, the periodogram is often calculated only for N equidistant points in the frequency domain from zero until $(N-1)/N$ Hz:

$$S(2\pi\frac{k}{N}) = \frac{1}{N}\left|\sum_{n=1}^{N} x_n e^{-2j\pi k\frac{n}{N}}\right|^2 , \quad k = 0,1,\cdots,N-1 \tag{3.45}$$

The inverse Fourier transform of this N point periodogram

$$\vec{r}(k) = \sum_{n=0}^{N-1} S\left(\frac{2\pi n}{N}\right) e^{2j\pi k\frac{n}{N}} \tag{3.46}$$

also has length N and cannot contain the full two-sided covariance of length $2N-1$. It has an aliased combined value, consisting of two terms as defined in (3.30):

$$\vec{r}(k) = \hat{r}(k) + \hat{r}(N-k) \tag{3.47}$$

If the autocovariance function were shorter than $N/2$, this could be an acceptable estimate for the autocovariance function because the two contributions on the right-hand side do not overlap. However, many processes have an infinitely long autocovariance function, like a declining power series, and the inverse transform of the periodogram will already be aliased with the nonzero true autocovariance at lags greater than N. Moreover, due to estimation variance, the estimated covariance $\hat{r}(k)$ for lags k between $N/2$ and N will not have estimates equal to zero, even if the

true expectation of the autocovariance function were zero there. That is a consequence of the variance in (3.37) that does not vanish at high lags. The inverse
transform (3.46) will contain only $\hat{r}(k)$ at lag k if the periodogram of N observed
data points is computed for $2N - 1$ or more frequencies by adding at least $N - 1$
zeros to the observations prior to the Fourier transform. In that case, the inverse
transform is long enough to contain the whole two-sided autocovariance estimate;
for details, see Priestley (1981).

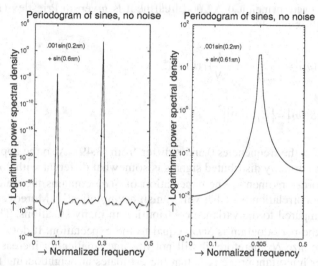

Figure 3.10. Estimation of the periodogram of $N = 100$ pure sinusoidal data without noise.
In the left-hand figure, the measurement interval contains exactly a whole number of periods
for both components of the signal; in the right-hand figure, the number of periods for the
highest frequency is 30.5.

The first example of a periodogram is given in Figure 3.10 for periodic
functions for which the fast Fourier transform is a powerful computational tool.
The periodogram from 100 observations is standard computed for 100 equidistant
frequencies, from 0.00 – 0.99 with steps of 0.01. Taking higher frequencies gives a
periodic continuation; at $f = 1$, the result for $f = 0$ returns. Furthermore, the result is
symmetrical around $f = 0.5$. Therefore, only the part with frequencies between 0
and 0.5 is shown; it can be mirrored to negative frequencies or also to higher
frequencies. The plot shows connecting lines, but only 51 points in this figure
follow from the computation. The left-hand figure gives the results for a sum of
two sinusoids that fit exactly in the interval with a number of complete periods.
The right-hand side has 30.5 periods of the stronger sine in the observation interval
as well as the same weaker component. The triangular bias of the periodogram
prevents the detection of the weak sine then.

Many tricks and procedures have been described to improve this behaviour for
periodic processes or for stochastic processes. That includes a data taper, which
multiplies the measured signal by some symmetrical bell-shaped function before it
is transformed. Another method is to truncate the autocorrelation function and to

multiply it by a lag window. The same result can be obtained with a spectral window convolved with the periodogram. Harris (1978) gives an extensive survey of periodic processes. More details are found in many places; see Priestley (1981), Kay (1988), Marple (1987), Kay and Marple (1981) and references therein. They are not treated here in any detail because time series methods give a better solution for most practical random data.

For the application of a raw periodogram to random data, the statistical properties are given here. The expectation and the variance of the periodogram can be derived from basic principles. A good treatment is given in Priestley (1981). The results are approximately

$$E[S(\omega)] = \sum_{k=-(N-1)}^{N-1} (1-\frac{k}{N}) r(k) e^{-j\omega k} \qquad (3.48)$$

$$\mathrm{var}[S(\omega)] = \{E[S(\omega)]\}^2 \qquad (3.49)$$

The variances at the frequencies 0 and π differ from (3.49). Also the approximation for other than normally distributed signals is somewhat different with contributions from higher order moments. The expectation of $S(\omega)$ contains the transform of a triangular autocorrelation bias that will vanish asymptotically. Moreover, it will be negligible compared to the variance contribution in many situations. The standard deviation of the periodogram is $S(\omega)$, equal to the expectation. It does not depend on the sample size N; therefore, it will not become smaller for increasing sample sizes. Another important property is that the estimates at neighbouring frequencies q/N and $(q+1)/N$ are almost completely independent. If the periodogram is computed at more than N frequencies, the range between the frequency points at q/N and $(q+1)/N$ can be considered an interpolation between those independent points; see Priestley (1981).

Figure 3.11 shows the estimated periodogram for 50 and 500 observations, on linear and logarithmic scales. Positive estimates for which the standard deviation is equal to the expectation will vary from almost zero to three or four times the expectation. The rough estimates show that the estimates at the frequencies q/N, q = 0, 1, 2, ... are independent. Taking 10 times more observations gives 10 times more independent estimates, which are all equally inaccurate. From the periodogram, it can be concluded that there is more power in the low-frequency range than in the higher range. However, it is impossible to conclude something about spectral details. If the periodogram at two neighbouring frequency points differs by a factor of 100, it can just be a consequence of the estimation uncertainty.

Autocovariance estimation is connected with spectral analysis by the Fourier transform. It is remarkable that raw periodograms have never been considered useful spectral estimates for random processes, whereas their Fourier transforms, the mean lagged products, are considered natural estimates for the autocovariance. The introduction of the FFT algorithm for fast Fourier transforms by Cooley and Tukey (1965) was a computational support for the nonparametric approach with periodograms and lagged products. From that moment on, the reduced computer

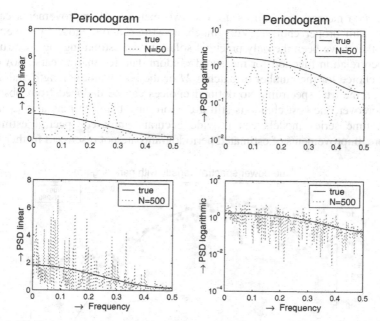

Figure 3.11. Estimation of the periodogram of $N = 50$ and 500 observations of a stationary stochastic process with the true spectrum $h(\omega) = 0.8$ $(1.25+\cos\omega)$. Taking more observations gives more independent estimates all of which have variances equal to the square of the expectation. In the logarithmic plots, the width of the inaccuracy band is a constant over the frequency range, which demonstrates that the standard deviation is proportional to the expectation.

effort enabled the routine Fourier analysis of extensive sets of random data by the nonparametric method. This was the reason that analysis with tapered and windowed periodograms has been the main practical tool for spectral analysis and autocovariance estimation for a long time.

Several improvements of this inaccurate raw periodogram of a random data estimate have been reported in the literature on spectral estimation; see Priestley (1981) and references:

- dividing the N observations in M batches, estimating the periodogram of each batch separately, and using the average of the M periodograms as the spectral estimator that contains N/M independent frequency points now.
- taking the average of M neighbouring points of the full periodogram, which gives the same estimate as above. This is called the Daniell window or rectangular spectral window.
- using only a small truncated part of the estimated autocorrelation function, say, less than 10% of the data length or shorter, multiply that remaining part by a bell-shaped lag window, and transform it to find an estimate for the spectrum.
- convolving the periodogram with a spectral window or smoothing it over a number of neighbouring frequencies.

The first two methods as well as the last two methods of improvements can be identical with proper choices. Very much is written about those improvements because they have been the only practical solutions for estimating the spectrum or the autocorrelation function for measured random data for many years. However, the best choice for the number of batches M or the best window length and shape depend on the true spectrum. No optimal choices can be deduced from observed data. Moreover, the best choice is still a compromise between bias and variance, whereas time series models can provide accurate unbiased spectral estimates. Therefore, improvements of estimated periodograms are not discussed further here.

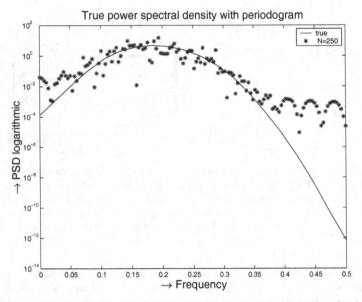

Figure 3.12. Estimation of the periodogram of the same 1000 observations that have been used to estimate the autocorrelation function in Figure 3.7. The triangular bias of Figure 3.4 is also found in the estimated periodogram.

Figure 3.12 shows the estimated periodogram from 250 observations of the process for which the autocorrelation function of Figure 3.7 has been estimated. The triangular bias that is present in the transform of the true correlation in Figure 3.4 is also present in the estimated spectrum. A remarkable feature in this figure is that the variance at high frequencies is almost invisible in the periodogram that looks smooth there. Above $f = 0.4$, the variance in the logarithmic plot is much smaller than the variance for lower frequencies, because the bias error is dominant. The variance of periodogram estimates is given by (3.49) with the true value of $E[S(\omega)]$, also for those frequencies. Hence, the standard deviation is much smaller than the bias in this frequency area.

One peculiar property of estimated autocovariances should be mentioned. The sum of all lagged product autocovariances equals the spectrum at zero frequency, which is identically zero if the mean of the data is subtracted. With subtraction of the mean, it follows that

$$\sum_{k=-(N-1)}^{N-1} \hat{r}(k) = S(0) = \frac{1}{N} \left| \sum_{n=1}^{N} \left(x_n - \frac{1}{N} \sum_{m=1}^{N} x_m \right) e^{-j0n} \right|^2 = 0 \qquad (3.50)$$

Suppose that the true covariance is positive for all shifts, *e.g.*, a power series. Then it is obvious that an estimate which adds up to zero must have a strange appearance in comparison with the true covariance function, as demonstrated in Figure 3.6.

3.7 Summary of Nonparametric Methods

The estimation variance of a raw periodogram is about the square of its expectation. The quality does not improve if more data are available. Only the number of frequencies at which an independent estimate can be obtained increases proportionally with the sample size. Therefore, the periodogram gives only limited information about the true power spectral density of stochastic data.

The periodogram has independent estimates at frequencies p/N and has a very irregular appearance. In contrast, the estimated autocovariance function will look very smooth with a high covariance between neighbouring lags. This is the reason that estimated lagged product autocovariance functions lead to an artificial and misleading confidence in completely accidental details. For stationary stochastic processes, the major problems in using periodograms for spectral analysis are easily recognized because of the rough appearance and the very irregular shape of estimates. The same problems should also be present in the associated lagged product autocovariance estimate, which is the inverse Fourier transform of the periodogram.

The autocovariance estimates for large lags can be seen as the cause of many problems in the accuracy of periodograms. Taking more observations introduces more lags in the estimated lagged product autocovariance function. The total length is $N - 1$ and it increases with sample size. This increased length is also the reason that the estimated periodogram contains more independent frequencies for more observations, all of which keep the same inaccuracy. A stationary stochastic process is no sum of harmonics. Trying to model it like that with a periodogram is erroneous.

Dividing $N - k$ lagged covariance products by N introduces a triangular bias in the autocovariance estimate that is used. Dividing by $N - k$ would produce unbiased autocovariance estimates which are not positive-semidefinite and could lead to negative spectral estimates.

A popular practical solution to diminish this bias problem is putting a taper over the data before transforming them. This leaves the middle 80% of the observations unchanged, but it multiplies the final 10% on both sides by a function going from 1 to 0, *e.g.*, a half raised cosine. This diminishes the bias at the cost of distorting the data. Because the finite Fourier transform treats stochastic data as one period of a periodic signal, the taper also removes the artifact of the apparent jump by treating the first and the last observations as if they were two contiguous observations.

The introduction of a multitaper approximation by Percival and Walden (1993) does not give a satisfactory solution. Broersen (1998) used simulations to show that the quality of this approach is not better than that obtained with traditional windows.

Summarizing, no proper solutions have been found for estimating accurate spectra or autocovariance functions for stationary stochastic observations with periodogram analysis. Moreover, the choice of the taper and/or the window depends on the experimenter. It has been practice to judge results by eye because no firm statistical rules can be given independently of the measured data. It is very easy to generate data synthetically as examples where judging by eye is misleading and leads to wrong conclusions. Periodograms are an "easy, quick, and dirty" method of spectral estimation for stochastic processes. Likewise, lagged product autocorrelation estimates are "easy, quick, and dirty." The popularity of periodograms for spectra and lagged product estimators for the autocovariance might largely be that this has been the only algorithm available to evaluate practical data on the computer for a long time. Other solutions were not feasible for routine computation of spectra and correlations of stationary stochastic observations without further detailed knowledge about the properties of the data.

Of course, periodograms remain an excellent solution for spectral estimation if the observations are periodic or if they consist of a sine wave with very little additive noise. The frequency content of deterministic measurements can also be studied with the periodogram. Furthermore, integrated periodograms over frequency subbands give a rough idea of the distribution of power over different frequency ranges.

Finally, the variance of autocorrelations diminishes with $1/N$ for increasing sample sizes. If the process is stationary and if enough observations can be made, the estimate can have any desired accuracy for every individual lag.

3.8 Exercises

3.1   Prove that the autocorrelation function of a stationary stochastic process is symmetrical around zero lag.

3.2   Prove that the power spectral density function of a stationary stochastic process is symmetrical around zero frequency.

3.3   Prove the transition from the second to the third line of Equation 3.18.

3.4   Suppose we would like to predict a single stationary time series x_n with zero mean and autocovariance function $r(k)$ at some future time $n+k$, with $k > 0$. We use only the present value x_n and an unknown constant α. Show that the mean square prediction error $\text{MSE}(\alpha) = \text{E}(x_{n+k} - \alpha x_n)^2$ is minimised by the value $\alpha = \rho(k)$.

3.5   The autocovariance function $r(k)$ of a stationary stochastic process is only nonzero for the lags 0, 1, and − 1. It is characterized by $r(0) = 5$ and $r(1) = 2$. N observations of the process are available to estimate the autocovariance function with a positive-semidefinite lagged product estimator.
Calculate the asymptotic approximations for the variances of the estimators

$$\hat{r}(0),\ \hat{r}(1),\ \hat{r}(2),\ \hat{r}(10).$$

Calculate the asymptotic approximations for the covariances between the autocovariance estimators for different lags:

$$\text{cov}\big[\hat{r}(0),\hat{r}(1)\big],\ \text{cov}\big[\hat{r}(0),\hat{r}(2)\big],\ \text{cov}\big[\hat{r}(0),\hat{r}(10)\big].$$

Calculate the asymptotic approximations for the variances of the estimators

$$\hat{\rho}(0),\ \hat{\rho}(1),\ \hat{\rho}(2),\ \hat{\rho}(10).$$

3.6   Show that the autocovariance function $r(k)$ of a stationary stochastic process with mean μ is given by $r(k) = E[x_n x_{n+k}] - \mu^2$.

3.7   Derive Equation (3.33). Why is the bias term with k/N present only in the first part of the expectation?

3.8   Calculate the variance of the mean if that is estimated from N observations and if the autocovariance function of the data is given by

$$r(k) = \sigma_x^2 a^{|k|},\ \forall k.$$

3.9   Calculate the variance of the variance that can be estimated from N observations if the autocovariance function of the data is given by

$$r(k) = \sigma_x^2 a^{|k|},\ \forall k.$$

4

ARMA Theory

4.1 Time Series Models

A major problem of nonparametric estimation of the lagged product autocovariance function and the periodogram is that the size of the estimated function increases with the number of observations. The periodogram gets more independent estimates in the frequency region from $0 - 0.5$, and the autocovariance function becomes longer. The nonzero part of the estimated lagged product autocovariance function is always equal to the sample size. The first lags of the estimated autocorrelation function become more accurate for increasing sample size. But at the high lag end, the estimates for the final lags until $N - 1$ are always very inaccurate. Furthermore, it is not clear how to cut off the autocorrelation function beyond an objectively chosen limit. Neither is it clear in how many independent batches the data should be divided to find an optimal average periodogram or, equivalently, how many neighbouring frequencies in the full data periodogram have to be averaged for an optimal result.

It would be interesting to estimate functions from the data that become more accurate only if more observations are available and do not grow in length. Therefore, several classes of functions or models will be defined. It is the intention that a model from one of those classes can be assigned to every possible stationary stochastic process. The classes that will be defined are

- white noise
- autoregressive process, known as AR
- moving average process, known as MA
- autoregressive moving average process, known as ARMA.

The words "process" and "model" are often used to describe more or less the same system. Process is generally used to describe the actual process of generating the data, which are also called the observations or signal. Each stationary stochastic data set has one generating process. The word model is more associated with the estimation based on given observations. Different types of models can be calculated for measured data, for which the generating process might be unknown. In the literature, the difference between the words process and model is not always used very strictly. This does not lead to much confusion because the actual meaning follows from the context. This chapter describes how those models can describe the autocorrelation function and the power spectral density.

4.2 White Noise

The white noise process type is the basic element of time series analysis in practice. It is often called a *purely random process*. It is defined as a sequence of uncorrelated random variables. It may or may not be stationary, independent, or normally distributed. The observed variable x_n is modeled as a sequence of uncorrelated random variables ε_n with the mean and the variance as a function of n if the process is not stationary:

$$x_n = \varepsilon_n \; . \tag{4.1}$$

For stationary (or wide sense stationary) processes, the mean μ_ε and the variance σ_ε^2 of ε_n and x_n are constants. The autocorrelation function $\rho(k)$ is given by

$$\begin{aligned}\rho(k) &= 1, \quad k = 0 \\ &= 0, \quad k \neq 0\end{aligned} \tag{4.2}$$

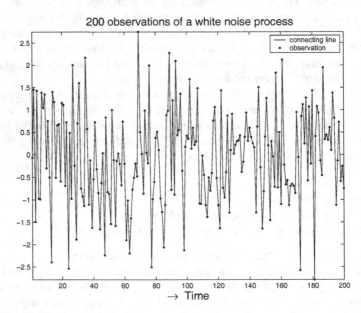

Figure 4.1. Realisation of 200 white noise observations. In many figures, connecting lines are drawn without warning, but all discrete time signals are defined only for $t = 0, 1, 2,...$ The dots represent the actual observations; the lines have no significance for discrete time signals.

4.3 Moving Average Processes

A moving average process of order q, a MA(q) process, is a weighted combination of $q+1$ shifted or lagged white noise data:

$$x_n = \varepsilon_n + b_1\varepsilon_{n-1} + \cdots + b_q\varepsilon_{n-q} \; . \tag{4.3}$$

The first parameter b_0 is generally taken as one and ε_n is a stationary, purely random process with mean μ_ε and variance σ_ε^2. In a more concise way, the MA equation can also be written as

$$x_n = B(z)\,\varepsilon_n \tag{4.4}$$

with

$$B(z) = 1 + b_1 z^{-1} + \cdots + b_q z^{-q} \tag{4.5}$$

and z^{-1} is the backward difference operator defined by

$$z^{-1}\varepsilon_n = \varepsilon_{n-1}$$
$$z^{-k}\varepsilon_n = \varepsilon_{n-k}$$
$$z\varepsilon_n = \varepsilon_{n+1} \tag{4.6}$$

The roots of $B(z)$ are called the zeros of the MA model. An MA process is called invertible if the zeros are inside the unit circle. The expression $B(z)$ is a transform in the time domain with q backward shifts in this example.
   The expectation of x_n is

$$\begin{aligned} E[x_n] &= E[\varepsilon_n + b_1\varepsilon_{n-1} + \cdots + b_q\varepsilon_{n-q}] \\ &= E[\varepsilon_n] + b_1 E[\varepsilon_{n-1}] + \cdots + b_q E[\varepsilon_{n-q}] \\ &= \mu_\varepsilon + b_1\mu_\varepsilon + \cdots + b_q\mu_\varepsilon = \mu_\varepsilon \sum_{i=0}^{q} b_i \end{aligned} \tag{4.7}$$

It is usual to assume that the mean μ_ε equals zero. The reason is that it is a good practice to subtract the average of random signals before further examination takes place. It would be quite useless to mix models for long-term average temperature or pressure with models for quick variations. Now, assuming $\mu_\varepsilon = 0$, it can be derived that

$$r(k) = \sigma_x^2 \rho(k) = E[x_n x_{n+k}]$$
$$= E[\{\varepsilon_n + b_1\varepsilon_{n-1} + \cdots + b_q\varepsilon_{n-q}\}\{\varepsilon_{n+k} + b_1\varepsilon_{n+k-1} + \cdots + b_q\varepsilon_{n+k-q}\}]$$
$$= \sigma_\varepsilon^2 \sum_{i=0}^{q-k} b_i b_{i+k}, \quad 0 \le k \le q$$
$$= 0, \qquad\qquad k > q \tag{4.8}$$

$$r_x(k) = r_x(-k), \qquad k < 0 \tag{4.9}$$

It follows elementarily that

$$\sigma_x^2 = r(0) = E[x_n x_n]$$
$$= E\left[\left(\varepsilon_n + b_1\varepsilon_{n-1} + \cdots + b_q\varepsilon_{n-q}\right)\left(\varepsilon_n + b_1\varepsilon_{n-1} + \cdots + b_q\varepsilon_{n-q}\right)\right]$$
$$= \sigma_\varepsilon^2 \sum_{i=0}^{q} b_i^2 \tag{4.10}$$

The normalized autocorrelation function is given by

$$\rho(k) = E[x_n x_{n+k}]/\sigma_x^2$$
$$= \sum_{i=0}^{q-k} b_i b_{i+k} \Big/ \sum_{i=0}^{k} b_i^2, \quad 0 \le k \le q$$
$$= 0, \qquad\qquad k > q \tag{4.11}$$

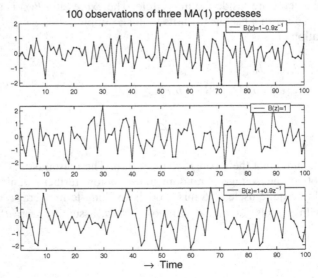

Figure 4.2. Two realisations of 100 MA(1) observations and a white noise signal, showing the difference between a negative correlation above and a positive correlation below

The autocorrelation function is zero for $k > q$. Hence, the MA model is for finite length correlations. An example of the autocorrelation function of a MA(13) process has been given in Figure 3.2. With the sequence ε_n as an input signal, a MA model can be seen as a filtered white noise signal with a finite impulse response, or FIR filter. Figure 4.2 gives some observations of MA processes. A negative correlation for lag 1 gives mostly an opposite sign for the next observation, whereas a positive correlation will more often give the same sign.

4.3.1 MA(1) Processes with Zero Outside the Unit Circle

The *autocorrelation* function remains the same the processes with parameter b and parameter $1/b$. This is demonstrated by an example of a MA(1) process with parameter b and $1/b$. The parameter b gives as autocorrelation for lag 1

$$x_n = \varepsilon_n + b\varepsilon_{n-1}$$
$$r_x(0) = \sigma_\varepsilon^2(1+b^2), \ \ r_x(1) = \sigma_\varepsilon^2(b)$$
$$\rho_x(1) = b/(1+b^2)$$

The parameter $1/b$ gives different outcomes for the autocovariance function because the variance is different. However, the autocorrelation function is the same:

$$y_n = \varepsilon_n + b^{-1}\varepsilon_{n-1}$$
$$r_y(0) = \sigma_\varepsilon^2(1+b^{-2}), \ \ r_y(1) = \sigma_\varepsilon^2(b^{-1})$$
$$\rho_y(1) = b^{-1}/(1+b^{-2}) = b/(1+b^2) \tag{4.12}$$

For a given MA(1) autocorrelation function, it is impossible to decide whether it belongs to an invertible process with the parameter inside the unit circle or to a process with a parameter greater than one. The value of $\rho(1)$ is limited by ± 0.5 for a MA(1) process, as can be derived from (4.12). Otherwise, the autocorrelation function with only zeros beyond lag 1 would not be positive-semidefinite. The limiting value is attained if the parameter b is equal to $+1$ or -1.

Furthermore, it can be proved that mirroring a zero of a MA(q) process with respect to the unit circle leaves the complete autocorrelation function undisturbed. Therefore, the autocorrelation function of a MA(q) process has 2^q different possible combinations of MA(q) models, with each zero inside or outside the unit circle. The invertible MA(q) model is the only one with all zeros inside the unit circle.

4.4 Autoregressive Processes

In deterministic filtering, the complement of FIR filters is IIR filters with infinite impulse response. In difference equations, MA models have only a right-hand side part. Models with only a left-hand side or with both sides can also be constructed. An autoregressive process of order p, or briefly an AR(p) process, is defined as

$$x_n + a_1 x_{n-1} + \cdots a_p x_{n-p} = \varepsilon_n \qquad (4.13)$$

The first parameter a_0 is generally taken as one and ε_n is a purely random stationary process with mean μ_ε and variance σ_ε^2. Taking the first parameters a_0 and b_0 equal to one is no loss of generality. The scale factor is manipulated with the variance of the random input. This gives enough freedom because multiplying all parameters including a_0 by a factor c and taking $c^2 \sigma_\varepsilon^2$ as the variance of ε_n gives exactly the same signal x_n. More concisely, the equation can also be written as

$$A(z) x_n = \varepsilon_n \qquad (4.14)$$

with

$$A(z) = 1 + a_1 z^{-1} + \cdots a_p z^{-p} \qquad (4.15)$$

The roots of $A(z)$ are called the poles of the AR(p) processes. Processes are called stationary if all poles are within the unit circle.

4.4.1 AR(1) Processes

It is a good idea to start with a derivation of the properties of AR processes for an AR(1) process:

$$x_n + a x_{n-1} = \varepsilon_n \qquad (4.16)$$

where the index 1 for the first and only parameter a has been left out in this first-order section. The equation can be rewritten as

$$x_n = -a x_{n-1} + \varepsilon_n \qquad (4.17)$$

In this equation, we can substitute x_{n-1} by its AR(1) model and so on, yielding

$$
\begin{aligned}
x_n &= \varepsilon_n - a x_{n-1} \\
&= \varepsilon_n - a(\varepsilon_{n-1} - a x_{n-2}) \\
&= \varepsilon_n - a \varepsilon_{n-1} + (-a)^2 \varepsilon_{n-2} + \cdots + (-a)^{n-1}(\varepsilon_1 - a x_0) .
\end{aligned}
\qquad (4.18)
$$

Using the $A(z)$ notation of (4.14) in computations gives the same result by considering the AR polynomial as the sum of an infinite series

$$x_n = \frac{1}{A(z^{-1})}\varepsilon_n = \frac{1}{1+az^{-1}}\varepsilon_n$$

$$= [1+(-az^{-1})+(-az^{-1})^2+\cdots+(-az^{-1})^N+\cdots]\varepsilon_n$$

$$= \varepsilon_n - a\varepsilon_{n-1}+(-a)^2\varepsilon_{n-2}+\cdots+(-a)^N(\varepsilon_{n-N})+\cdots \qquad (4.19)$$

Now, suppose that the initial value x_0 equals zero and that the signal was also zero for negative times. In fact, this is artificial because it is in contrast with the notion of stationary stochastic processes. However, this will be interpreted so that the difference from the present time n is so large that the initial values have been forgotten at time n. It is shown with (4.19) that the AR observation at time n can be decomposed and becomes a linear combination of all previous input variables.

The expectation of x_n becomes

$$E[x_n] = E[\varepsilon_n - a\varepsilon_{n-1}+(-a)^2\varepsilon_{n-2}+\cdots+(-a)^{n-1}(\varepsilon_1-ax_0)]$$

$$= \mu_\varepsilon\{1-a+(-a)^2+\cdots+(-a)^{n-1}\}$$

$$= \mu_\varepsilon\frac{1-(-a)^n}{1+a} \quad \text{for } a \neq 1. \qquad (4.20)$$

If $\mu_\varepsilon = 0$, then $E[x_n] = 0$ for all n and x_n is stationary up to order one. If μ_ε is not zero, then

$$E[x_n] \approx \mu_\varepsilon\frac{1}{1+a} \qquad (4.21)$$

for large n. This is mathematically denoted as x_n being asymptotically stationary to order one. A straightforward derivation of the mean of a stationary AR(1) process is given by

$$E[x_n]+aE[x_{n-1}] = E[\varepsilon_n]$$

$$E[x_n]+aE[x_n] = \mu_\varepsilon$$

$$E[x_n] = \frac{\mu_\varepsilon}{1+a}. \qquad (4.22)$$

In this derivation, the stationary property means that the expectation of x_{n-i} does not depend on the time index n and can be written as the expectation of x_n. Figure 4.3 gives some realisations of AR(1) processes with moderate and strongly positive and negative correlations. If the correlation is strongly positive, as in the left-hand figures, it is possible that the signal remains above or below the long-term average for a long time. Analysing a very short signal can give odd results then. Looking only at the observations from 15 to 45 in the lower left figure would suggest variations around an average of -1.50. Therefore, a stochastic signal should be much longer than the correlation length to obtain a reliable impression of the character of the signal.

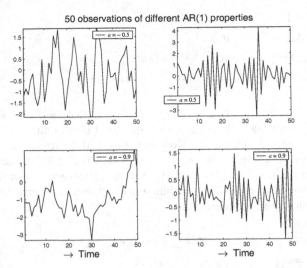

Figure 4.3. Four realisations of 50 AR(1) observations for different parameter values of a

From now on, assume that $\mu_\varepsilon = 0$ to simplify derivations. Then ε_n is a stochastic variable with mean $\mu_\varepsilon = 0$, variance σ_ε^2, and autocovariance $\mathrm{cov}(\varepsilon_n, \varepsilon_m) = 0$ for $m \neq n$. The variance of the signal x_n becomes

$$\mathrm{var}[x_n] = \sigma_x^2 = E[\varepsilon_n - a\varepsilon_{n-1} + (-a)^2 \varepsilon_{n-2} + \cdots + (-a)^{n-1} \varepsilon_1]^2$$
$$= \sigma_\varepsilon^2 \{1 + a^2 + a^4 + \cdots + a^{2n-2}\}$$
$$= \sigma_\varepsilon^2 \frac{1 - a^{2n}}{1 - a^2} \quad for\ a \neq 1.$$

For $|a| < 1$, this can be simplified to the asymptotic expression

$$\mathrm{var}[x_n] = \sigma_x^2 = \frac{\sigma_\varepsilon^2}{1 - a^2} \quad for\ |a| < 1. \tag{4.23}$$

Finally, multiplying x_n by x_{n+k} and taking the expectation, one can derive the autocovariance function:

$$r(k) = E[x_n x_{n+k}]$$
$$= E\left\{ \left[\varepsilon_n - a\varepsilon_{n-1} + \cdots + (-a)^{n-1} \varepsilon_1 \right] \times \left[\varepsilon_{n+k} - a\varepsilon_{n+k-1} + \cdots + (-a)^{n+k-1} \varepsilon_1 \right] \right\}$$

$$r(k) = \sigma_\varepsilon^2 \left[(-a)^k + (-a)^{k+2} + \cdots + (-a)^{k+2n-2} \right]$$
$$\approx \frac{\sigma_\varepsilon^2}{1 - a^2} (-a)^k, \quad for\ k \geq 0. \tag{4.24}$$

For negative k,

$$r(k) = r(-k) \qquad (4.25)$$

The same result can be derived by multiplying the AR(1) equation by x_{n-k} and taking the expectations. First, multiplying (4.16) by x_n and x_{n-1} gives

$$E[x_n x_n] + aE[x_{n-1} x_n] = E[\varepsilon_n x_n]$$
$$E[x_n x_{n-1}] + aE[x_{n-1} x_{n-1}] = E[c_n x_{n-1}] . \qquad (4.26)$$

The present value of x_n is a linear combination of the present value of ε_n plus previous input values ε_{n-k}. Moreover, the ε_n are uncorrelated, so only multiplication of ε_{n-i} by itself will contribute to the expectation. The expectations in the equations become the true autocovariances. Hence,

$$\sigma_x^2 + a\rho(1)\sigma_x^2 = \sigma_\varepsilon^2$$
$$\rho(1)\sigma_x^2 + a\sigma_x^2 = 0 \qquad (4.27)$$

for which the solution is

$$\rho(1) = -a$$
$$\sigma_x^2 = \frac{\sigma_\varepsilon^2}{1-a^2} \qquad (4.28)$$

Further,

$$E[x_n x_{n-k}] + aE[x_{n-1} x_{n-k}] = E[\varepsilon_n x_{n-k}]$$

$$\rho(k) + a\rho(k-1) = 0, \qquad \text{for } k \geq 1$$
$$\rho(k) = -a\rho(k-1) = (-a)^2 \rho(k-2) = \cdots = (-a)^k \rho(0) = (-a)^k$$
$$\rho(|k|) = (-a)^{|k|} \qquad (4.29)$$

It turns out that the autocorrelation function is a power series with its pole as the root of $1 + az^{-1} = 0$ or of $z + a = 0$. For $|a| < 1$, the root is within the unit circle and the autocorrelation is a decaying sequence of infinite length. Figure 4.4 gives two decaying power series as autocorrelation functions. Where finite-order MA processes have a finite correlation length, the AR correlation never dies out completely. The single value of the AR(1) parameter defines the whole autocorrelation function. Autoregressive models are an important class of time series models. Therefore, some specific properties will be derived.

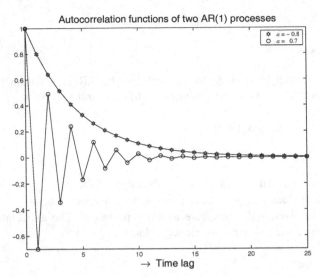

Figure 4.4. Autocorrelation functions of two AR(1) processes, with positive and negative parameter a

4.4.2 AR(1) Processes with a Pole Outside the Unit Circle

The question what happens with poles outside the unit circle is simple. If one tries to generate such a signal, the values will explode. The process is not stationary. Mathematically, an interesting description is possible if the direction of time is reversed. This will be illustrated for an AR(1) process with $|a| > 1$.

$$x_n + ax_{n-1} = \varepsilon_n \tag{4.30}$$

This can be rewritten formally as

$$\frac{x_{n+1}}{a} + x_n = \frac{\varepsilon_{n+1}}{a} \tag{4.31}$$

and it is easily seen that x_n can be written as a convergent sum of future excitations with parameter $1/a$

$$x_n = \frac{\varepsilon_{n+1}}{a} - \frac{x_{n+1}}{a} = \frac{\varepsilon_{n+1}}{a} - \frac{\varepsilon_{n+2}}{a^2} + \cdots \frac{(-1)^{k+1}\varepsilon_{n+k}}{a^k} + \cdots \tag{4.32}$$

The variance becomes

$$\text{var}[x_n] = \sigma_\varepsilon^2 \left(\frac{1}{a^2} + \frac{1}{a^4} + \frac{1}{a^6} + \cdots \right)$$

$$= \frac{\sigma_\varepsilon^2}{a^2} \frac{1}{1 - 1/a^2} = \frac{\sigma_\varepsilon^2}{a^2 - 1} \tag{4.33}$$

Hence, the variance of this signal has an expression other than (4.28) for the AR(1) process with parameter inside the unit circle. The autocorrelation follows as

$$\rho(1) = \frac{r(1)}{\sigma_x^2} = \frac{E[x_n x_{n+1}]}{\sigma_x^2}$$

$$= \frac{1}{\sigma_x^2} E \left\{ \left[\frac{\varepsilon_{n+1}}{a} - \frac{\varepsilon_{n+2}}{a^2} + \cdots \frac{(-1)^{i+1}\varepsilon_{n+i}}{a^i} - \cdots \right] \right.$$

$$\left. \times \left[\frac{\varepsilon_{n+2}}{a} - \frac{\varepsilon_{n+3}}{a^2} + \cdots \frac{(-1)^i \varepsilon_{n+1+i}}{a^i} - \cdots \right] \right\}$$

$$= (a^2 - 1) \left(-\frac{1}{a^3} - \frac{1}{a^5} - \frac{1}{a^7} - \cdots \right)$$

$$= -\frac{a^2 - 1}{a^3} \frac{1}{1 - 1/a^2} = -\frac{1}{a} \tag{4.34}$$

The autocorrelation function is the same as the autocorrelation of the process with the reciprocal root inside the unit circle. Mirroring a root with respect to the unit circle has influence on the variance but not on the normalized autocorrelation function.

This mathematical exercise has no practical meaning, but it may assist in understanding why reversing the time index in a signal will have no consequences for the autocorrelation function, the spectrum, and the time series model that describe it. Reversing the time or taking backward residuals will play a role in several well-known AR estimation algorithms.

4.4.3 AR(2) Processes

An AR(2) process is given by

$$x_n + a_1 x_{n-1} + a_2 x_{n-2} = \varepsilon_n \tag{4.35}$$

The expectation is found with $E[x_n](1 + a_1 + a_2) = \mu_\varepsilon$, and it will be assumed that $\mu_\varepsilon = 0$; this can always be realised in practice by subtracting the sample mean from the data. The AR polynomial is

$$A(z) = 1 + a_1 z^{-1} + a_2 z^{-2} \tag{4.36}$$

The roots of $A(z) = 0$ are the poles of this process.

$$x_n(1 + a_1 z^{-1} + a_2 z^{-2}) = \varepsilon_n$$
$$x_n(1 - p_1 z^{-1})(1 - p_2 z^{-1}) = \varepsilon_n \qquad (4.37)$$

where $p_{1,2}$ are the solutions of

$$z^2 + a_1 z + a_2 = 0 \qquad (4.38)$$

Hence, the process can be written as

$$x_n = \frac{1}{\left(1 - p_1 z^{-1}\right)\left(1 - p_2 z^{-1}\right)} \varepsilon_n$$

$$= \frac{1}{p_1 - p_2}\left(\frac{p_1}{1 - p_1 z^{-1}} - \frac{p_2}{1 - p_2 z^{-1}}\right)\varepsilon_n \qquad (4.39)$$

This can be written as an infinite summation

$$x_n = \frac{1}{p_1 - p_2}\left\{ p_1\left[1 + \sum_{k=1}^{\infty}\left(p_1 z^{-1}\right)^k\right] - p_2\left[1 + \sum_{k=1}^{\infty}\left(p_2 z^{-1}\right)^k\right]\right\} \varepsilon_n$$

$$= \sum_{k=0}^{\infty}\left(\frac{p_1^{k+1} - p_2^{k+1}}{p_1 - p_2}\right)\varepsilon_{n-k} \cdot \qquad (4.40)$$

Clearly, the process is asymptotically stationary if both poles are less then one in absolute value, hence if the poles are located inside the unit circle. This equation also shows how an AR(2) process can be written explicitly as an infinite sum of present and previous excitations with one as coefficient for the present excitation. Therefore,

$$x_n = \sum_{i=0}^{\infty} \gamma_i \varepsilon_{n-i}, \ \gamma_0 = 1 \qquad (4.41)$$

This can be interpreted as a MA(∞) process. It also follows that

$$E[\varepsilon_n x_n] = \sigma_\varepsilon^2$$
$$E[\varepsilon_n x_{n-k}] = 0, \ k \geq 1 \qquad (4.42)$$

Multiplying both sides of the AR(2) equation (4.35) by x_{n-k} and taking expectations gives

$$E[x_n x_{n-k}] + a_1 E[x_{n-1} x_{n-k}] + a_2 E[x_{n-2} x_{n-k}] = E[\varepsilon_n x_{n-k}] \cdot \qquad (4.43)$$

This can be solved easily for $k \geq 0$. Using negative values for k in this equation becomes more cumbersome because the right-hand side does not vanish then. Cross correlations are not symmetrical around zero lag. Eventually, the outcome should be the same, with more manipulation. For $k = 0$, this equation becomes

$$\sigma_x^2 \left[1 + a_1 \rho(1) + a_2 \rho(2) \right] = \sigma_\varepsilon^2 \tag{4.44}$$

For $k = 1$ and 2,

$$\rho(1) + a_1 + a_2 \rho(1) = 0 \tag{4.45}$$

$$\rho(2) + a_1 \rho(1) + a_2 \rho(0) = 0 \tag{4.46}$$

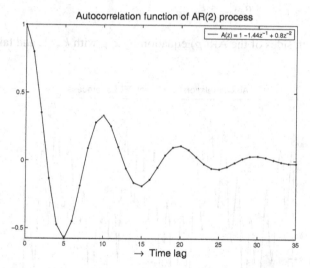

Figure 4.5. Autocorrelation function of AR(2) process, complex poles $p_{1,2}$ at $0.72 \pm j \, 0.53$

Solving those three equations gives

$$\rho(1) = -\frac{a_1}{1 + a_2}$$

$$\rho(2) = \frac{a_1^2}{1 + a_2} - a_2$$

$$\sigma_x^2 = \frac{(1 + a_2)\sigma^2}{(1 - a_2)(1 - a_1 + a_2)(1 + a_1 + a_2)} \tag{4.47}$$

For k arbitrary but greater than one

$$\rho(k) + a_1 \rho(k-1) + a_2 \rho(k-2) = 0 \tag{4.48}$$

The solution is

$$\rho(k) = C_1 p_1^k + C_2 p_2^k, \; k \geq 0 \tag{4.49}$$

where $p_{1,2}$ are the poles of (4.37) and C_1 and C_2 are constants whose values are determined by the boundary conditions $\rho(0)$ and $\rho(1)$. With real AR parameters, the poles can both be real or they can be complex conjugated. Figure 4.5 gives an example of an AR(2) autocorrelation function with complex poles.

4.4.4 AR(p) Processes

An AR(p) process is given by

$$x_n + a_1 x_{n-1} + \cdots + a_p x_{n-p} = \varepsilon_n \tag{4.50}$$

Multiplying both sides of the AR(p) equation by x_{n-k} with $k \geq 1$, and taking expectations gives

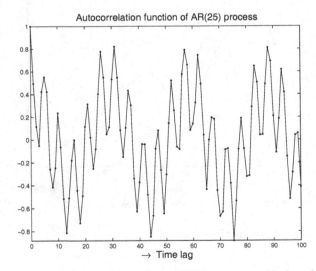

Figure 4.6. Autocorrelation function of an AR(25) process with 3 real poles and 11 complex pole pairs. Regularly shaped as well as quite irregular autocorrelation functions are possible.

$$\rho(k) + a_1 \rho(k-1) + \cdots + a_p \rho(k-p) = 0 \tag{4.51}$$

whose solution is

$$\rho(k) = C_1 p_1^k + C_2 p_2^k + \cdots C_p p_p^k, \; k \geq 0 \tag{4.52}$$

where p_i are the poles and C_i are constants whose values are determined by the boundary conditions which are determined by $\rho(0)$, $\rho(1)$,..., $\rho(p-1)$. In the same way as for the AR(2) process, an AR(p) process can be expressed as a MA(∞) process, with sums of power series of the poles as coefficients, if all poles are within the unit circle. The autocorrelation can then be expressed as a function of the autocorrelation of infinitely many MA parameters, as in a MA(q) process. The autocorrelation has infinite length, but it dies out when all poles are inside the unit circle. Finally, the pole or complex conjugated pair of poles with the largest radius will dominate the autocorrelation for larger time lags. Figures 4.6 and 4.7 show that the autocorrelation function of higher order AR processes can have a regular as well as a rather irregular shape.

In terms of conditional probability density functions, (4.50) can be expressed as

$$f(x_n \mid x_{n-1}, x_{n-2}, \cdots, x_1) = f(x_n \mid x_{n-1}, x_{n-2}, \cdots, x_{n-p})$$

In words, the present observation of an AR(p) process depends only on its p preceding observations, not on how those preceding observations originated.

Furthermore, it can be proved that mirroring a pole of an AR process with respect to the unit circle leaves the complete autocorrelation function undisturbed. Therefore, an AR(p) process has, in principle, 2^p different combinations of poles that would produce exactly the same autocorrelation function. A stationary AR(p) process is a unique process with that autocorrelation function and with all poles inside the unit circle. The autocorrelation function is insensitive for mirroring a pole. The autocovariance function, however, includes the variance of the process and will not remain unchanged by mirroring poles or zeros with respect to the unit circle.

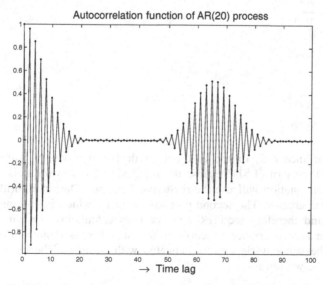

Figure 4.7. Regularly shaped autocorrelation function of AR(20) process, 10 complex conjugate pole pairs with radius 0.99

4.5 ARMA(p,q) Processes

The generating equation of an ARMA(p,q) process is

$$x_n + a_1 x_{n-1} + \cdots + a_p x_{n-p} = \varepsilon_n + b_1 \varepsilon_{n-1} + \cdots + b_q \varepsilon_{n-q} \tag{4.53}$$

with shorthand notation

$$A(z)x_n = B(z)\varepsilon_n \tag{4.54}$$

This process has p poles and q zeros. A simple method for extracting results for ARMA processes comes from linear filter theory. An ARMA(p,q) process is equivalent to a series of an AR(p) and a MA(q) process. An AR(p) process with white noise as input is followed by a MA(q) process. The linear combination of correlated variables in the MA process gives no problem in calculating the autocorrelation function.

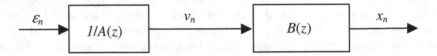

Figure 4.8. Representation of an ARMA process as a series of an AR and a MA process

The formulas for the representation in Figure 4.8 are

$$x_n = \frac{B(z)}{A(z)} \varepsilon_n \tag{4.55}$$

which gives

$$v_n = \frac{1}{A(z)} \varepsilon_n$$
$$x_n = B(z)v_n \tag{4.56}$$

The autocovariance $r_v(k)$ of the first part with $1/A(z)$ is easily found with the standard AR theory of (4.51) because the input signal is a white noise sequence. The actual computation will use the recursive Levinson-Durbin algorithm described in the next chapter. The second part has input v_n, which is no longer a white noise input and therefore requires some care in calculating the autocovariance function. The autocovariance function of the signal x_n is denoted $r(k)$ without index. All other autocovariances are indicated with an index. The second equation in (4.56) can be written as

$$x_n = B(z)\, v_n = v_n + b_1 v_{n-1} + \cdots + b_q v_{n-q} \tag{4.57}$$

For $r(k)$,

$$r(k) = E[x_n x_{n-k}]$$
$$= E\left[\left(v_n + b_1 v_{n-1} + \cdots + b_q v_{n-q}\right)\left(v_{n-k} + b_1 v_{n-k-1} + \cdots + b_q v_{n-k-q}\right)\right] \qquad (4.58)$$

This can further be written as

$$r(k) = \begin{pmatrix} 1 & b_1 & \cdots & b_q \end{pmatrix} \begin{pmatrix} r_v(k) & r_v(k+1) & \cdots r_v(k+q) \\ r_v(k-1) & r_v(k) & \\ \vdots & & \vdots & \ddots \\ r_v(k-q) & r_v(k-q+1) & & r_v(k) \end{pmatrix} \begin{pmatrix} 1 \\ b_1 \\ \vdots \\ b_q \end{pmatrix} \qquad (4.59)$$

or in a more compact style as

$$r(k) = \sum_{m=-q}^{q} \left[r_v\left(k+m\right) \sum_{i=0}^{q} b_i b_{i+|m|} \right], \quad \forall k \qquad (4.60)$$

The whole autocovariance function $r(k)$ for an interval of lag k can be computed conveniently as a convolution of the autoregressive autocovariance function $r_v(k)$ with the summation of products of the MA parameters, which is a normalized version of the autocovariance function of the MA part in (4.8). The parameters b_i are taken as zero beyond the interval $0 \leq i \leq q$. With this arrangement, the definition

$$r_{MA}(k) = \sum_{i=0}^{q} b_i b_{i+|k|}, \quad -q \leq k \leq q \qquad (4.61)$$

can be given, and the ARMA autocovariance function can be written compactly as

$$r(k) = \sum_{m=-q}^{q} \left[r_v\left(k+m\right) r_{MA}(m) \right], \quad \forall k. \qquad (4.62)$$

The ARMA autocovariance function is a convolution of the separate autocovariances of the AR and the MA parts. The AR part is computed with the innovation variance σ_ε^2, and the MA part in (4.61) uses a unit input variance. It is a simple method for actual computation of ARMA autocovariance functions. This autocovariance function also gives an expression for the variance of an ARMA process by taking $k = 0$. The autocorrelation function $\rho(k)$ is found by dividing (4.62) by $r(0)$. Figure 4.9 gives an ARMA(1,2) autocorrelation function. The MA order two determines, together with the AR parameter, an initial part of length two. The further autocorrelation function is completely determined by the AR part.

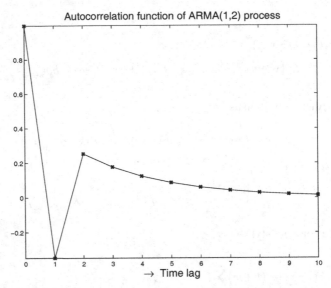

Figure 4.9. Autocorrelation function of an ARMA(1,2) process. The regular power series extension as in an AR(1) process is visible for k > 2.

Another solution for describing the autocovariance function uses the long MA representation of the process that is given by

$$x_n = \frac{B(z)}{A(z)} \varepsilon_n = G(z)\varepsilon_n = \sum_{i=0}^{\infty} g_i \varepsilon_{n-i} \qquad (4.63)$$

For given $A(z)$ and $B(z)$, the polynomial $G(z)$ can be obtained by long polynomial division. The same result is obtained by multiplying $B(z)$ by the infinitely long MA representation $1/A(z)$. A direct recursion can also be given with

$$G(z) = \frac{B(z)}{A(z)} \Rightarrow A(z)G(z) = B(z) \qquad (4.64)$$

The coefficients of z^{-m} are given by the equation,

$$a_0 g_m + a_1 g_{m-1} + \cdots + a_p g_{m-p} = b_m \qquad (4.65)$$

This gives a solution for computing the g_i coefficients for increasing orders from $A(z)$ and $B(z)$:

$$
\begin{aligned}
g_i &= 0,\, i < 0,\ \ g_0 = 1 \\
g_m &= -a_1 g_{m-1} - \cdots - a_p g_{m-p} + b_m, \quad m = 1, \cdots, q \\
g_m &= -a_1 g_{m-1} - \cdots - a_p g_{m-p}\ \ , \qquad m > q
\end{aligned}
\qquad (4.66)
$$

The inverse problem, computing $A(z)$ and $B(z)$ for an ARMA(p,q) process with given orders p and q from the infinitely long $G(z)$, is most conveniently solved by using p of the relations in (4.66) for $m > q$ to solve the a_i. For an ARMA(p,q) process, the values of the b_i coefficients are known to be zero for $m > q$. Afterward, computing b_i from the first q equations is straightforward when $G(z)$ and a_i are known.

The cross covariance between a single input and output lag of an ARMA(p,q) process is given by

$$E[x_{n-k}\varepsilon_{n-m}] = E\left[\sum_{i=0}^{\infty} g_i \varepsilon_{n-k-i}\varepsilon_{n-m}\right]$$

$$= \sigma_\varepsilon^2 g_{m-k} \tag{4.67}$$

This follows because the expectation in (4.67) is nonzero only for $k + i = m$. Furthermore, g_{m-k} is nonzero for nonnegative arguments, and hence for $m \geq k$ in (4.67).

By multiplying the ARMA(p,q) equation (4.53) by x_{n-k} and taking expectations it follows that

$$E[x_n x_{n-k} + a_1 x_{n-1} x_{n-k} + \cdots + a_p x_{n-p} x_{n-k} = \varepsilon_n x_{n-k} + b_1 \varepsilon_{n-1} x_{n-k} + \cdots + b_q \varepsilon_{n-q} x_{n-k}]. \tag{4.68}$$

For $k \geq 0$, it can be simplified to

$$r(k) + a_1 r(k-1) + \cdots + a_p r(k-p) = Q_k \tag{4.69}$$

$$Q_k = \sigma_\varepsilon^2 \sum_{m=0}^{q-k} g_m b_{m+k}, \quad 0 < k \leq q$$

$$= 0, \qquad k > q \tag{4.70}$$

Of course $r(-k) = r(k)$ for negative time shifts. This result shows that after q disturbed terms, which also deliver the boundary conditions, only the AR parameters influence the further development of the autocorrelation function. Therefore, at lags greater than the MA order q, the autocorrelation behaves as the autocorrelation of an AR(p) process with $Q_k = 0$ in (4.69).

Finally, the long MA representation $G(z)$ of (4.64) can be used to derive

$$r(k) = E\{x_n x_{n+k}\} = E\left\{\sum_{i=0}^{\infty} g_i \varepsilon_{n-i} \sum_{m=0}^{\infty} g_m \varepsilon_{n+k-m}\right\} = \sigma_\varepsilon^2 \sum_{i=0}^{\infty} g_i g_{i+k}, \quad k \geq 0. \tag{4.71}$$

The autocovariance function of arbitrary ARMA(p,q) processes can also be written as an infinite summation of lagged MA parameters. By defining $g_i = 0$ for $i < 0$, the lower limit in the summation in (4.71) can also be taken as minus infinity and the limitation to positive values of k can be released.

Priestley (1981) defines an autocovariance generating function $R(z)$ as

$$R(z) = \sum_{k=-\infty}^{\infty} r(k) z^{-k} \tag{4.72}$$

With (4.71) and $g_i = 0$ for $i < 0$, it is easily derived that

$$R(z) = \sum_{k=-\infty}^{\infty} \left(\sigma_\varepsilon^2 \sum_{i=-\infty}^{\infty} g_i g_{i+k} \right) z^{-k} = \sigma_\varepsilon^2 \sum_{k=-\infty}^{\infty} \sum_{i=-\infty}^{\infty} g_i z^i g_{i+k} z^{-(i+k)}$$

$$= \sigma_\varepsilon^2 G(z) G(z^{-1}) \tag{4.73}$$

$G(z)$ follows like the definitions of $A(z)$ and $B(z)$ in (4.15) and (4.5). The condition

$$\sum_{i=0}^{\infty} g_i z^{-i} < \infty \tag{4.74}$$

implies that $G(z)$ is analytic inside the unit circle. It follows that if the poles of $A(z)$ and the zeros of $B(z)$ are less than one in absolute value, the sum in (4.74) is finite. Therefore, the condition (4.74) requires AR processes to be stationary and MA processes to be invertible. The substitution $z = e^{j\omega}$ transforms the autocorrelation generating function into the power spectral density.

4.6 Harmonic Processes with Poles on the Unit Circle

So far, only poles and zeros inside the unit circle have been considered. With (4.73), the same autocorrelation function is generated with poles and zeros mirrored with respect to the unit circle because both z and z^{-1} are arguments in the autocovariance generating function. The definition of the shift operator z in (4.6) shows that using both z and z^{-1} give a two-sided transform. Hence, 2^{p+q} different combinations of poles and zeros yield the same autocovariance of an ARMA(p,q) process.

The model with its poles exactly on the unit circle represents a special class of its own: the harmonic function. This cannot be described as an ordinary AR process because the output variance would become infinite for a nonzero input signal. Taking the two complex conjugate poles, they describe the process

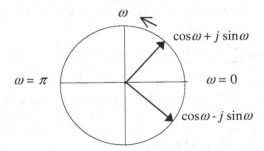

Figure 4.10. Position of two complex conjugated poles of harmonic process

$$x_n \left[1-(\cos\omega + j\sin\omega)z^{-1}\right]\left[1-(\cos\omega - j\sin\omega)z^{-1}\right] = 0$$

$$x_n \left(1-2\cos\omega z^{-1} + z^{-2}\right) = 0$$

$$x_n - 2\cos\omega x_{n-1} + x_{n-2} = 0 \tag{4.75}$$

and the solution is

$$x_n = C\cos(\omega n + \phi) \cdot \tag{4.76}$$

Hence, an AR(2) process *without any input signal* and with the second parameter equal to one represents a sinusoidal solution with a frequency determined by the first parameter, leaving free values of the amplitude and for the phase. It is very important that the input is identically zero.

Priestley (1981) defined this sine wave solution as a stochastic variable by choosing a random variable for ϕ, with uniform distribution between $-\pi$ and π and with probability density function $1/(2\pi)$ between those boundaries. If ϕ is a constant, the solution x_n is a deterministic variable, but it can be made random by taking a random variable for ϕ. The special property of the random harmonic process is that a complete realisation from $-\infty$ to ∞ becomes deterministic by choosing one specific value for ϕ. It is clear that AR processes with poles on the unit circle are a separate class of processes.

For random ϕ with uniform probability density function, it is possible to calculate the expectation and the autocorrelation function. The expectation becomes

$$E[x_n] = E\left[C\cos(\omega n + \phi)\right]$$

$$= \int_{-\pi}^{\pi} C\cos(\omega n + \phi)\frac{1}{2\pi}d\varphi = \frac{C}{2\pi}\left[\sin(\omega n + \phi)\right]_{-\pi}^{\pi}$$

$$= 0 \tag{4.77}$$

This result is valid for all n. It just says that the expectation of a sine wave with known frequency but unknown phase equals zero. The autocovariance function is found as follows:

$$r(k) = E[x_n x_{n-k}] = \frac{C^2}{2\pi}\int_{-\pi}^{\pi}\cos(\omega n + \phi)\cos(\omega n - \omega k + \phi)\,d\varphi$$

$$= \frac{C^2}{2\pi}\int_{-\pi}^{\pi}[\cos^2(\omega n + \phi)\cos(\omega k) + \cos(\omega n + \phi)\sin(\omega n + \phi)\sin(\omega k)]d\varphi$$

$$= \frac{C^2}{4\pi}\int_{-\pi}^{\pi}\left\{[1+\cos(2\omega n + 2\phi)]\cos(\omega k) + \sin(2\omega n + 2\phi)\sin(\omega k)\right\}d\varphi$$

$$= \frac{C^2}{4\pi}\int_{-\pi}^{\pi}\cos(\omega k)\,d\varphi = \tfrac{1}{2}C^2\cos(\omega k) \tag{4.78}$$

The autocorrelation function found by dividing by the variance of the signal becomes

$$\rho(k) = \cos(\omega k) \tag{4.79}$$

The autocorrelation is infinitely wide.

With this basic ingredient, a **harmonic process** is defined as

$$x_n = \sum_{i=1}^{M} C_i \cos(\omega_i n + \phi_i) \tag{4.80}$$

The phases ϕ_i are independent, uniformly distributed, random variables, each between $-\pi$ and $+\pi$. As the frequencies can have arbitrary values, the process x_n will generally not be periodic. The expectation of independent random variables is the sum of individual expectations, hence,

$$E[x_n] = \sum_{i=1}^{M} \frac{C_i}{2\pi} \int_{-\pi}^{\pi} \cos(\omega_i n + \phi_i) \, d\varphi_i = 0 \tag{4.81}$$

Likewise, the autocovariance is the sum of individual covariances because of the independence of the phases that makes the expectation of all cross products zero, or

$$r(k) = E[x_n x_{n-k}] = \frac{1}{2} \sum_{i=1}^{M} C_i^2 \cos(\omega_i k) \tag{4.82}$$

Furthermore, it is also possible to take a random variable as the amplitude.

All stationary AR processes and invertible MA processes have a decaying autocorrelation function. The power series of (4.52) decays if the radius of the poles is less than one and the autocovariance of MA in (4.8) is finite. Therefore, the harmonic process is the only stochastic model with an autocorrelation function that never dies out because (4.82) is a periodic function.

4.7 Spectra of Time Series Models

The spectral properties will be derived for the infinite representation (4.63) that includes all AR, MA, and ARMA models. This is an elegant way to derive formulas for the spectra, but it is not advisable for a practical computation due to the infinite length. The general time series model is given in (4.63) by

$$x_n = \frac{B(z)}{A(z)} \varepsilon_n = G(z)\varepsilon_n = \sum_{i=0}^{\infty} g_i \varepsilon_{n-i}$$

The autocovariance function is given in (4.71) as

$$r(k) = E[x_n x_{n+k}] = \sigma_\varepsilon^2 \sum_{i=0}^{\infty} g_i g_{i+k} , \; \forall k$$

where $g_i = 0$ for $i < 0$. The power spectral density is given by (3.8):

$$h(\omega) = \frac{1}{2\pi} \sum_{k=-\infty}^{\infty} r(k) e^{-j\omega k}$$

$$= \frac{\sigma_\varepsilon^2}{2\pi} \sum_{k=-\infty}^{\infty} \sum_{i=0}^{\infty} g_i g_{i+k} e^{-j\omega k}$$

$$= \frac{\sigma_\varepsilon^2}{2\pi} \sum_{k=-\infty}^{\infty} \sum_{i=0}^{\infty} g_i e^{j\omega i} g_{i+k} e^{-j\omega(i+k)} \tag{4.83}$$

This can further be written as

$$h(\omega) = \frac{\sigma_\varepsilon^2}{2\pi} G(e^{-j\omega}) \, G(e^{j\omega}) = \frac{\sigma_\varepsilon^2}{2\pi} \left| G(e^{j\omega}) \right|^2 \tag{4.84}$$

A comparison with (4.73) shows that this outcome is also found by substituting $z = e^{j\omega}$ in the autocovariance generating function. With (4.63), the spectrum can also be written as

$$h(\omega) = \frac{\sigma_\varepsilon^2}{2\pi} \frac{\left| B(e^{j\omega}) \right|^2}{\left| A(e^{j\omega}) \right|^2} , \quad -\pi < \omega \le \pi \tag{4.85}$$

This general ARMA spectrum immediately gives also the spectra of pure AR and pure MA processes by setting $B(e^{j\omega})$ and $A(e^{j\omega})$ equal to one, respectively. The figures with spectral densities will show the normalized spectrum $\varphi(f)$ that is given by

$$\phi(f) = \frac{\sigma_\varepsilon^2}{\sigma_x^2} \frac{\left| B(e^{j2\pi f}) \right|^2}{\left| A(e^{j2\pi f}) \right|^2} , \quad -0.5 < f \le 0.5 \tag{4.86}$$

4.7.1 Some Examples

The spectrum of white noise becomes

$$h(\omega) = \frac{\sigma_\varepsilon^2}{2\pi} \qquad (4.87)$$

An AR(1) process with parameter a yields

$$h(\omega) = \frac{\sigma_\varepsilon^2}{2\pi} \frac{1}{\left| A(e^{j\omega}) \right|^2}$$

$$= \frac{\sigma_\varepsilon^2}{2\pi} \frac{1}{\left(1 + ae^{-j\omega}\right)\left(1 + ae^{j\omega}\right)}$$

$$= \frac{\sigma_\varepsilon^2}{2\pi} \frac{1}{1 + 2a\cos\omega + a^2} \qquad (4.88)$$

For the normalized spectrum $\varphi(f)$,

$$\phi(f) = \frac{1 - a^2}{1 + 2a\cos(2\pi f) + a^2} \qquad (4.89)$$

The poles of an AR(1) process are always real. In the complex plane of Figure 4.10, the angle of the poles is zero for negative a and π for $a > 0$.

Figure 4.11 gives the spectra of an AR(1) process with a positive and a negative parameter. The maximum of the spectrum is found at the frequency of the pole in Figure 4.11.

The power spectral density of a MA(1) process becomes

$$h(\omega) = \frac{\sigma_\varepsilon^2}{2\pi}\left| B(e^{j\omega}) \right|^2 = \frac{\sigma_\varepsilon^2}{2\pi}\left(1 + 2b\cos\omega + b^2\right) \qquad (4.90)$$

It is given in Figure 4.12 for a positive and a negative value of the MA(1) parameter. With those examples, it is clear how the spectrum of higher order MA models and ARMA models can be calculated.

An AR(2) process has the spectrum

$$h(\omega) = \frac{\sigma_\varepsilon^2}{2\pi} \frac{1}{\left(1 + a_1 e^{-j\omega} + a_2 e^{-j2\omega}\right)\left(1 + a_1 e^{j\omega} + a_2 e^{j2\omega}\right)}$$

$$= \frac{\sigma_\varepsilon^2}{2\pi} \frac{1}{\left(1 - a_2\right)^2 + a_1^2 + 2a_1\left(1 + a_2\right)\cos\omega + 4a_2\cos^2\omega} \qquad (4.91)$$

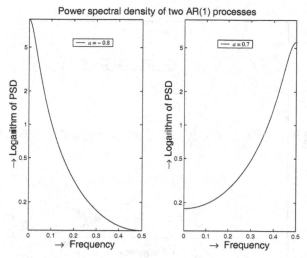

Figure 4.11. Normalized power spectral density of the two AR(1) processes of Figure 4.4. The slowly varying autocorrelation has a concentration of power in the lowest frequencies.

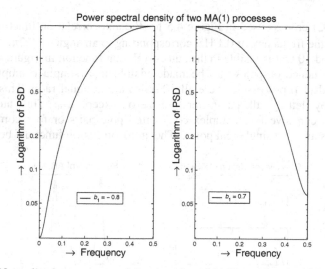

Figure 4.12. Normalized power spectral density of two MA(1) processes with real zeros at $\omega = 0$ and at $\omega = \pi$, respectively. The spectra show a valley at the frequency of a zero.

For the normalized spectrum $\varphi(f)$

$$\phi(f) = \frac{1-a_2}{1+a_2} \frac{(1+a_2)^2 - a_1^2}{(1-a_2)^2 + a_1^2 + 2a_1(1+a_2)\cos(2\pi f) + 4a_2 \cos^2(2\pi f)} \tag{4.92}$$

Poles are recognized as spectral peaks, when they are close to the unit circle. Likewise, zeros are seen as spectral valleys. Figure 4.13 gives the power spectrum

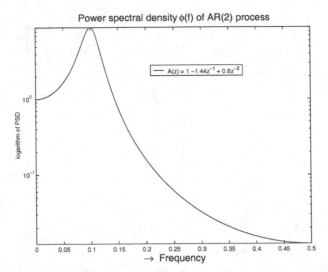

Figure 4.13. Normalized power spectral density of the AR(2) processes of Figure 4.5. The angle of the poles is at $\omega \approx \pi / 5$ or $f \approx 0.1$.

of the AR(2) process used in Figure 4.5 for the autocorrelation function. The pole is at about the frequency of 0.1 Hz, corresponding to an angle of 0.2π. This belongs to the period 10 that is visible in the autocorrelation function in Figure 4.5.

The influence of poles can be made visible in a simple example with two different AR(10) processes where all 10 poles are at equal radii. This example is obtained by letting all parameters be zero, except a_{10}. This autoregressive polynomial can give five complex conjugated pole pairs or four complex conjugated pairs and two single real poles. The autocorrelation function becomes zero

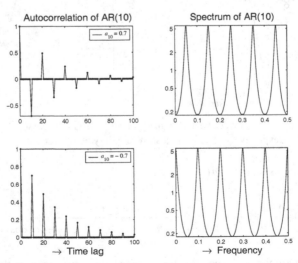

Figure 4.14. Autocorrelation function and normalized power spectral density of two AR(10) processes with all parameters zero except a_{10}

for all lags that are not multiples of 10. At those lags, however, the autocorrelation function is a power series, as in an AR(1) process. The spectra have constant heights for every equidistant peak.

Figure 4.14 shows what happens with a single positive or negative parameter a_{10} for this special AR(10) process. The positive parameter gives five complex conjugated pole pairs, which can be seen in the power spectrum that has five peaks in the interval. The negative parameter gives four complex conjugated pole pairs and two real poles at the frequencies of 0 and 0.5.

There is a strong relation between the width of a single peak and its height. The angle of a pole determines the frequency; the distance of the pole to the unit circle determines the width and the height of a spectral peak. Narrow AR(2) peaks are also high. It will generally require an AR process with many close poles to obtain a spectral peak where the width of the peak and the shape of the peak do not belong to the natural AR(2) combination of height and width. Zeros close to the unit circle give deep valleys. The pole exactly on the unit circle is the limiting case, with a single spectral line as a peak with vanishing width.

Figure 4.15 gives an ARMA(10,10) example where the 10 poles are the same as in the upper half of Figure 4.14 and the 10 zeros are located at the positions that were poles of the lower half. At the frequencies of the zeros, six valleys are seen; each has the same depth because the equidistant zeros all had the same radius. This chapter deals with time series theory. It describes the relation between the parameters, on one hand, and the autocorrelation function and the power spectral density on the other hand. It should be stressed that applying the relations to estimated models is not always allowed. Most relations for the AR theory can be applied to estimated parameters and autocorrelations without a problem. However, the MA and ARMA relations create a problem. Estimating q MA parameters from q autocorrelations is not efficient and produces no useful model. Conversely, estimating q autocorrelations from q estimated MA parameters is allowed.

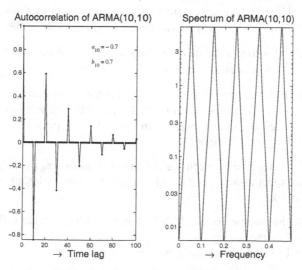

Figure 4.15. Autocorrelation function and normalized power spectral density of an ARMA(10,10) process with all parameters zero except a_{10} and b_{10}

4.8 Exercises

4.1   For a MA(1) process, show that $|\rho(1)| \leq 0.5$ for any parameter value. For which parameter value is $\rho(1)$ maximal and for which minimal.

4.2   Find the first five parameters of the long MA representation $G(z)$ of the process

$$x_n + 0.5x_{n-1} = \varepsilon_n.$$

4.3   Find the first five parameters of the long MA representation $G(z)$ of the process

$$x_n - 0.5x_{n-1} = \varepsilon_n + 0.9\varepsilon_{n-1} + 0.5\varepsilon_{n-2}.$$

4.4   Derive the autocovariance function $r(k)$, as given in Equation (4.60).

4.5   The first parameters g_i of a long MA model of an ARMA(2,2) process are:

$$g_0 = 1, g_1 = -1.3, g_2 = -0.03, g_3 = 0.937, g_4 = -0.8223, \ldots\ldots$$

Find the parameters of the ARMA(2,2) process.

4.6   Use ARMASA to plot the autocorrelation function of an AR(1) process.

4.7   Use ARMASA to plot the power spectral density function of an AR(1) process.

4.8   Use ARMASA to plot the autocorrelation function of a MA(1) process.

4.9   Use ARMASA to plot the power spectral density function of a MA(1) process.

4.10   Compute the parameters of an AR(2) process with a spectral peak at $\omega = \pi/3$. Use the complex conjugated poles at $R(\cos\omega + j\sin\omega)$ and $R(\cos\omega - j\sin\omega)$ with $R < 1$ to find poles within the unit circle.

4.11   Use ARMASA to plot the autocorrelation function of an AR(2) process with a spectral peak at $\omega = \pi/3$.

4.12   Use ARMASA to plot the power spectral density function of an AR(2) process with a spectral peak at $\omega = \pi/3$. Take different values of the radius of the complex conjugated poles. Can the width of a peak be varied independently of the height?

4.13   Study the difference between logarithmic and linear amplitude scaling for the power spectral density of an AR(2) process if the poles are approaching the unit circle.

4.14   Compute the parameters of a MA(2) process with a spectral valley at $\omega = \pi/4$. Use the conjugated poles at $R(\cos\omega + j\sin\omega)$ and $R(\cos\omega - j\sin\omega)$ with $R < 1$ to find zeros within the unit circle.

4.15 Use ARMASA to plot the power spectral density function of a MA(2) process with a spectral valley at $\omega = \pi/4$.

4.16 Use ARMASA to plot the power spectral density function of an ARMA(2,2) process with a spectral peak at $\omega = \pi/4$ and a valley at $\omega = \pi/3$.

4.17 Verify that the same AR(1) autocorrelation function is also the solution of (4.29) if the negative value $k = -1$ is substituted for x_{n-k}. Hence, x_n is multiplied by x_{n+1}. This requires the computation of a right-hand side term.

5

Relations for Time Series Models

5.1 Time Series Estimation

Almost all stationary stochastic processes can be modeled by a unique, stationary, and invertible ARMA process. It does not matter how the observed process had been generated, and it is not necessary to know whether it is noise or the sum of many simple processes plus coloured noise. Any observed signal that can be modeled as a stationary stochastic process has a unique ARMA representation. This can be ARMA(p,q), AR(p), or MA(q). AR(∞) and MA(∞) are always a good description, but it is important in practice to find the model with the fewest parameters. That model can generally be estimated with good accuracy. Any continuous spectral density can mathematically be approximated arbitrarily closely with a rational model. The important conclusion, which required much mathematical skill to prove, is that the ARMA models provide a complete model set for arbitrary stationary stochastic processes. Therefore, using a time series model is not equivalent to forcing a model upon the data from a limited class of candidates. Time series models offer a complete description without restrictions. The model type and the model order are generally supposed to be known in the theory.

With a suitable computer program ARMAsel (Broersen 2000, 2002), it is possible now to detect automatically the best model type, AR, MA, or ARMA, and the best model order for a given set of observations. The selection criteria use the notion of parsimony: they add a penalty for each additional parameter. In this way, they look for an accurate model with a minimum number of parameters.

Before treating the estimation and order-selection algorithms, some theoretical relations for time series models are given.

5.2 Yule-Walker Relations and the Levinson-Durbin Recursion

The Yule-Walker relations describe the relation between the autocovariances and the parameters of AR processes (Priestley, 1981). Multiplying both sides of the AR(p) equation (4.50) by x_n and taking expectations gives

$$r(0) + a_1 r(1) + \cdots + a_p r(p) = \sigma_\varepsilon^2 \tag{5.1}$$

Multiplying both sides of the AR(p) equation by x_{n-k} with $k > 0$, and taking the expectations gives the Yule-Walker relations

$$r(k) + a_1 r(k-1) + \cdots + a_p r(k-p) = 0 \qquad (5.2)$$

The variance of the innovations appears only in (5.1) with $k = 0$; all other equations with $k > 0$ have zero on the right-hand side. The shape of the equation is regular for k greater than p. If k is between 1 and p, negative indexes of the lag $k - i$ are mostly replaced by the autocovariance at the positive lag because the autocorrelation function is symmetrical. That gives the possibility of obtaining $p + 1$ equations containing the autocovariances $r(0)$, ..., $r(p)$, on the one hand, and the p parameters together with the innovation variance on the other hand.

The normalized version of (5.2) can also be written as

$$\rho(k) = -a_1 \rho(k-1) - \cdots - a_p \rho(k-p) \qquad (5.3)$$

This equation shows how the AR(p) autocorrelation function for the lags 1 to p is extrapolated for $k > p$.

The autocorrelation function is a measure of the degree of correlation between x_n and x_{n-k}, as a function of lag k. However, x_n and x_{n-k} are not only directly related, but also through the intermediate observations. For example, for an AR(1) process with parameter a, $E\{x_n x_{n-2}\}$ equals $a^2 r(0)$, which is exactly the extrapolation of the autocovariance of order one with (5.2). Box and Jenkins (1976) and Priestley (1981) use a partial autocorrelation function to describe the new contribution of an additional AR order to the autocorrelation. The partial correlation is loosely defined as the difference between the autocorrelation coefficient at a certain lag and its extrapolation from lower order correlations. For true AR(p) processes, all *partial* correlations above order p are zero, whereas the autocorrelations remain nonzero for any time shift. The negative of the partial correlation is often called the reflection coefficient. By the nature of correlations, reflection coefficients also are always between −1 and 1. AR(p) models can be characterized by their p nonzero parameters or by their p reflection coefficients.

The first $p + 1$ Yule-Walker equations will be used to solve the parameters and the variance σ_ε^2 of the innovations from the autocovariances of x_n. Together, for $k = 0$ until $k = p$, Equations (5.1) and (5.2) can be written in the matrix form

$$\begin{bmatrix} r(0) & r(-1) & \cdots & r(-p) \\ r(1) & r(0) & & \\ \vdots & & \ddots & r(-1) \\ r(p) & \cdots & r(1) & r(0) \end{bmatrix} \begin{bmatrix} 1 \\ a_1 \\ \vdots \\ a_p \end{bmatrix} = \begin{bmatrix} \sigma_\varepsilon^2 \\ 0 \\ \vdots \\ 0 \end{bmatrix} \qquad (5.4)$$

Here, the negative lags are still present in the matrix. Due to the symmetry of the autocovariance function, it is possible to replace the negative lags by their positive equivalents. Leaving the first equation out, the last p equations become

$$\begin{bmatrix} r(0) & r(1) & \cdots & r(p-1) \\ r(1) & r(0) & & \\ \vdots & & \ddots & r(1) \\ r(p-1) & \cdots & r(1) & r(0) \end{bmatrix} \begin{bmatrix} a_1 \\ \vdots \\ \vdots \\ a_p \end{bmatrix} = - \begin{bmatrix} r(1) \\ r(2) \\ \vdots \\ r(p) \end{bmatrix} \qquad (5.5)$$

Without loss of information, this latter matrix formula can be divided by $r(0)$ to remove the influence of the variance completely from those equations. With some new matrix and vector variables, the first equation of (5.4) and the matrix of (5.5) become

$$r(0) + \alpha_p^T r_p = \sigma_\varepsilon^2$$
$$R_p \alpha_p = -r_p \qquad (5.6)$$

R_p is the doubly symmetrical $p \times p$ Toeplitz matrix in (5.5). The parameter vector is defined without the first parameter a_0 as

$$\alpha_p = \begin{bmatrix} a_1 & a_2 & \cdots & a_p \end{bmatrix}^T \qquad (5.7)$$

and the autocovariance vector as

$$r_p = \begin{bmatrix} r(1) & r(2) & \cdots & r(p) \end{bmatrix}^T \qquad (5.8)$$

The solution is expressed mathematically by

$$\alpha_p = -R_p^{-1} r_p \qquad (5.9)$$

Inverting the autocovariance matrix is never used to determine the solution in practice. Much faster and numerically more stable algorithms are available. They use the very special character of R_p in (5.5). That matrix is symmetrical about both the main diagonal and the main antidiagonal, and it is a Toeplitz matrix. Toeplitz means that all diagonal lines have equal elements.

The solution of (5.9) is generally found recursively with an algorithm called the Levinson-Durbin algorithm or sometimes the Levinson recursion. To have a convenient derivation, a symbol ~ is introduced to indicate the reversed version of a vector, with the elements in reverse order. The reversal of the parameter vector (5.7) gives

$$\tilde{\alpha}_p = \begin{bmatrix} a_p & a_{p-1} & \cdots & a_1 \end{bmatrix}^T$$

and the reversed autocovariance vector (5.8) becomes

$$\tilde{r}_p = \begin{bmatrix} r(p) & r(p-1) & \cdots & r(1) \end{bmatrix}^T \qquad (5.10)$$

It follows elementarily by writing out the terms that

$$\tilde{\alpha}_p^T r_p = \tilde{r}_p^T \alpha_p$$
$$\tilde{\alpha}_p^T \tilde{r}_p = \alpha_p^T r_p \qquad (5.11)$$

The Levinson-Durbin algorithm starts with the solution of (5.9) by taking $p = 1$ and recursively determines the solutions for increasing orders.

The individual parameters of lower order models in a recursive solution are given for an intermediate order K as

$$\alpha^{[K]} = \left(a_1^K \quad a_2^K \quad \cdots \quad a_K^K \right)^T \qquad (5.12)$$

The solution is based on partial autocorrelations and has very good numerical performance. Suppose that a solution of order K is available for Equations (5.6). The true variance of the innovations σ_ε^2 will follow only for the true order p. Therefore, some intermediate variance measure is introduced. The solution for the intermediate order K can be written symbolically as

$$r(0) + \alpha^{[K]T} r_K = s_K^2$$
$$R_K \alpha^{[K]} = -r_K$$
$$\Rightarrow \alpha^{[K]} = -R_K^{-1} r_K \qquad (5.13)$$

where s_K^2 denotes the residual variance after calculating K parameters. The final line is a mathematically closed form expression for the solution, not the way the solution has been obtained.

The solution for order $K + 1$ has to be expressed in the variables of (5.13) to obtain a recursive solution. The equations for order $K + 1$ in the notation of (5.13) become

$$r(0) + \alpha^{[K+1]T} r_{K+1} = s_{K+1}^2$$
$$R_{K+1} \alpha^{[K+1]} = -r_{K+1} \qquad (5.14)$$

Now define the new parameter vector for order $K + 1$, that is one longer than the previous one, as the old parameter vector with an extra addition for each vector element and one new element to increase the order from K to $K + 1$. In a formula, this becomes

$$\alpha^{[K+1]} = \begin{bmatrix} \alpha^{[K]} + \Delta^{[K]} \\ k_{K+1} \end{bmatrix} \qquad (5.15)$$

The coefficients k_i are the reflection coefficients of order i. Furthermore $-k_i$ is also the partial correlation coefficient. The new reflection coefficient k_{K+1} would be zero if the extrapolation with (5.2) of the autocovariance function of the AR(K)

model beyond order K were equal to $r(K+1)$. The difference determines the value of k_{K+1}.

Introduce a partitioned Toeplitz covariance matrix of order $K + 1$:

$$R_{K+1} = \begin{bmatrix} R_K & \tilde{r}_K \\ \tilde{r}_K^T & r_x(0) \end{bmatrix} \tag{5.16}$$

For the matrix equation in (5.14),

$$\begin{bmatrix} R_K & \tilde{r}_K \\ \tilde{r}_K^T & r(0) \end{bmatrix} \begin{bmatrix} \alpha^{[K]} + \Delta^{[K]} \\ k_{K+1} \end{bmatrix} = -\begin{bmatrix} r_K \\ r(K+1) \end{bmatrix} \tag{5.17}$$

which can be written as two separate equations

$$\begin{aligned} R_K \left[\alpha^{[K]} + \Delta^{[K]} \right] + k_{K+1}\tilde{r}_K &= -r_K \\ \tilde{r}_K^T \left[\alpha^{[K]} + \Delta^{[K]} \right] + k_{K+1}r(0) &= -r(K+1) \end{aligned} \tag{5.18}$$

Substituting the solution for order K of (5.13) in the equation for stage $K + 1$ gives

$$\begin{aligned} R_K \Delta^{[K]} &= -k_{K+1}\tilde{r}_K \\ \tilde{r}_K^T \alpha^{[K]} + \tilde{r}_K^T \Delta^{[K]} + k_{K+1}r(0) &= -r(K+1) \end{aligned} \tag{5.19}$$

The solution of the first equation is

$$\begin{aligned} \Delta^{[K]} &= -k_{K+1}R_K^{-1}\tilde{r}_K \\ &= k_{K+1}\tilde{\alpha}^{[K]} \end{aligned} \tag{5.20}$$

The solution for the new parameter vector (5.15) is

$$\alpha^{[K+1]} = \begin{bmatrix} \alpha^{[K]} + k_{K+1}\tilde{\alpha}^{[K]} \\ k_{K+1} \end{bmatrix} \tag{5.21}$$

The new parameter vector consists of the old parameter vector and as an addition, the reversed old parameter vector, multiplied by the additional parameter of order $K + 1$. Furthermore, the parameter of order $K + 1$ is equal to the reflection coefficient of that order.

The second Equation (5.18) is used to derive the value of the reflection coefficient:

$$\tilde{r}_K^T \alpha^{[K]} + \tilde{r}_K^T k_{K+1} \tilde{\alpha}^{[K]} + k_{K+1} r(0) = -r(K+1)$$

$$k_{K+1} \left[r_K^T \alpha^{[K]} + r(0) \right] = -r(K+1) - \tilde{r}_K^T \alpha^{[K]}$$

$$k_{K+1} = -\frac{r(K+1) + \tilde{r}_K^T \alpha^{[K]}}{s_K^2} \qquad (5.22)$$

where s_K^2 has been substituted from (5.13). The first line contains the product of two reversed vectors, which equals the product of the original vectors. Further, using partitioned matrices, the new residual expression in (5.14) becomes

$$
\begin{aligned}
s_{K+1}^2 &= r(0) + \alpha^{[K+1]T} r_{K+1} \\
&= r(0) + \left(\alpha^{[K]T} + k_{K+1} \tilde{\alpha}^{[K]T} \right) r_K + k_{K+1} r(K+1) \\
&= r(0) + \alpha^{[K]T} r_K + k_{K+1} \left[r(K+1) + \tilde{\alpha}^{[K]T} r_K \right] \\
&= s_K^2 + k_{K+1} \left[-k_{K+1} s_K^2 \right] \\
&= s_K^2 (1 - k_{K+1}^2)
\end{aligned}
\qquad (5.23)
$$

where the last line of (5.22) has been substituted. The Levinson-Durbin recursion is the solution to determine the AR parameters from the first p given true autocovariances:

LEVINSON-DURBIN RECURSION

Starting values

$$s_0^2 = r(0)$$

$$k_1 = a^{[1]} = -\frac{r(1)}{s_0^2}$$

$$s_1^2 = s_0^2 (1 - k_1^2)$$

for $K = 1, 2, ..., p - 1$

$$k_{K+1} = -\frac{r(K+1) + \tilde{r}_K^T \alpha^{[K]}}{s_K^2}$$

$$s_{K+1}^2 = s_K^2 (1 - k_{K+1}^2)$$

$$\alpha^{[K+1]} = \begin{bmatrix} \alpha^{[K]} + k_{K+1} \tilde{\alpha}^{[K]} \\ k_{K+1} \end{bmatrix} \qquad (5.24)$$

5.3 Additional AR Representations

Each of the representations with parameters and with reflection coefficients has its own advantages. It is common that a relation that is simply formulated in one representation can be very cumbersome in an other representation. As an example, the successive residual variances in (5.24) are very easily given as a function of reflection coefficients of increasing order. However, they would look very complicated if expressed in the parameters. On the contrary, the definition (4.13) of an AR(p) process would have a complicated appearance if formulated with reflection coefficients.

Several equivalent representations of AR(p) processes are known:

- $p + 1$ autocovariances
- p parameters and the variance of the innovations
- p reflection coefficients and the variance of the innovations
- p poles and the variance of the innovations
- p autocorrelations and the variance of the innovations
- p cepstral coefficients and the variance of the innovations (Markel and Gray, 1976)
- In speech coding where autoregressive modeling is often called linear prediction, many more representations are known (Viswanathan and Makhoul, 1975; Markel and Gray, 1976; Erkelens, 1996), *e.g.*,
  - line spectral pairs
  - log area ratios
  - arcsine of reflection coefficient
  - immittance spectral pairs.

An important property of reflection coefficients is that an AR model is stationary if all reflection coefficients are smaller than one in amplitude. Checking the magnitude of the reflection coefficients is the usual way to verify the stationarity or invertibility of AR or MA polynomials. Reflection coefficients are also known as the negative partial correlation coefficients of each order. Reflection coefficients with magnitudes less than one guarantee that

- the AR(p) process is stationary
- the poles of the AR(p) polynomial $A(z)$ are within the unit circle
- the matrix R_p of (5.6) is positive-semidefinite.

Stationary processes have all reflection coefficients between -1 and 1, and all estimated reflection coefficients have to be less than one to obtain a useful model.

Relations for the other representations have also been given. However, their use is less general. Log area ratios, line spectral pairs, cepstra, and other dedicated transformations are typical representations in speech analysis and they are treated in dedicated literature; see Erkelens (1996) or Markel and Gray (1976).

It is an enormous advantage that the given AR relations are also useful, when they are applied to *estimated* parameters and to *estimated* autocorrelation functions. That will be an incentive to search for recurrent AR estimation methods. The main reason is that p estimated autocorrelations are asymptotically efficient estimates (Porat, 1994). They are also asymptotically sufficient (Arato, 1961) to calculate the parameters or the reflection coefficients or any other representation of an AR(p) model.

5.4 Additional AR Relations

It is often important to use the proper representation to facilitate derivations and expressions. Therefore, relations between autocorrelations, parameters, and reflection coefficients will be treated in some detail.

5.4.1 The Relation Between the Variances of x_n and of ε_n for an AR(p) Process

This is rather difficult and complex if expressed in parameters, but it is simply a product of $(1 - k_i^2)$ when expressed in reflection coefficients. With (5.24), it follows that

$$\sigma_x^2 = \sigma_\varepsilon^2 / \prod_{i=1}^{p}\left(1 - k_i^2\right) \tag{5.25}$$

The reduction of the residual variance with increasing order K is given by

$$s_K^2 = \sigma_x^2 \prod_{i=1}^{K}\left(1 - k_i^2\right) \tag{5.26}$$

which yields

$$s_K^2 = \sigma_\varepsilon^2, \ K = p, p+1, \cdots \tag{5.27}$$

All reflection coefficients for orders greater than the true AR order p are zero. The relation (5.25) can look very unattractive if any representation other than reflection coefficients is used. Expressions for the reduction of the residual variance as a function of the model order are much longer and much less informative if they are written in terms of parameters, although the numerical results will coincide.

5.4.2 Parameters from Reflection Coefficients

The Yule-Walker relations give the relation between parameters and autocorrelations. They have been solved with the recursive Levinson-Durbin algorithm to find reflection coefficients and parameters from autocovariances. The last relation in (5.24)

$$\alpha^{[K+1]} = \begin{bmatrix} \alpha^{[K]} + k_{K+1}\tilde{\alpha}^{[K]} \\ k_{K+1} \end{bmatrix}, \qquad K = 2, \cdots, p-1 \tag{5.28}$$

is used recursively to transform reflection coefficients into parameters, starting with the first $a^{[1]} = k_1$. The final vector $\alpha^{[p]}$ contains the parameters of the AR(p) process.

5.4.3 Reflection Coefficients from Parameters

In this case, the parameter vector with a_1, a_2, ..., a_p is given. The start is found by writing the complete parameter vector in the final state that is obtained with (5.28) as

$$\alpha^{[p]} = \begin{bmatrix} a_1 \\ \vdots \\ a_p \end{bmatrix} = \begin{bmatrix} a_1^p \\ \vdots \\ a_p^p \end{bmatrix}, \qquad k_p = a_p^p \tag{5.29}$$

and then recursively the reflection coefficients of lower are found with

$$\alpha^{[K-1]} = \frac{\begin{bmatrix} a_1^K \\ \vdots \\ a_{K-1}^K \end{bmatrix} - k_K \begin{bmatrix} a_{K-1}^K \\ \vdots \\ a_1^K \end{bmatrix}}{1 - k_K^2}, \qquad K = p, p-1, \cdots, 2$$

$$k_{K-1} = a_{K-1}^{K-1} \tag{5.30}$$

5.4.4 Autocorrelations from Reflection Coefficients

This starts with $\rho(0) = 1$, $\rho(1) = -k_1$ and continues with building the parameter vector for increasing orders and computing the next autocorrelation

$$\rho(K+1) = -\tilde{\rho}_K^T \left[\alpha^{[K]} + k_{K+1}\tilde{\alpha}^{[K]} \right] - k_{K+1}, \qquad K = 2, \cdots, p-1$$

$$\rho(K) = -a_1\rho(K-1) - \cdots - a_p\rho(K-p), \qquad K \geq p \tag{5.31}$$

Multiplying the autocorrelations by $r(0)$, the variance of x_n, the autocovariances are found.

5.4.5 Autocorrelations from Parameters

This starts by transforming the parameters into reflection coefficients with (5.29) and using (5.31) to determine the autocorrelation from those reflection coefficients.

The reverse, parameters and reflection coefficients from autocovariances or autocorrelations, is carried out with the Levinson-Durbin recursion (5.24). It is clear that the construction of intermediate parameter vectors from reflection coefficients plays a role in all transformations that are given here.

5.5 Relation for MA Parameters

A number of convenient relations have been derived for AR models. Those relations for relating the true process parameters to the spectrum, autocorrelation function, and all other AR representations will also be important for estimation algorithms. The relation between MA parameters and the autocorrelation function that belongs to them has been given in (4.11). It is simple and straightforward to compute the autocorrelation function for given MA parameters.

However, the inverse is much less simple because those relations are non-linear. Wilson (1969) developed a nonlinear search algorithm for a solution.

To demonstrate the problems with the relation between autocorrelation and MA parameters, a MA(1) process with $b_1 = 0.5$ is treated here. With (4.11), it follows easily that $\rho(1) = 0.4$ if b_1 is given. The other way around, solving b_1 from a given value $\rho(1)$ uses

$$\rho(1) = b_1 / 1 + b_1^2$$

with the solution

$$b_1 = \frac{1}{2\rho(1)} \pm \sqrt{\frac{1}{4\rho(1)^2} - 1} \qquad (5.32)$$

The solutions for b_1 are 0.5 and 2, which is just $1/b_1$ and which gives the same autocorrelation according to (4.12).

Higher order processes require iterative computer calculations to find the true parameters from the true autocorrelations. For a MA(1) process, $\rho(1)$ is between -0.5 and 0.5; these limiting values are obtained for $b_1 = -1$, $b_1 = 1$, respectively. A real solution of (5.32) requires that $-0.5 < \rho(1) < 0.5$. That is no problem for true autocorrelation functions because they obey automatically the positive-semidefinite requirement if the zeros of the MA polynomial of (4.5) are within the unit circle. But it would give difficulties if it were applied to estimated lagged product autocorrelations. If the true autocorrelation $\rho(1)$ is 0.5, a slightly higher estimate will no longer produce a real solution with (5.32).

This relation between q MA estimated autocorrelations and q MA parameters will not be treated any further because it is no longer valid or efficient for *estimated* autocorrelation functions. Porat (1994) has shown that estimated autocorrelations of a MA process are not efficient. This means in practice that it is not statistically efficient to compute q MA parameters from q estimated autocorrelations. Theoretically, efficient estimates for MA parameters from estimated autocorrelations, with the minimum variance from the given observations, can be computed only from all estimated autocorrelations (Godolphin and de Gooijer, 1982). This property is the reason that MA and also ARMA estimation is much more difficult in practice than AR estimation.

5.6 Accuracy Measures for Time Series Models

Several methods have been described for estimating parameters of time series models. A comparative presentation of the spectral results of Kay and Marple (1981) showed that different estimators may yield completely different spectra from the same data. Before discussing the estimation methods, an objective criterion will be derived to compare the quality of estimated models with the true process. This prevents preferring a certain estimation method as a matter of taste, which would be quite unacceptable in a scientific environment. Therefore, an objective measure for the comparison of estimated models has to be derived.

Gray and Markel (1976) have compared different distance measures from the point of view of practical applications, especially speech processing. They found that using differences between parameters or reflection coefficients did not give reliable measures. Taking the logarithm of spectra was much better. Martin (2000) concluded also that it is not advisable to use the difference between true and estimated parameters as an accuracy measure. The difference between an AR(1) process $a_1 = 0.98$ and a model $a_1 = 1.02$ would be the same as the difference between $a_1 = 0.02$ and $a_1 = -0.02$. In practice, the first difference is very important for a stationary process and a nonstationary model; the second difference is very small. The difference between $a_1 = 0.99$ and $a_1 = 0.95$ is also much more important than the difference between $a_1 = 0.19$ and $a_1 = 0.15$. This would follow from a sensible accuracy measure. Therefore, the unweighted squared difference between parameters is not considered a useful measure.

5.6.1 Prediction Error

It is possible to use basic principles to define a general accuracy measure for an estimated model. That measure is the squared error of the one step ahead prediction that can be found with the model in fresh data that have not been used to estimate the model. It should be realised that the fit to the observations that are used to estimate the model is minimised in some sense and the statistical properties of the estimated model depend on those observations. Above the true order, the true values of the parameters are zero, but estimated values will have small values that reduce the least-square fit to the given observations, known as the residual variance. Therefore, the prediction error in new and independent observations of the same process is the basis for an accuracy measure.

Suppose that an ARMA(p,q) process (4.54) gives the true description of the data:

$$A(z)x_n = B(z)\varepsilon_n \tag{5.33}$$

Also suppose that an *estimated* model ARMA(p',q') is given with the estimated parameter polynomials $\hat{A}(z)$ and $\hat{B}(z)$. True or estimated polynomials may have different numbers of parameters and they may also be equal to one. Therefore, the prediction error of a MA model $\hat{B}(z)$ to an AR process $A(z)$ is also defined by this definition. The model may be estimated from data that have been observed previously, or it may just be an arbitrary model for which one wants to know how close it is to the true ARMA(p,q) process. It is well known that a model with more parameters, estimated by the least-squares principle, always seems to fit better to the data from which it was estimated. Every extra estimated parameter reduces the residual variance. This phenomenon is also found in polynomial regression. To avoid this artifact in fitting properties, the quality of a model is defined as the output of that model applied to *fresh* or *new* data of the ARMA(p,q) process. Those new data may not have been used for estimating of the model because that would create dependence between the estimated parameters and the data. In this derivation, it is essential that the parameters in $\hat{A}(z)$ and $\hat{B}(z)$ are independent of the data x_n that are used in Figure 5.1 as the input of a series of two linear filters.

Figure 5.1. Fresh and new data filtered by an estimated ARMA model

Figure 5.1 gives a relation between new data x_n as input signal and the output $\hat{\varepsilon}_n$ that gives the prediction error or the part of x_n which cannot be explained by the model

$$\hat{B}(z)\hat{\varepsilon}_n = \hat{A}(z)x_n \tag{5.34}$$

The output is no longer the true ε_n from (5.33) unless the estimated model and the true process are identical. The prediction error (PE) is defined as variance $\sigma_{\hat{\varepsilon}}^2$ of $\hat{\varepsilon}_n$ for new data as input in Figure 5.1. The squared error of one step ahead prediction is often called simply the prediction error (PE) or the one step ahead prediction error. PE is the variance of a new "product ARMA" process, which is found by substitution of x_n from the true process Equation (5.33),

$$\hat{\varepsilon}_n = \frac{\hat{A}(z)}{\hat{B}(z)}x_n = \frac{\hat{A}(z)B(z)}{A(z)\hat{B}(z)}\varepsilon_n \tag{5.35}$$

The variance $\sigma_{\hat{\varepsilon}}^2$ of $\hat{\varepsilon}_n$ is the integral of the power spectral density of $\hat{\varepsilon}_n$. Therefore, the prediction error can also be defined in the frequency domain with (4.85) as

$$\text{PE} = \sigma_{\hat{\varepsilon}}^2 = \frac{\sigma_{\varepsilon}^2}{2\pi} \int_{-\pi}^{\pi} \left| \frac{\hat{A}(e^{j\omega})B(e^{j\omega})}{A(e^{j\omega})\hat{B}(e^{j\omega})} \right|^2 d\omega \qquad (5.36)$$

This can also be written as

$$\text{PE} = \sigma_{\hat{\varepsilon}}^2 = \frac{\sigma_{\varepsilon}^2}{2\pi} \int_{-\pi}^{\pi} \frac{h(\omega)}{\hat{h}(\omega)} d\omega \qquad (5.37)$$

because the spectral density is a real number for every frequency. It is essential in this last equation that both the true and the estimated spectrum in (5.37) have been computed with the same innovation variance σ_{ε}^2. The integrals in (5.36) and in (5.37) depend only on the true and on the estimated parameters and have no explicit contribution from the true or the estimated variance of the signal. All different estimated spectra can be normalized by the same estimate $\hat{\sigma}_x^2$. Therefore, a spectral quality measure should be independent of this scaling factor.

For AR processes, the MA polynomials in (5.36) are both equal to one. Then, the ratio $\text{PE}/\sigma_{\varepsilon}^2$ is known as the likelihood ratio LR (Erkelens, 1996; Gray and Markel, 1976)

$$\text{LR} = \frac{1}{2\pi} \int_{-\pi}^{\pi} \left| \frac{\hat{A}(e^{j\omega})}{A(e^{j\omega})} \right|^2 d\omega \qquad (5.38)$$

That ratio, closely related to the PE, is used in speech coding as a quality measure.

This PE formula (5.37) shows the relative character of the prediction error measure in the frequency domain. Without proof, it is claimed that the minimal obtainable value is given by

$$\text{PE} = \sigma_{\varepsilon}^2 \qquad (5.39)$$

which is found if the true and the estimated ARMA model are the same. The proof is based on properties of the monic polynomials that appear in (5.36). The situation in the frequency domain is then that numerator and the denominator are equal, which means that the spectrum of the minimal error signal is white. The minimal value of the integral is 2π. This can also be interpreted that the best fitting model is the model that gives perfect white residuals. The frequency domain formula for the PE shows the relative character of this measure. The best estimated model is equally characterized by

white residuals
minimum PE
the quotient in (5.37) equals one

The latter property guarantees that the minimum PE is given by (5.39).

5.6.2 Model Error

Finally, the model error (ME) is defined for models that are estimated from N observations (Broersen, 1998). This is a scaled version of the prediction error, multiplied by the number of observations N from which the models $\hat{A}(z)$ and $\hat{B}(z)$ have been estimated:

$$\mathrm{ME}\left[\frac{\hat{B}_{q'}(z)}{\hat{A}_{p'}(z)}, \frac{B_q(z)}{A_p(z)}\right] = N([\mathrm{PE} - \sigma_\varepsilon^2] / \sigma_\varepsilon^2)$$

$$= N(\sigma_{\hat{\varepsilon}}^2 / \sigma_\varepsilon^2 - 1) \tag{5.40}$$

Akaike (1970, 1970a) has shown that the expectation of PE for efficiently estimated unbiased models of AR(p) models is $1 + p/N$ times the variance of ε_n, giving the value p to the ME. Likewise, the PE for "*unbiased*" ARMA(p',q') models with the model orders $p' \geq p$ and $q' \geq q$ has a lower limit,

$$E(\mathrm{ME}) \geq p' + q' \tag{5.41}$$

The minimal expectation of the model error is equal to the number of estimated parameters and is independent of the number of observations. It is the Cramér-Rao lower bound (Stoica and Moses, 1997) for achievable accuracy with unbiased models. The ME gives an easy measure for judging the quality of estimated models because its expectation has been made independent of the sample size from which the model was estimated.

The numerical value of the PE, and hence of the ME, is found with the variance of the ARMA process,

$$A(z)\hat{B}(z)\hat{\varepsilon}_n = \hat{A}(z)B(z)\varepsilon_n \tag{5.42}$$

These products of AR and MA polynomials can be multiplied to single new ARMA polynomials to give

$$C_{p+q'}(z)\hat{\varepsilon}_n = D_{p'+q}(z)\varepsilon_n \tag{5.43}$$

The variance of such an ARMA process is computed by the standard theory of Section 4.5. That separates the computations into an AR and a MA part, with the AR part first:

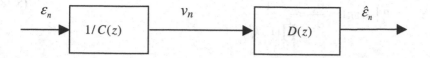

Figure 5.2. Calculation of model PE as the variance of an ARMA model

$$\hat{\varepsilon}_n = \frac{D(z)}{C(z)} \varepsilon_n \tag{5.44}$$

By computing first the AR model as a filter and afterward the MA part as a linear combination, the AR and the MA process are

$$v_n = \frac{1}{C(z)} \varepsilon_n$$
$$\hat{\varepsilon}_n = D(z)v_n$$

The MA process written out yields

$$\hat{\varepsilon}_n = D(z)v_n = v_n + d_1 v_{n-1} + \cdots + d_{p'+q} v_{n-p'-q} \tag{5.45}$$

The PE, the variance of the prediction error, is found as in (4.59) with

$$\sigma_{\hat{\varepsilon}}^2 = r_{\hat{\varepsilon}}(0) = \left(1 \ d_1 \cdots d_{p'+q}\right) \begin{bmatrix} r_v(0) & r_v(1) & \cdots & r_v(p'+q) \\ r_v(1) & r_v(0) & & \\ \vdots & \vdots & \ddots & \\ r_v(p'+q) & r_v(p'+q-1) & & r_v(0) \end{bmatrix} \begin{pmatrix} 1 \\ d_1 \\ \vdots \\ d_{p'+q} \end{pmatrix} \tag{5.46}$$

The PE is proportional to the innovation variance σ_ε^2. The model error ME follows with (5.40) and is made independent of this scaling factor.

5.6.3 Power Gain

The power gain $P_g[A(z), B(z)]$ of an ARMA process with the AR and MA polynomials $A(z)$ and $B(z)$ is defined as the ratio of the output and input variance of the ARMA process:

$$P_g[A(z), B(z)] = \frac{\sigma_x^2}{\sigma_\varepsilon^2} \tag{5.47}$$

For AR processes, it follows with (5.25) that

$$P_g\left[A(z),1\right] = \frac{\sigma_x^2}{\sigma_\varepsilon^2} = \frac{1}{\displaystyle\prod_{i=1}^{p}\left(1-k_i^2\right)} \tag{5.48}$$

This ratio is the only characteristic that is required for the ME; the absolute variance level plays no role in the ME. The relation between the ME and the power gain is

$$\mathrm{ME}\left[\frac{\hat{B}_{q'}(z)}{\hat{A}_{p'}(z)}, \frac{B_q(z)}{A_p(z)}\right] = N\left\{P_g\left[A_p(z)\hat{B}_{q'}(z), \hat{A}_{p'}(z)B_q(z)\right]-1\right\} \tag{5.49}$$

The power gain is computed as $\sigma_{\hat{\varepsilon}}^2/\sigma_\varepsilon^2$ of the ARMA process (5.43). Calculation of the variance of an ARMA process gives the numerical value for the difference between two models or between the true process and an arbitrary estimated model.

5.6.4   Spectral Distortion

The model error is one measure in a class of strongly related relative measures (de Waele, 2003; Broersen, 2001). It is a scaled prediction error. Other relative measures are the spectral distortion (SD) defined as

$$\begin{aligned}
\mathrm{SD} &= \frac{0.5}{2\pi}\int_{-\pi}^{\pi}\left[\ln h(\omega)-\ln\hat{h}(\omega)\right]^2 d\omega \\
&= \frac{0.5}{2\pi}\int_{-\pi}^{\pi}\left[\ln\frac{h(\omega)}{\hat{h}(\omega)}\right]^2 d\omega
\end{aligned} \tag{5.50}$$

where the quantity with ^ denotes the model estimate of the spectrum, as in (5.37). A comparison with (5.39) shows the asymptotic equivalence of N times the SD and the ME. The square root of the SD is also sometimes used. Gray and Markel (1976) have concluded that log spectral distance measures make the best reference for a comparison in speech coding.

5.6.5   More Relative Measures

The Kullback-Leibler discrepancy is a measure that has been developed in information theory. De Waele (2003) showed the close relationship between the Kullback-Leibler discrepancy and the model error for small differences between model and process. It is a general property of all relative measures that they are close for small differences between the true and the estimated quantities.

Stoica and Moses (1997) defined the spectral flatness as

$$SF = \frac{\exp\left[\dfrac{1}{2\pi}\displaystyle\int_{-\pi}^{\pi}\ln\dfrac{h(\omega)}{\hat{h}(\omega)}d\omega\right]}{\dfrac{1}{2\pi}\displaystyle\int_{-\pi}^{\pi}\dfrac{h(\omega)}{\hat{h}(\omega)}d\omega} \qquad (5.51)$$

The integrated ratio of squared spectral densities (IRSE) is

$$IRSE = \frac{0.5}{2\pi}\int_{-\pi}^{\pi}\left[\frac{h(\omega)-\hat{h}(\omega)}{\hat{h}(\omega)}\right]^2 d\omega \qquad (5.52)$$

The measures have different scaling. Because the performance of all relative measures is similar for small differences, a favourite measure can be chosen freely. The usual measure in this book will be the model error (ME) of (5.40). The numerical values of the different relative measures may not be the same due to scaling. However, in a comparison between a true process and several estimated models, relative measures will generally reach their minimum value for the same model.

5.6.6 Absolute and Squared Measures

Absolute accuracy measures compute the absolute value of the difference between spectra,

$$\frac{1}{2\pi}\int_{-\pi}^{\pi}\left|h(\omega)-\hat{h}(\omega)\right| d\omega \qquad (5.53)$$

or the integrated square error (ISE) of the difference between the spectra

$$ISE = \frac{1}{2\pi}\int_{-\pi}^{\pi}\left[h(\omega)-\hat{h}(\omega)\right]^2 d\omega$$

$$= \sum_{k=-\infty}^{\infty}\left[r(k)-\hat{r}(k)\right]^2 \qquad (5.54)$$

Absolute measures have no firm theoretical ground. Parseval's law applied to the Fourier transform pair of the spectrum and the autocovariance in (3.8) gives equality with the sum of squared differences of the autocovariance function in the ISE measure of (5.54). One problem of this measure follows from the estimation variance (3.37) of the autocovariance function at great lags. That has a limiting constant value for all lags where the true autocovariance is zero. That means that the estimated lagged product autocovariance is always far from the true autocovariance function. Actual differences between true and estimated autocorrelations are seen in Figure 3.6.

Of course, if this measure approaches zero, the estimated spectrum with ^ and the true spectrum are very close. However, some models with a small value for such a measure may be acceptable, and others with the same value may be completely wrong. A difference of 0.01 between the true and the estimated spectrum for a certain frequency has the same influence in absolute measures when the true spectrum is 0.001 as when it is one. Broersen (2001) has shown by simulations in an AR(1) process that the behaviour of absolute errors as a function of the model order of the estimated AR model is very irregular. Some high-order estimated AR models seemed to be the best choice because they had the smallest values of an absolute error measure. However, those high-order AR models are not good in any other useful respect. The estimated AR(1) model should be the most accurate model with any sensible quality measure. That is found with all relative measures and not with absolute measures. Therefore, absolute measures are not suitable for establishing the quality of time series models.

The influence of scaling with the estimated variance of the signal has also been studied (Broersen, 2001). It influences the absolute and squared measures, but it does not lead to a better or more useful quality measure.

The same arguments against ISE in (5.54) apply to the integral of the squared difference between the normalized spectra. With Parseval's relation, this gives an expression in both the time and the frequency domain for the integrated mean square error measure (IMSE):

$$\text{IMSE} = \frac{1}{2\pi} \int_{-\pi}^{\pi} \left[\phi(\omega) - \hat{\phi}(\omega) \right]^2 d\omega$$

$$= \sum_{k=-\infty}^{\infty} \left[\rho(k) - \hat{\rho}(k) \right]^2 \tag{5.55}$$

Figure 5.3 shows the behaviour of the PE and of three different accuracy measures, as a function of the AR order in simulations with an AR(2) process. The PE and the ME have clear minima at the true order. The absolute difference and IMSE have a minimum at order eight; order two is not even a local minimum with those measures.

Absolute and squared measures pay practically no attention to weak parts of the spectrum. Relative measures are generally preferable. IMSE may in many examples show good behaviour, but not always; sometimes completely wrong models have the smallest IMSE. The behaviour of the ME and of the other relative measures is always good. On the other hand, a small value of IMSE is no guarantee of a good or acceptable model. Taking the true autocorrelation in Figure 3.6 and subtracting 0.06 from $\rho(4)$, leaving the rest of the true function undisturbed, gives a small IMSE but makes the function no longer positive-semidefinite. Its Fourier transform is no longer positive everywhere and is no spectral estimate. Therefore, the quality of this artificial example is much worse than the quality of the estimate in Figure 3.6 for $N = 100$, but the IMSE is much smaller. Good models may generally have good quality with low values of ISE or IMSE, but the reverse is not always true. Models with a small value of ISE or IMSE are not always good or even acceptable.

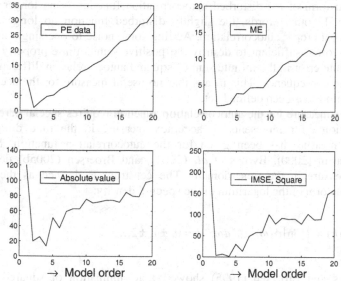

Figure 5.3. PE estimated from the observations and ME, the sum of the differences of the absolute value, and IMSE, the sum of the squared differences between true and estimated spectrum, all as a function of the AR model order. Data are generated with an AR(2) process with $A(z)=1 - 1.12\,z^{-1} + 0.4\,z^{-2}$, $N = 100$.

The relation between the ME (5.40) and the SD (5.50) is very strong. Apart from scaling, the SD and the ME are the sum of squared differences in the log spectrum, whereas the IMSE and the ISE are given by the sum of squares in the linear spectrum. Figure 3.3 demonstrates the difference. Large relative errors are important only for IMSE and ISE if they occur in the range $0.1 < f < 0.3$ in that example. Large relative errors have hardly any influence if they occur in the range where the true spectrum is less than 0.01.

5.6.7 Cepstrum as a Measure for Autocorrelation Functions

Normally distributed variables are completely characterized by their auto-correlation function or by their spectrum. Therefore, an accuracy measure for the autocorrelation function should favour the same characteristics as the spectral measure. That is not ISE of (5.54) or IMSE of (5.55), for the reasons given above. Also the argument that a small artificial error at one lag can already change the positive-semidefinite condition is a good reason not to use ISE and IMSE for autocorrelations.

The visual appearance of autocorrelation functions can also be very misleading. Large relative errors in the biased spectra of Figure 3.4 are invisible in the linear spectrum at the left-hand side of the same figure. Hence, they are also invisible in the plot of the autocorrelation function. Reversing the argument, autocorrelation functions can look identical in a plot, but at the same time have very large differences in the weak parts of the spectra belonging to them. A specific small absolute error in the time domain can have as a consequence that the inverse Fourier transform of the disturbed autocorrelation function is no longer positive everywhere. In other words, the slightly disturbed function no longer has the properties of a proper autocorrelation. Adding some noise to one single lag of the true $r(k)$ is often sufficient to destroy the positive-semidefinite property. A very small absolute error in the infinite sum of squared autocorrelation differences may have a large consequence. This means that no useful measure for the accuracy of autocorrelations has been defined yet.

An error measure for the autocorrelation function requires special care. So far, the prediction error represents an accuracy measure in the time domain. No equivalent measure has been given for the autocorrelation function. Makhoul (1976), Martin (2000), Byrnes et $al.$ (2001), and Broersen (2005b) describe a cepstral measure in the time domain. The cepstrum is defined as the inverse Fourier transform of the logarithm of the spectral density:

$$c(k) = \int_{-\pi}^{\pi} \ln[\phi(\omega)]e^{j\omega k}\,d\omega, \quad k = 0, \pm 1, \pm 2, \ldots \tag{5.56}$$

Erkelens and Broersen (1995) showed that summation of squared cepstral differences has properties similar to the spectral distortion (SD) of (5.50). Parseval's relation gives the equivalence in the time and the frequency domain between the cepstrum and the logarithm of the normalized spectrum.

$$\sum_{k=-\infty}^{\infty}\left[c(k)-\hat{c}(k)\right]^2 = \frac{1}{2\pi}\int_{-\pi}^{\pi}\left\{\ln[\phi(\omega)] - \ln\left[\hat{\phi}(\omega)\right]\right\}^2 d\omega \tag{5.57}$$

A comparison with the spectral distortion (SD) of (5.50) shows that the cepstral distance is just the same as the SD, apart from a constant two and some variance normalization. Hence, the ME can also be seen as a measure of the quality of autocorrelation functions. The literature does not describe other useful accuracy measures based on the autocorrelation function itself. Spectral measures will be used instead. An accurate spectrum belongs to an accurate autocorrelation function and vice versa.

5.7   ME and the Triangular Bias

This book treats AR models as a parametric estimator for spectrum and autocorrelation. In speech processing, AR estimation is often indicated as linear prediction. The parameters are used for efficient coding of the speech signal. It is also possible to consider an AR model as an infinite impulse response filter. Therefore, many different descriptions in various scientific disciplines treat a similar theoretical problem. An example is bias propagation in linear prediction. If lagged product estimates for the autocorrelation function are used in the Yule-Walker relations, the triangular bias of (3.33) plays a role.

Erkelens and Broersen (1997) showed that triangular bias becomes very noticeable if an AR(p) process has a reflection coefficient close to the unit circle for an order lower than p. If only the final reflection coefficient of order p is close to the unit circle, this has no strong influence. Using the expectation of the unbiased and of the biased autocorrelation function of (3.32), the expectations of the parameters of the biased and of the unbiased model can be calculated with the Levinson-Durbin recursion (5.24). The unbiased expectations are the true AR(p) parameters. The difference can be evaluated with the ME. It can become large if reflection coefficients of order lower than p are close to $+1$ or -1. It turns out that the ME becomes very large if one of the intermediate reflection coefficients, of an order lower than the true order, has a radius of about $1 - 1/N$. That means that the ME becomes much greater than the number of estimated parameters that gives the expected variance contribution in (5.41) to the ME for the smallest possible inaccuracy of unbiased estimation.

Table 5.1. The model error ME caused by triangular bias applied to the true autocorrelation function for two AR(2) processes as a function of N

N	$A_1(z)$	$A_2(z)$
100	161	175
1,000	447	19
10,000	147	1.9
100,000	17	0.2
1,000,000	1.8	0.02

Table 5.1 gives an example of two AR(2) processes with the same values for the reflection coefficients but in a different sequence:

$A_1(z)$ with $k_1 = 0.999$ and $k_2 = 0.8$, poles at radius 0.8 and 0.99
$A_2(z)$ with $k_1 = 0.8$ and $k_2 = 0.999$, two poles at radius 0.9995

The high ME value for $N = 100$ shows that 100 observations will not be enough to analyse those AR processes accurately. Both examples have serious bias problems. The autocorrelation length is much longer than 100, and the influence of the triangular bias is not well predicted for finite N by the asymptotic formulas, where it disappears. For more observations, the ME becomes much greater for the first process. That shows that reflection coefficients with an absolute value close to one give a high bias to the reflection coefficients of higher orders; see also Erkelens and Broersen (1997). If only the highest order reflection coefficient is close to ± 1, such a bias will not develop. In contrast with a popular belief, this example shows that it is not the radius of the poles that determines the influence of triangular bias on the accuracy of AR models.

A constant bias would give a constant PE and hence a ME value that is proportional to N. The ME decreases in Table 5.1 for increasing N, which shows that the influence of the bias decreases. The triangular bias will vanish asymptotically but can have a strong influence in finite-sample sizes. This example shows that asymptotic unbiased models can still have serious bias problems in very large sample sizes.

The difference between absolute and relative error measures has already been shown in Figure 3.4, where the influence of triangular bias was demonstrated in linear and logarithmic plots of the spectrum. That example had an autocorrelation function of length 13, which can be generated by a MA(13) process. Take the true and the biased true autocorrelations of (3.32) as an example to compute the difference between error measures. The absolute squared error in the autocorrelation with the triangular bias of this MA(13) process becomes with (5.55)

$$
\text{IMSE} = \sum_{k=-13}^{13} \left[\rho(k) - \hat{\rho}(k) \right]^2
$$

$$
= \sum_{k=-13}^{13} \left[\frac{k}{N} \rho(k) \right]^2 \approx \frac{25.1}{N^2} \tag{5.58}
$$

This gives a very small number for $N > 1000$, which explains why absolute figures like the left-hand side in Figure 3.4 with a linear scale for the spectral density do not show this bias. The IMSE value becomes very small and the differences disappear within the line thickness. The difference of the logarithms is not small on the right-hand side of Figure 3.4. That difference defines the spectral distortion in (5.50). The values for the model error (ME), the spectral distortion (SD) and the integrated squared difference (IMSE) of (5.58) are given in Table 5.2, as a function of sample size N.

Table 5.2. The model error (ME), N times the spectral distortion (SD), and N times IMSE, caused by the triangular bias applied to the true MA(13) autocorrelation function of Figure 3.2, as a function of N.

N	ME	SD*N	IMSE*N
10^3	$1.2 \ 10^4$	$1.9 \ 10^4$	$2.5 \ 10^{-2}$
10^6	$1.8 \ 10^6$	$5.7 \ 10^6$	$2.5 \ 10^{-5}$
10^9	$2.8 \ 10^8$	$7.9 \ 10^8$	$2.5 \ 10^{-8}$
10^{12}	$2.6 \ 10^9$	$3.2 \ 10^9$	$2.5 \ 10^{-11}$

The ME is defined by multiplication by N; the spectral distortion (SD) and IMSE have been defined without. To facilitate a comparison, the last two measures have been multiplied by N in Table 5.2. The ME and the SD show a similar tendency. For infinitesimally small errors, they can coincide completely with proper scaling; for larger errors, some differences are present. The ME values for the triangular bias in this example are very much greater than the expectation of the ME due to estimation variance that will be only about 13 for all values of N. The absolute IMSE measure of (5.58) decreases with N^2 and has completely different behaviour. This example shows once more that models with a very small IMSE value can have a large PE or ME.

5.8 Computational Rules for the ME

So far, several equivalent formulas for the model error (ME) have been given in (5.40) and in (5.49). The two equivalent expressions for the ME are

$$
\mathrm{ME}\left[\frac{\hat{B}_{q'}(z)}{\hat{A}_{p'}(z)}, \frac{B_q(z)}{A_p(z)}\right] = N\left(\frac{\mathrm{PE}}{\sigma_\varepsilon^2} - 1\right)
$$

$$
= N\left\{ P_g\left[A_p(z)\hat{B}_{q'}(z), \hat{A}_{p'}(z)B_q(z)\right] - 1 \right\} \tag{5.59}
$$

The last formulation especially shows that only the two *products* in the numerator and in the denominator play a role. Therefore, some rules for the ME computation can be derived and proved easily with the final representation in (5.59). All equivalent representations are found by moving the true and estimated polynomials in the first representation of (5.59) from the numerator on the left-hand side to the denominator of the right-hand side. Some equivalent ME results are

$$
\mathrm{ME}\left[\frac{\hat{B}_{q'}(z)}{\hat{A}_{p'}(z)}, \frac{B_q(z)}{A_p(z)}\right] = \mathrm{ME}\left[\frac{A_p(z)\hat{B}_{q'}(z)}{1}, \frac{\hat{A}_{p'}(z)B_q(z)}{1}\right]
$$

$$
= \mathrm{ME}\left[\frac{1}{1}, \frac{\hat{A}_{p'}(z)B_q(z)}{A_p(z)\hat{B}_{q'}(z)}\right]
$$

$$
= \mathrm{ME}\left[\frac{1}{\hat{A}_{p'}(z)}, \frac{B_q(z)}{A_p(z)\hat{B}_{q'}(z)}\right]
$$

$$
= \mathrm{ME}\left[\frac{\hat{B}_{q'}(z)}{1}, \frac{\hat{A}_{p'}(z)B_q(z)}{A_p(z)}\right]
$$

$$
= \mathrm{ME}\left[\frac{A_p(z)}{\hat{A}_{p'}(z)B_q(z)}, \frac{1}{\hat{B}_{q'}(z)}\right] \tag{5.60}
$$

The variance of the ARMA process $B_q(z)/A_p(z)$ is not the same as the variance of the inverse process $A_p(z)/B_q(z)$. Interchanging the numerator and denominator in the ME is not allowed theoretically. However, interchanging the sequence in P_g expressions will sometimes be a close approximation if the ME is small. That means that the prediction error is close to the variance of the innovations and that the resulting ARMA process is close to white noise or the estimated model is close to the true process. For accurate models, it follows approximately that

$$
\mathrm{ME}\left[\frac{B_q(z)}{A_p(z)}, \frac{\hat{B}_{q'}(z)}{\hat{A}_{p'}(z)}\right] \approx \mathrm{ME}\left[\frac{\hat{B}_{q'}(z)}{\hat{A}_{p'}(z)}, \frac{B_q(z)}{A_p(z)}\right] \approx \mathrm{ME}\left[\frac{\hat{B}_{q'}(z)}{1}, \frac{\hat{A}_{p'}(z)B_q(z)}{A_p(z)}\right]
$$

$$
\approx \mathrm{ME}\left[\frac{1}{\hat{A}_{p'}(z)}, \frac{B_q(z)}{A_p(z)\hat{B}_{q'}(z)}\right] \tag{5.61}
$$

The first line of (5.61) changes the sequence of the arguments in the ME, which is the same as interchanging the numerator and the denominator in the power gain P_g. The second line gives the difference between a MA(q') process with q' parameters on the left-hand side and an ARMA(p, $p'+q$) process on the left-hand side. The final equation compares an AR(p') model with an ARMA($p+q'$, q) process.

The Yule-Walker relations and the Levinson-Durbin recursion describe the relation of AR models of increasing orders with a given true autocorrelation function. The best lower order approximation uses the lower order reflection coefficients, derived from the lower order autocorrelation function. Therefore, the best approximating AR(m) model to a true ARMA($p+q'$, q) process is the model with the first m AR autocorrelations equated to the first m autocorrelations of the ARMA($p+q'$, q) process.

The best AR(m) model is also the AR(m) model that minimises the model error (ME). Hence, the best AR(m) model of an arbitrary true ARMA(p,q) process can be written as

$$\hat{A}_m(z) = \arg\min_{\tilde{A}_m(z)} \left\{ ME\left[\frac{1}{\tilde{A}_m(z)}, \frac{B_q(z)}{A_p(z)} \right] \right\}, \quad \forall m \tag{5.62}$$

This can be solved with the Yule-Walker relations applied to the autocovariance generated by the right-hand side ARMA process $B_q(z) / A_p(z)$.

It is easy and straightforward to find the best AR(m) model with arbitrary values of m that approximates an ARMA process. It is the AR model that is fitted to the first m points of the autocorrelation function. It is always stationary.

However, it is much more difficult to find a MA(m) model that is best fitting to a higher order ARMA process. This has been demonstrated in (5.32), where one MA parameter is found from one true autocorrelation. Finding the best fitting MA model of arbitrary order to a process requires an extensive nonlinear search for the minimum of the second line in (5.61). That is not attractive. The calculation of those MA models is nonlinear and requires initial conditions. This is a good reason to formulate estimation problems, preferably looking for the best AR model by interchanging polynomials, as in Equation (5.61). That interchange may not be accurate for models with poor quality, but it is a good approximation for those models where the residuals are close to white noise.

The different operations with the ME in this section will assist in finding the best formulation for MA and ARMA estimation. The approximate interchange of the sequence of the arguments in ME is especially useful in deriving estimation algorithms for MA and ARMA models.

5.9 Exercises

5.1   Vectors in Matlab® are represented as $[a\ b\ c\ d\ e\ ...]$. This notation is also used in the exercises. The reflection coefficients of an AR(2) process are given by the vector $[1,\ 0.8,\ 0.5]$. It is usual to include the zero-order parameter or reflection coefficient that should always be equal to one. Use ARMASA to determine the parameters of the process.

5.2   The first two autocorrelations of a process are $\rho(1)$ and $\rho(2)$. What is the best AR(1) model?

5.3   The first two reflection coefficients of a process are k_1 and k_2. What is the best AR(1) model?

5.4   The parameters of an AR(2) process are a_1 and a_2. What is the best AR(1) model?

5.5   The reflection coefficients of an AR(2) process are given by $[1,\ 0.8,\ 0.5]$. Use ARMASA to determine the autocorrelation function until lag 20.

5.6   The parameters of an AR(3) process are given by $[1,\ 0.08,\ -0.472,\ 0.4]$. Use ARMASA to determine the reflection coefficients of the process.

5.7   The parameters of an AR(3) process are given by $[1,\ 0.08,\ -0.472,\ 0.4]$. Use ARMASA to determine the autocorrelation function until lag 20.

5.8   The parameters of an AR(3) process are given by $[1,\ 0.08,\ -0.472,\ 0.4]$. Use ARMASA to determine the parameters of the best AR(2) process that is an intermediate model in the Levinson-Durbin recursion.

5.9   The autocovariance function of an AR process is given by $[5,\ -4.5,\ 3.29,\ -1.9692,\ 0.9753,\ -0.4045,\ 0.1422,\ -0.0428,\ 0.024,\ -0.0796,\ 0.2252]$. Use ARMASA to determine the parameters of the AR process.

5.10   Use ARMASA to determine the reflection coefficients of the AR process with the autocovariance function given in Exercise 5.6.

5.11   Show that the second reflection coefficient of an AR(p) process is given by

$$k_2 = \frac{r^2(1) - r(2)r(0)}{r^2(0) - r^2(1)}.$$

5.12   Use ARMASA to determine the power spectral density of the AR process with the autocovariance function given in Exercise 5.6.

5.13   Use ARMASA to determine whether the parameter vector
$$[1, -1.5, 2, -0.9]$$
belongs to a stationary process. Hint: Use the fact that processes are stationary with roots inside the unit circle if all reflection coefficients are not greater than one in amplitude.

5.14   Use ARMASA to determine whether the parameter vector
$$[1, -2.6, 2.66, -1.15, 0.1]$$
belongs to an invertible process. Hint: Use the fact that the polynomials of an AR and a MA process are similar.

5.15   The variance of an ARMA(2,2) process given by

$$x_n - 1.2x_{n-1} + 0.8x_{n-2} = \varepsilon_n + 0.2\varepsilon_{n-1} + 0.7\varepsilon_{n-2}$$

is known to be 10. Use ARMASA to determine the variance of the innovations ε_n.

5.16   The influence of triangular bias on the true autocorrelation function can be evaluated for a given AR(p) process by computing the true autocorrelation function, multiplying $r(m)$ by $1 - m/N$. Then, the new parameters belonging to the biased autocovariance are calculated and the difference is evaluated with the model error.
Find an AR(3) process where the model error caused by this bias is greater than 1000 for $N = 1000$ and a second example where the model error is less than 0.01. Use ARMASA to determine the model errors and the biased parameter values.

5.17   Use ARMASA to verify that the model error is not symmetrical with respect to AR and MA parameters.

5.18   Use ARMASA to verify that interchanging the true AR and the approximating MA polynomials does not influence the model error. The same is valid for the true MA and the approximating AR polynomial.

5.19   Use ARMASA and trial and error to find the best approximating AR(1) model parameter for the AR(2) process

$$x_n + 0.8x_{n-1} + 0.6x_{n-2} = \varepsilon_n$$

by looking for the smallest model error. Find an explanation for the resulting parameter value.

5.20   Use ARMASA and trial and error to find the best approximating MA(1) model parameter for the MA(2) process

$$x_n = \varepsilon_n + 0.8\varepsilon_{n-1} + 0.6\varepsilon_{n-2}$$

by looking for the smallest model error.

5.21   Use ARMASA and trial and error to find the best approximating AR(1) model parameter for the AR(3) process

$$x_n + 0.8x_{n-1} + 0.6x_{n-2} + 0.4x_{n-3} = \varepsilon_n$$

by looking for the smallest model error. Find an explanation for the resulting parameter value.

5.22   Use ARMASA and trial and error to find the best approximating MA(1) model parameter for the MA(3) process

$$x_n = \varepsilon_n + 0.8\varepsilon_{n-1} + 0.6\varepsilon_{n-2} + 0.4\varepsilon_{n-2}$$

by looking for the smallest model error.

5.23   Use ARMASA and trial and error to find the best approximating AR(1) model parameter for the MA(1) process

$$x_n = \varepsilon_n + 0.5\varepsilon_{n-1}$$

by looking for the smallest model error. Find an explanation for the resulting parameter value.

5.24   Use ARMASA and trial and error to find the best approximating MA(1) model parameter for AR(1) process

$$x_n - 0.4x_{n-1} = \varepsilon_n$$

by looking for the smallest model error.

5.25   Use ARMASA to verify with an example that a lower order approximating AR model for a higher order true AR process is the best, with the smallest model error, if the lower order reflection coefficients are made equal to the corresponding true process reflection coefficients. Verify also that the same result is obtained by using the only first autocorrelation lags to estimate the lower order model.

5.26   The reflection coefficients of an AR(3) process are k_1, k_2, and k_3. Find an expression for the parameters of the process.

6

Estimation of Time Series Models

6.1 Historical Remarks About Spectral Estimation

Spectral estimation has a long history, in which the progress has been influenced by theoretical and by computational developments. Only the studies on *stationary* stochastic processes are followed here. This seems to be a severe mathematical restriction for measured random data. In practice, however, the definition of stationarity can be treated very loosely. For example, data like speech often can be considered stationary enough over small intervals, and autocorrelations and spectra have useful interpretations for each interval.

A clear computational influence was the use of the FFT (fast Fourier transform) algorithm of Cooley and Tukey (1965) for Fourier transforms. The reduced computer effort of the FFT algorithm enabled routine analysis of extensive sets of data with periodogram analysis. Therefore, nonparametric spectral analysis with tapered and windowed periodograms has been the main practical tool for spectral analysis for a long time.

If one knew the model order and the model type, a single time series model of a low order could also be estimated from practical data. However, the best order and the best model type are almost never known for a finite number of measured observations. Therefore, time series models were not practical tools in the past.

Some historical developments in spectral estimation show the combined growth of both parametric and nonparametric methods. More than a century ago, Schuster (1898) used periodogram analysis to find hidden periodicities. Yule (1927) published an article about autoregressive models. Throwing peas on a pendulum, thus giving a physical introduction to autoregressive modeling, supposedly causes deviations from a pure harmonic motion in a pendulum. The first description of real data with moving average models is attributed to Slutsky; it was translated in 1937 but written 10 years earlier. Time and frequency domain considerations were united for stochastic processes by the independent contributions of Wiener (1930) and Khintchine (1934).

Maximum likelihood is a reliable principle for deriving efficient estimators in ordinary linear regression problems. Mann and Wald (1943) proved that *for large N*, it is justified to use the same AR data both as dependent and as independent variables in regression theory and to apply the maximum likelihood principle.

Their reasoning relied heavily on the fact that the exact parameters relate the data x_n to the exact innovations ε_n and the properties of the true white noise signal can be exploited. The theoretical properties of maximum likelihood estimators are derived when parameters are estimated without any error, which might happen if the sample size is infinite. The theory itself does not give any indication how many observations are at least required before the theory may be applied.

What happens for a finite value of N or what is the minimal value of N to apply this asymptotic theory has not been defined. This was a reason for Whittle (1953) to state, "*it is by no means obvious that these* (ML) *properties are conserved when the sample variates are no longer independent, e.g., in the case of a time series.*"

Whittle (1953) also showed that using only the first two estimated sample autocovariances to calculate a MA(1) model is very inefficient, although the expectations of all higher true autocovariances are zero.

The invariance property of maximum likelihood estimates is very important for the theoretical background of spectral estimation. It states that, under rather mild mathematical conditions, the maximum likelihood estimator of a function of a stochastic variable is equal to the function of the maximum likelihood estimate of the variable itself (Zacks, 1971; Porat, 1994). This invariance property of maximum likelihood estimates is the theoretical underpinning of the use of time series analysis for the estimation of the autocorrelation function and the spectrum of measured observations. If it is possible to obtain efficient estimates for the parameters of a time series model, the autocorrelation is estimated efficiently with (4.62) as a function of those parameters, and an efficient estimate of the spectrum follows with (4.85).

The spectrum of a MA(1) process, obtained from only one lagged product autocorrelation estimate is not efficient. The autocovariance and the spectrum can be estimated much more accurately from the estimated MA(1) time series model (Porat, 1994). Porat (1994) proved that $0 \le k \le p - q$ sample autocovariances are asymptotically efficient in an ARMA(p,q) process. Hence, lagged products are not generally efficient autocovariance estimates. The periodogram as an inverse Fourier transform of the lagged product autocovariance can be nothing but an inaccurate estimator of the spectral density for random signals.

Arato (1961) showed that the first $p + 1$ lagged products are asymptotically a sufficient estimator if the data are from an AR(p) process. This is no longer true for other types of processes.

Maximum likelihood estimation is a nonlinear operation for all time series models. That gives problems with convergence and with stationarity or invertibility of models, especially in small samples. Different computationally efficient algorithms have been derived by using approximations of the likelihood. Durbin (1959) introduced an algorithm for MA estimation from a long estimated AR model. A year later, Durbin (1960) used the long AR model to reconstruct estimated residuals to be used in ARMA estimation and an additional alternating update of the MA and the AR parameters.

Melton and Karr (1957) used only the sign of stochastic processes for the detection of signals in noise with the polarity coincidence correlator. Wolff *et al.* (1962) concluded that for Gaussian processes, the polarity coincidence principle is inferior to lagged product correlation. Blackman and Tukey (1959) showed how a few lagged product autocovariance estimates could be transformed into a valid spectral estimate by applying lag windows.

Burg (1967) described a very robust AR estimator that estimates one reflection coefficient at a time from forward and backward residuals. Meanwhile, after the publication of Cooley and Tukey (1965), the FFT caused a revival of periodogram based spectral estimation and related lagged product autocovariance estimation. Box and Jenkins (1970) showed how time series models could be estimated from practical data. Pioneering work on order selection has been done by Akaike (1970, 1974, 1978), who introduced the celebrated selection criterion AIC. Parzen (1974) discussed the relation between time series models and prediction and the physical and econometrical modeling of a truly AR(∞) process with finite-order AR approximations. Priestley (1981) described parametric and non-parametric spectral estimators in a mathematically accessible style.

Kay and Marple (1981) conclude after an extensive survey of different algorithms for time series models, that, "When the model is an accurate representation of the data, spectral estimates can be obtained whose performance exceeds that of the classical periodogram." In other words, if the model type and model order are known *a priori*, time series models give the best solution.

Two problems remained for the standard application of time series models:

- order selection
- the availability of practical fast algorithms.

Selection of the model order could give peculiar results in finite samples with a preference for the highest candidate orders. Ulrych and Bisshop (1975) came up with the approach to use *sometimes* the first local minimum of a selection criterion. The theoretical asymptotic answer of consistent criteria of Akaike (1978), Rissanen (1978), and Hannan and Quinn (1979) could not solve this selection problem in practice. It turned out that finite-sample statistics of the estimated AR reflection coefficients is so much different from asymptotic properties that it has to be incorporated in order-selection criteria (Broersen, 1985, 2000). Moreover, finite-sample AR order selection has to be adapted to the AR estimation method (Broersen and Wensink, 1993).

The second problem was the variety of estimation algorithms with similar asymptotic properties. No choice could be made on the basis of asymptotic theory, although experience taught that the statistical accuracy of algorithms may be poor in certain finite-sample examples. Moreover, the mathematical theory is almost exclusively valid for models that have the true model type and model order. This knowledge is not available for newly measured data. In practice, a large number of candidate models of the three types, AR, MA, and ARMA, and of different orders have to be estimated. Almost all candidates are from the wrong model type and model order. From all candidates, a single estimated model has to be selected that best represents the statistical properties of the measured data. This requires robust algorithms, fast computers, and reliable order-selection criteria.

A successful and robust attempt was reported to select the model type and order from stochastic data with unknown characteristics by Broersen (2000). The increased computer speed gives the opportunity of estimating hundreds or thousands of candidate models and selecting with reliable statistical criteria one of the best among the candidates, and often the very best.

6.2 Are Time Series Models Generally Applicable?

The spectral density of an ARMA process is a rational function of $e^{j\omega}$ because it is the quotient of polynomials in $e^{j\omega}$. Conversely, all processes with rational spectra can be written as ARMA models. The mathematical background follows from (4.85). The question has two sides: which functions can be described and how can the parameters for some given spectral function be determined?

Which polynomials $A(e^{j\omega})$ and $B(e^{j\omega})$ can describe an arbitrary continuous positive symmetrical function with the quotient

$$h(\omega) = \frac{\sigma^2}{2\pi} \frac{B(e^{j\omega})B(e^{-j\omega})}{A(e^{j\omega})A(e^{-j\omega})}$$ (6.1)

in the interval $-\pi < \omega \le \pi$? The solution can be unique only if an additional requirement is used that all poles and zeros must be inside the unit circle. This is valid for all continuous spectra. Therefore, harmonic processes that have discrete spectral components are excluded.

Mathematical arguments or simple reasoning will show that a function that is identically zero over some interval cannot be represented by a function like (6.1). Historically, this has been formulated as follows (Priestley, 1981). If

$$\int_{-\pi}^{\pi} \log[h(\omega)] d\omega > -\infty$$ (6.2)

then a unique one-sided $G(z)$ exists with the sequence $g_0, g_1, \dots$ which has all zeros inside the unit circle. The integral of the logarithm will become $-\infty$ only if the power spectral density is exactly zero over an interval or over more than one consecutive point. The spectrum may touch zero at a single point, not at an interval. This is a mild requirement for spectra in practice. Therefore, almost all stationary stochastic processes can be modeled by a unique, stationary, and invertible ARMA process.

The use of time series models for spectra does not imply more assumptions about the data than the use of modified periodograms.

Only a small selection of all algorithms that have been proposed will be mentioned. Mainly those estimation algorithms that are used in ARMAsel will be described in some detail in this chapter. More information about time series estimation can be found in Priestley (1981), Kay (1988), Marple (1987), Kay and Marple (1981), Box and Jenkins (1976), Brockwell and Davis (1987), Porat (1994), Stoica and Moses (1997), Hamilton (1994), Makhoul (1976), Pollock (1999), and in many other books and journal papers.

6.3 Maximum Likelihood Estimation

In estimation theory, maximum likelihood estimators are known to have nice asymptotic properties. The invariance property transfers the properties of estimated parameters to spectral analysis. Therefore, the first estimators that will be investigated are the maximum likelihood (ML) estimators for AR, MA, and ARMA models. As ML estimators require the knowledge of the probability density function of the observations, the density has to be chosen. A logical choice is the normal distribution, and the joint normal density (2.22) will be the basis. The model parameters enter in that equation because they determine the covariance matrix R_{XX}. ML estimation maximises the likelihood or the probability density function (2.22) by seeing which parameters maximise the likelihood if the given observations are substituted. Often, minus the logarithm of the likelihood is minimised, which gives the same parameter vector as the solution.

ML estimation requires a numerical search procedure to find the minimum. It turns out that often precautions are necessary to make sure that the search algorithm will not end with a nonstationary or a noninvertible model. An arbitrary monic polynomial can be expressed in reflection coefficients k_i with (5.30). All reflection coefficients have an absolute value less than one if the roots of the polynomial are within the unit circle. Unfortunately, optimisation without constraints will often find a minimum of the log likelihood with roots outside the unit circle. Minimisation with constraints will give answers that depend on the prescribed constraints rather than on the data. By using the unconstrained optimisation of tangent $\theta_i = \tan(\pi/2 * k_i)$, it can be ascertained that for each solution $-\infty < \theta_i < \infty$, a solution k_i is found as $2/\pi * \arctan(\theta_i)$ with the property that $-1 < k_i < 1$.

ML estimation is a method that should be efficient asymptotically. That means that the asymptotically expected value for the model error (ME) is equal to the number of estimated parameters. That is the Cramér-Rao lower bound for achievable accuracy. Firstly, the quality of ML estimators is investigated and compared with this best attainable accuracy. Afterward, other estimation algorithms will be compared with the same Cramér-Rao benchmark.

6.3.1 AR ML Estimation

Low-order AR models are estimated without any problem by the maximum likelihood estimator. The theory promises favourable asymptotic properties, and finite-sample performance can be studied with Monte Carlo simulations. De Waele (2003) has given a comparison of ML estimates for 20 observations with estimates obtained by the method of Burg (1967). He concluded that the model error (ME) of ML estimates is generally some percent greater than the ME of the Burg estimates. But other examples have also been examined, e.g., an AR(6) process with three complex conjugated pole pairs at radius β. De Waele (2003) found that models estimated from $N = 20$ observations are better with Burg's method for $\beta < 0.85$ and ML was better for $\beta > 0.85$. For $\beta < 0.8$, the accuracy of the ME was close to the Cramér-Rao lower bound, which is six for the AR(6) process. For higher values of β, the Cramér-Rao lower bound for the model error (ME) was not attained by any estimation method. For $\beta = 0.95$, the ME of ML estimates is about 3.5 times higher than the lower bound, and the ME of Burg estimates is about 6.5 times. The differences in ME values are always small if Burg's method is better; they can become greater if all poles are at a radius greater than $1-1/N$. However, that is a rather extreme example.

One aspect, the required computing time, has not been considered so far. Without going into detail, it can be said that ML requires much more time. Whereas one or two ML parameters can be computed within a second, it will require about several hours to optimise the likelihood of an AR(30) model with the present computer software and hardware. Moreover, in trying to optimise AR models of orders still higher than 20 or 30, often numerical problems appeared that prevented convergence. It was not possible to compute AR models of orders higher than 50 on a regular basis. Often, convergence problems appear already for models of orders less than 20. The Burg algorithm can compute an AR(1000) model in less time than ML requires to estimate an AR(5) model. For those practical reasons, ML is not recommended as the standard solution for estimating AR models.

6.3.2 MA ML Estimation

Maximum likelihood estimation of MA parameters is a nonlinear problem with possible difficulties in convergence and invertibility of solutions. Nonlinear methods sometimes don't converge to invertible models for small sample sizes N, unless the algorithm uses constrained minimisation or takes reciprocals of estimated non-invertible roots. If estimated poles and zeros can be inside or outside the unit circle, they can also be close to it. The single realisations where invertible solutions do have zeros very close to the unit circle, with distance less than $1/N$, dominate the simulation average of quality measures for estimated models. Therefore, ML estimation for MA models has never been popular.

Davidson (1981) has described the many problems that occur in finite-sample estimation for MA(1) processes. A problem is the symmetry of the likelihood function with respect to the unit circle. Mirrored zeros have identical autocorrelation functions in Equation (4.12). The symmetry of the likelihood function always causes a local extreme value on the unit circle, which can be either a maximum or minimum. As a result, the global maximum of the likelihood function is often found exactly on the unit circle, even if the true process has a zero within. Approximations of the likelihood also suffer from the same problem. Godolphin and de Gooijer (1982) have studied many approximations of the exact likelihood, but they encountered convergence problems in some cases for simple MA(1) processes. It has been shown in (5.32) that no real solution for b_1 exists if the estimated $|\rho(1)|$ is greater than 0.5. In practice, the estimation variance of the first autocorrelation coefficient will often give estimates greater than 0.5, and the ML solution will always be found exactly on the unit circle then.

In simulations, de Waele (2003) determined that the local minimum on the unit circle is often also the global minimum. In a MA(1) example, he found the ML estimate in 12% of the simulation runs exactly on the unit circle, although the true MA(1) parameter was 0.5. Using constraints to force the estimated zeros within the unit circle will often produce that imposed constraint as a solution. ML does not seem to be a good estimation method for MA models in finite samples. For low-order MA models from finite samples, the average model error of the ML solution in simulations is never close to the Cramér-Rao lower bound.

6.3.3 ARMA ML Estimation

ARMA models combine the properties of poles and zeros. With respect to estimated MA parameters, the peculiar ML properties of pure MA models are found in some simulations. There is a tendency to estimate zeros exactly on the unit circle. For a true ARMA(p,q) process with $A(z)$ and $B(z)$ in (4.54), a perfectly fitting ARMA($p+1, q+1$) model is found by multiplying the ARMA(p,q) process by an ARMA(1,1) process with an identical pole and zero. In estimating an over-complete ARMA($p+1, q+1$) model, all additive, canceling, pole-zero combinations represent the same solution. Therefore, convergence of the ML estimator will be a problem for overcomplete models. The same problem can also occur in the estimation of true ARMA(p, q) models, if a close pole-zero pair is present. That pair can wander over the complex plane without significant influence on the power spectral density of the resulting model. If parameters are rather small, poles and zeros are found close to the origin of the complex plane and automatically become close pole-zero pairs.

Some authors (Kay, 1988) impose the restriction that the ARMA model orders are not greater than the true process order. That eliminates the problem of over-complete models, but it requires *a priori* knowledge of the true order that is almost never available for practical data. De Waele (2003) determined in simulations that the finite-sample likelihood of overcomplete estimated models seems to be somewhat better than expected with the asymptotic theory, whereas the model error (ME) of those models was worse than expected. The explanation may be that freely selecting a position for a nearly canceling pole-zero pair gives an unexpected profit for the fit of the likelihood that is optimised for the given data and at the same time creates an extra loss of quality in an objective measure like the ME.

The performance of true order ML estimation in simulated ARMA processes with very significant parameters was acceptable. The performance of estimated models above the true order was poor. The performance of ML estimates for true order models with moderate true parameter values has also been tested. A simulation experiment with an ARMA(3,2) process with all poles and zeros at a radius of 0.4 gives an average ME of 86 for $N = 200$, far above the Cramér-Rao lower bound that is five for this process. The same experiment with all poles and zeros at a radius of 0.8 gives 6.2 as the average ME value in simulations, very close to the Cramér-Rao bound. Small true parameter values in the AR and in the MA part cause problems because they have poles and zeros with small radii. It is easily seen that the parameter of order p in an AR(p) model equals the product of the radii of all poles. In other words, small final parameters belong to a small radius of at least one pole.

The overall conclusion is that the performance of ML estimation in low-order AR models is acceptable, but the computation time is rather high. MA models converge too often to zeros exactly on the unit circle, and ML is not suitable for

MA models. Finally, the ML estimates for ARMA models are acceptable only for model orders that are not above the true order with the extra condition that all true parameters have to be very significant. Therefore, other estimators will be tried for all model types. The accuracy of those estimators will be compared to the Cramér-Rao bound. Only estimators that are close to that bound for all examples will be accepted.

6.4 AR Estimation Methods

The AR model type is the backbone of time series analysis in practice. In theory, it is a model with simple relations between the parameters and the autocorrelation function. Both computing parameters for a given autocorrelation function and the reverse problem are simple with Yule-Walker relations. Finding the parameters for a given MA autocorrelation function was much more difficult and is only easily obtained for a single MA(1) parameter with (5.32).

If a few observations of stochastic data are available, the best AR model order is often greater than $0.1N$, where N is the number of observations. If the model order is not very much smaller than the sample size N, say less than $0.1N$, Broersen (1985) showed that deviations from asymptotic theory become important. This requires some care because the outcome of AR estimation in finite samples depends on the algorithms used. The finite-sample theory empirically describes the different outcomes of those algorithms. The usual theory in the majority of the literature is asymptotic. It is the same for all AR estimation algorithms, and it gives no explanation for the observed differences among estimation methods (Ulrych and Bisshop, 1975).

Finite-sample effects are neglected, although they are completely responsible for the differences in performance of the various estimation methods. Therefore, the asymptotic literature has to remain vague on the subject of the differences among methods. Broersen and Wensink (1993) give a finite-sample comparison for AR estimation methods. They studied the bias and the variance of estimated parameters, as well as the average quality of estimated models. Several AR methods will be treated briefly here to indicate why they were not selected as the preferred algorithm. Detailed descriptions can be found elsewhere (Kay and Marple, 1981). Only the preferred algorithm will be treated in detail.

6.4.1 Yule-Walker Method

The *Yule-Walker method* of AR estimation uses the Yule-Walker Equations (5.4), with the lagged product autocovariance estimates defined in (3.30). The biased autocovariance estimates are substituted for the true autocovariances $r(k)$ in (5.4). Exactly the same solution can also be obtained with a least-squares algorithm. First, the signal x_n is made infinitely long by surrounding it with zero observations outside the interval 1 - N. A least-squares solution estimates all parameters simultaneously by minimising

$$\text{RSS}(p) = \sum_{n=-\infty}^{\infty} \left(x_n + \hat{a}_1 x_{n-1} + \cdots + \hat{a}_p x_{n-p} \right)^2 \tag{6.3}$$

A comparison with the ordinary least-squares regression solution (2.38) shows that this solution involves the estimated autocorrelation function of an infinitely long signal. That autocorrelation is just the lagged product biased estimate because all additional terms are zero.

This well-known Yule-Walker AR estimation method has a small variance, but it can have a very large bias. In Chapter 5.7, it has been demonstrated that the bias may become of magnitude one for true reflection coefficients with an absolute value close to one, instead of the bias of magnitude $1/N$ that applies for other AR estimation methods. Therefore, the Yule-Walker method cannot be advised for data with unknown characteristics. The small estimation variance of the parameters (Broersen and Wensink, 1993) will give favourable overall accuracy if the bias is not important for a given example. But this small variance advantage is completely lost for other processes with huge biases (Erkelens and Broersen, 1997).

6.4.2 Forward Least-squares Method

The forward least-squares method uses $N - K$ residuals to estimate an $AR\,(K)$ model. A least-squares solution estimates all parameters simultaneously by minimising

$$\text{RSS}(K) = \sum_{n=K+1}^{N} \left(x_n + \hat{a}_1 x_{n-1} + \cdots + \hat{a}_K x_{n-K} \right)^2 \tag{6.4}$$

The first K contributions would require observations from before the measurement interval and they are omitted in the residual sum of squares RSS(K) that is minimised in the forward least-squares estimation. Broersen and Wensink (1993) showed that the finite-sample variance of estimated parameters is approximately given by $1/(N + 2 - 2K)$ instead of by the asymptotic parameter variance $1/N$. The difference becomes noticeable for $K \approx N/10$ and becomes extreme for $K \approx N/2$. The finite-sample variance is explained approximately because K parameters are estimated by minimising $N - K$ residuals in (6.4). That leaves $N - 2K$ degrees of freedom; the correction two is used in the finite-sample variance approximation. A disadvantage of this least-squares estimation method is that the estimated model is not guaranteed to be stationary. It is possible that poles of the estimated polynomial are outside the unit circle.

6.4.3 Forward and Backward Least-squares Method

Reversing the observed measurements gives the same estimated lagged products autocovariance function. The reversed data sequence is also a possible realistion of the same AR process as the forward process. Therefore, least-squares minimisation can also be applied to backward residuals. One specific least-squares estimation method minimises the *sum* of forward and backward residuals.

$$\text{RSS}(K) = \sum_{n=K+1}^{N} \left(x_n + \hat{a}_1 x_{n-1} + \cdots + \hat{a}_K x_{n-K} \right)^2$$

$$+ \sum_{n=1}^{N-K} \left(x_n + \hat{a}_1 x_{n+1} + \cdots + \hat{a}_K x_{n+K} \right)^2 \tag{6.5}$$

Broersen and Wensink (1993) showed that the empirical finite-sample variance of estimated parameters is given approximately by $1/(N + 1.5 - 1.5K)$.

A problem is that none of the least-squares methods, forward and/or backward, can guarantee stationary models with all poles inside the unit circle. All K parameters are estimated simultaneously to find an AR(K) model. Examples can be found where the minimum of the residual variance is obtained with an estimated pole outside the unit circle. For that reason and also because Burg's method will have a smaller estimation variance, least-squares estimators are not preferred.

6.4.4 Burg's Method

Apart from the Yule-Walker method, there is a second estimation method for finding a single new reflection coefficient at each stage of the computations: *Burg's method* (Burg, 1967). The first $K - 1$ reflection coefficients are kept constant at stage K. The parameters for increasing model orders follow with the Levinson-Durbin recursion (5.24). This favourably influences the estimation variance. Burg's method estimates the reflection coefficients for increasing model orders, thus making sure that the model will be stationary with all roots of $A(z)$ within the unit circle. Of all methods that have been mentioned, this method of Burg gives the smallest expected mean square error for the parameters of the estimated AR model, and it is always stationary. Therefore, Burg's method is preferred as an estimator of AR parameters.

Each individual reflection coefficient is estimated from all available information for the reflection coefficient of that order. In the least-squares algorithms, only $N - K$ residuals contribute to the estimation of all K parameters simultaneously. In Burg's method, each reflection is estimated individually. $N - 1$ residuals contribute to the estimation of k_1, $N - 2$ residuals to k_2, and so on. It is difficult to give an explicit expression of parameter estimates from data. It is not possible to give an explicit expression for the residual variance that is minimised in Burg's method. The usual description of the method is in terms of the relation of estimates of order K, given the estimates at order $K - 1$. The Levinson-Durbin recursion is used to determine parameters of increasing model orders. It uses one new reflection coefficient for all parameters of one order higher and also for the decrease in the residual variance. This single new estimated reflection coefficient is a good measure of the improvements obtained by estimating a model of one order higher. Here, it is explained how k_K is found with Burg's method if an AR($K - 1$) model has already been estimated with the Burg algorithm. The basic idea is to filter the presently estimated model out of the data and to estimate one single reflection coefficient from the filtered data. Filtering out an AR model of order K

has the consequence that only $N - K$ residuals are available. Define forward and backward residuals of order zero up to intermediate order K as

$$f_0(n) = x_n$$
$$f_1(n) = x_n + \hat{k}_1 x_{n-1}$$
$$......$$
$$f_K(n) = x_n + \hat{a}_1^K x_{n-1} + \cdots + \hat{a}_K^K x_{n-K}, \quad n = K+1, \cdots, N \tag{6.6}$$

Starting with the measured data, the forward residuals represent the data from which the present model of order K has been filtered out. Likewise, backward residuals are found by filtering out the present model:

$$b_0(n) = x_n$$
$$b_1(n) = x_{n-1} + \hat{k}_1 x_n$$
$$......$$
$$b_K(n) = x_{n-K} + \hat{a}_1^K x_{n-K+1} + \cdots + \hat{a}_K^K x_n, \quad n = K+1, \cdots, N \tag{6.7}$$

The special way of indexing the reversed backward residual $b_K(n)$ of order n as the result for x_{n-K} gives computational advantages in the actual implementation of the filter operation. Both the forward and the backward residuals are defined for the same set of indexes n.

The filter results for the backward residuals at stage K can also be written with the reversal operator $\sim$ defined in (5.10):

$$f_K(n) = \begin{bmatrix} x_n & x_{n-1} & \cdots & x_{n-K} \end{bmatrix} \begin{bmatrix} 1 \\ \hat{\alpha}^{[K]} \end{bmatrix}$$
$$b_K(n) = \begin{bmatrix} x_n & x_{n-1} & \cdots & x_{n-K} \end{bmatrix} \begin{bmatrix} \tilde{\hat{\alpha}}^{[K]} \\ 1 \end{bmatrix} \tag{6.8}$$

The Levinson-Durbin recursion has been derived in (5.21) to relate the parameters of the AR($K-1$) model to the AR(K) parameters as

$$\hat{\alpha}^{[K]} = \begin{bmatrix} \hat{\alpha}^{[K-1]} + \hat{k}_K \tilde{\hat{\alpha}}^{[K-1]} \\ \hat{k}_K \end{bmatrix}. \tag{6.9}$$

With some elementary manipulation, the forward and backward residuals become

$$f_K(n) = f_{K-1}(n) + \hat{k}_K b_{K-1}(n-1), \quad n = K+1, \cdots, N$$
$$b_K(n) = b_{K-1}(n-1) + \hat{k}_K f_{K-1}(n) \tag{6.10}$$

From those equations, the single unknown $\hat{k}_K$ is estimated by minimising the sum of squares of the forward and backward residuals

$$\text{RSS}(K) = \sum_{n=K+1}^{N} \left[f_K^2(n) + b_K^2(n) \right] \tag{6.11}$$

which, by equating the derivative with respect to the reflection coefficient to zero, yields the following result for $\hat{k}_K$:

$$\hat{k}_K = -2 \frac{\sum_{n=K+1}^{N} \left\{ f_{K-1}(n) b_{K-1}(n-1) \right\}}{\sum_{n=K+1}^{N} \left\{ f_{K-1}^2(n) + b_{K-1}^2(n-1) \right\}} \tag{6.12}$$

This estimate $\hat{k}_K$ is always smaller than one in absolute value because

$$\sum_{n=K+1}^{N} \left[f_{K-1}(n) \pm b_{K-1}(n-1) \right]^2 \geq 0$$

$$\sum_{n=K+1}^{N} \left[f_{K-1}^2(n) \pm 2 f_{K-1}(n) b_{K-1}(n-1) + b_{K-1}^2(n-1) \right] \geq 0$$

$$\pm 2 \sum_{n=K+1}^{N} f_{K-1}(n) b_{K-1}(n-1) \leq \sum_{n=K+1}^{N} \left[f_{K-1}^2(n) + b_{K-1}^2(n-1) \right] \tag{6.13}$$

That all reflection coefficients are ≤ 1 guarantees that all poles of the AR parameter polynomial are within the unit circle. Hence, Burg's algorithm always produces stationary models. The residual variance of the Burg method could be related to the residual sum of squares (6.11), but it is better to define the fit of the AR(K) model as in the solution of the Yule-Walker relations (5.23), as

$$s_K^2 = s_{K-1}^2 (1 - \hat{k}_K^2) \tag{6.14}$$

Broersen and Wensink (1993) have shown that the average of (6.14) over many experiments is equal to the average of squared forward and backward residuals of (6.11):

$$s_K^2 \approx \frac{\text{RSS}(K)}{2(N-K)} = \frac{\sum_{n=K+1}^{N} \left[f_K^2(n) + b_K^2(n) \right]}{2(N-K)} \tag{6.15}$$

It will turn out later that using (6.14) is convenient for order selection.

6.4.5 Asymptotic AR Theory

Kay and Makhoul (1983) showed that asymptotic theory gives rather complicated formulas for the variance of the first $p - 1$ reflection coefficients of an AR(p) process. They will not be discussed here because they are not important for the problem of order selection that is concentrated on what happens at and around the true process order. The results for the true order p and for higher orders are much more simple:

$$\text{var}\left[\hat{k}_p\right] = \frac{1-k_p^2}{N}$$

$$\text{var}\left[\hat{k}_{p'}\right] = \frac{1}{N}, \quad p' > p \tag{6.16}$$

Moreover, all reflection coefficients for the true AR order p and higher are independent of the lower order estimates. The first $p - 1$ reflection coefficients of an AR(p) process, however, have a covariance matrix with strong correlation coefficients between the lower order estimates.

The covariance matrix of the reflection coefficients prevents easy analysis of the residual variance for model orders lower than the true process order. The theoretical asymptotic analysis of s_K^2 will be identical for all methods of estimation that have been discussed because K is neglected with respect to N in asymptotic theory.

Akaike (1969, 1970, 1970a) has derived the asymptotic formula:

$$E\left[s_K^2\right] = \sigma_\varepsilon^2\left(1-\frac{K}{N}\right), \quad K \geq p \tag{6.17}$$

for the decrease in the residual variance. He derived the result for $K = p$, but it is evident that every AR(p) process is also a true AR($p+k$) process with k additional parameters that are all equal to zero. Akaike (1970) also derived an asymptotic expression for the prediction error (PE) that has been defined for an ARMA process in (5.36). That definition can be applied to AR processes by taking $B(z) = 1$. The asymptotic expectation of the squared one step ahead prediction PE(K) becomes

$$E[\text{PE}(K)] = \sigma_\varepsilon^2\left(1+\frac{K}{N}\right), \quad K \geq p \tag{6.18}$$

The condition for those formulas is that $K << N$. In asymptotic theory, the multiplication $(1-1/N)(1-p/N)$ yields $1-(p+1)/N$, and all terms containing $1/N^2$ are neglected.

For white noise, the expectations of the residual variance and of the prediction error apply to all model orders. Estimating AR parameters from a white noise signal seems to decrease the residual variance, but the prediction error increases by the same amount. Using (6.14) and (6.16), the asymptotic expectation for a white noise signal is derived as

$$E\left[s_K^2\right] = E\left[\sigma_\varepsilon^2 \prod_{i=1}^{K}\left(1-k_i^2\right)\right] \approx \sigma_\varepsilon^2 \prod_{i=1}^{K}\left(1-\operatorname{var} k_i\right)$$

$$= \sigma_\varepsilon^2 \left(1-\frac{1}{N}\right)^K \approx \sigma_\varepsilon^2 \left(1-\frac{K}{N}\right) \tag{6.19}$$

by realising that the variance of all reflection coefficients is $1/N$ in (6.16) and their expectation is zero. The bias is proportional to $1/N$ and is neglected in the expectation of the squared reflection coefficient. Asymptotic theory also treats the last approximation as an equality. Often, $K + 1$ is found in the literature in (6.17) and (6.18). That indicates that the mean is subtracted and that an additional estimated quantity is explicitly included in the number of estimated parameters in those formulas.

6.4.6   Finite-sample Practice for Burg Estimates of White Noise

In finite-sample theory, the final approximation in (6.19), omitting all contributions with $1/N^2$ and higher, will not be allowed because it is not assumed that $K << N$. So far, an exact theoretical derivation of the properties of the reflection coefficients estimated from white noise by the Burg method is not available. Simulations will be used to establish the variance of reflection coefficients and the development of the residual variance and of the prediction error for increasing model orders. Jones (1976) studied the prediction error of the maximum entropy AR estimation method, which is just another name for the Burg algorithm. Broersen (1985, 1990) related the variance of estimated reflection coefficients to the residual variance and the prediction error. This exercise has been carried out for every AR estimation method, but only the Burg results are reported here. The bias of Burg estimates is much smaller than that of Yule-Walker estimates. Tjøstheim and Paulsen (1983) show that the bias of reflection coefficients is about $1/N$ in white noise experiments and much more important for even than for odd model orders. That is explained because the estimation of k_2 involves the squared estimate of $\rho(1)$ in the Levinson-Durbin recursion (5.24). The square of $\rho(1)$ is precisely the variance of $\rho(1)$ in white noise where the true expectation is zero. That variance is $1/N$ in (3.40) and the same reasoning applies to other even reflection coefficients. As a matter of convenience, bias is neglected in the finite-sample experiments as well as in finite-sample theory. Its contribution of $1/N^2$ to the mean square error is small in comparison with the variance contribution.

For the variance of reflection coefficients, Broersen (1985) used the finite-sample formula,

$$v_i = \operatorname{var}\left(k_i\right) = \frac{1}{N+1-i} \tag{6.20}$$

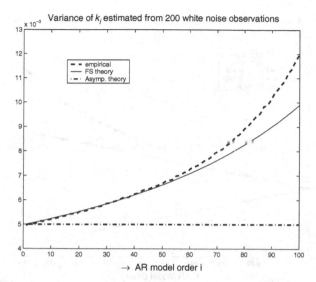

Figure 6.1. Measured variance of reflection coefficients, the asymptotic variance, and the finite-sample approximation $1/(N + 1 - i)$ of average Burg estimates of 200,000 simulation runs of 200 white noise observations.

as an approximation for the variance estimated by the Burg method in white noise. The relation has been found for different values of N. The variance for order $N/2$ is almost $2/N$ for all N. The v_i coefficient in (6.20) can be seen as the number of degrees of freedom that is available in estimating i parameters from N observations; the correction with one in the denominator improved the accuracy.

Figure 6.1 shows the average of simulations and the finite-sample relation of (6.20), compared with asymptotic theory. The v_i coefficients are much closer to the experimental data than $1/N$, but the approximation could perhaps be improved somewhat, especially for orders greater than $N/4$. However, the simple formula (6.20) has been considered sufficiently accurate until now. Many different white noise simulations with N between 4 and 1000 have been done. Always, v_i of (6.20) is a useful approximation for the variance of reflection coefficients. This shows one important feature of finite-sample behaviour: it is not the sample size N itself, but the ratio of the model order i and N that is decisive. If the model order is lower than about $N/10$, asymptotic theory is often applicable with sufficient accuracy. For higher orders, finite-sample behaviour may become important.

The empirical formula for the variance of reflection coefficients can be used to derive formulas for the finite-sample behaviour of the residual variance and the prediction error as a function of the model order. The approximation for the residual variance of white noise models is derived from the first line of (6.19) as

$$E_{FS}\left[s_K^2\right] = E_{FS}\left[\sigma_\varepsilon^2 \prod_{i=1}^{K}\left(1-k_i^2\right)\right] \approx \sigma_\varepsilon^2 \prod_{i=1}^{K}\left(1-\operatorname{var} k_i\right)$$

$$= \sigma_\varepsilon^2 \prod_{i=1}^{K}\left(1-v_i\right) \tag{6.21}$$

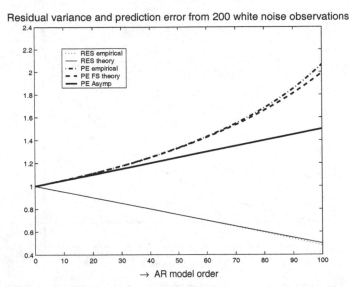

Figure 6.2. Measured residual variance and prediction error as the average Burg estimates of 200,000 simulation runs of 200 white noise observations with variance 1. Also the relations of asymptotic theory and of finite-sample theory are shown.

Substitution of v_i of (6.20) in (6.21) shows that the residual variance is exactly the same in asymptotic and in finite-sample theory. As in (6.17), it decreases linearly with the model order for white noise. This is accidentally the case only for finite-sample Burg estimates.

In white noise, every AR parameter that is used for prediction will deteriorate the result. In analogy with the residual variance, the prediction error in finite-sample theory is approximated by

$$E_{FS}\left[\mathrm{PE}(K)\right] = \sigma_\varepsilon^2 \prod_{i=1}^{K}\left(1+v_i\right) \qquad (6.22)$$

Broersen and Wensink (1993) have given a formal derivation of this finite-sample equation for the prediction error.

These finite sample formulas are compared with the average of simulations in Figure 6.2. The approximation for the residual variance is very good; the lines almost overlap. The approximation for the prediction error is excellent for orders less than $0.4N$ and still very good for higher orders. It starts to deviate from the asymptotic result for orders greater than $0.15N$. The good approximation in Figure 6.2 is the main reason that no better approximation of the variance in Figure 6.1 has been developed. Formulas (6.21) and (6.22) together will be referred to as finite-sample theory, although it is realised that the mathematical background is weak. They are empirical and simple approximations for the average result of simulations with white noise as a signal. Choosing (6.20) for the variance of reflection coefficients can be loosely connected with degrees of freedom.

6.4.7 Finite-sample Practice for Burg Estimates of an AR(2) Process

The knowledge of the behaviour of higher order estimates obtained from a white noise signal as a true process is not important in itself. However, asymptotic theory shows that all AR(p) processes show the same white noise behaviour for the true

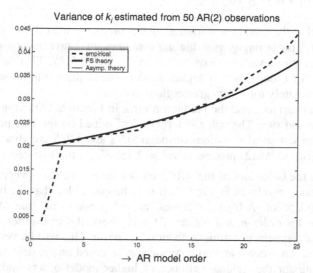

Figure 6.3. Measured variance of reflection coefficients, asymptotic white noise variance, and the finite-sample approximation $1/(N + 1 - i)$ of the average Burg estimates of 10,000 simulation runs of 50 AR(2) observations.

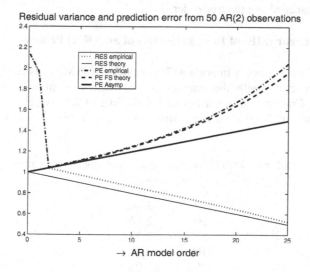

Figure 6.4. Measured residual variance and prediction error as the average Burg estimates of 10,000 simulation runs of 50 AR(2) observations. The white noise relations of asymptotic and finite-sample theory are also shown for $\sigma_\varepsilon^2 = 1$.

order p and for all higher orders. For lower orders, the behaviour depends on the true process parameters. To study finite-sample behaviour in some detail, simulations have been done. The example in Figure 6.3 shows the results for an AR(2) process with

$$A(z) = 1 + 0.51 \, z^{-1} + 0.7 \, z^{-2}.$$

It is evident that the empirical variance of the first two reflection coefficients in Figure 6.3 deviates strongly from the expectations for white noise signals. With (6.16), the asymptotic variance expression for k_2 gives $0.51/N$. That value is found in Figure 6.3 for order two. For higher model orders, the white noise behaviour returns approximately for orders greater than two.

The residual variance and the prediction error in Figure 6.4 are almost identical for orders one and two. The value is largely determined by the true process parameters, and the statistical variations of magnitudes about $1/N$ are almost invisible. The variance of the AR(2) process is $\sigma_x^2 = 2.155 \, \sigma_\varepsilon^2$. At the true order two and at higher orders, the behaviour of the AR(2) models in Figure 6.4 is very close again to the white noise results in Figure 6.2. It may be concluded that the behaviour of estimated models of AR(p) processes is independent of true AR process characteristics for order p and higher. At and above the true order, it becomes similar to what happens in estimates from white noise. This will be very helpful in order selection. An order-selection criterion can be based on the detection of white noise behaviour for the residual variance of higher model orders, independent of the true process parameters for low model orders. In biased models of orders lower than true order p, the bias contribution of the nonzero reflection coefficients to the residual variance will generally be much greater than the contribution of the estimation variance above the true order.

6.4.8 Model Error (ME) of Burg Estimates of an AR(2) Process

The accuracy of estimates is important. The model error (ME) has been introduced as measure in (5.40) with the number of parameters as minimal expectation. Therefore, the Cramér-Rao lower bound for the ME of the AR(2) process is two. The average ME of 1000 simulation runs is given in Table 6.1, as a function of sample size N.

Table 6.1. The model error (ME) of AR(2) estimates as a function of sample size N.

N	ME
10	2.86
20	2.56
50	2.23
100	2.11
200	2.08
500	2.00
1000	1.95

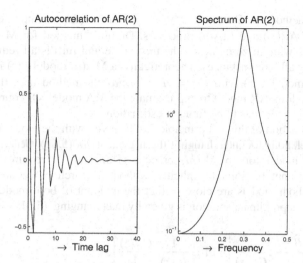

Figure 6.5. Autocorrelation function and normalized power spectral density of the AR(2) process with $A(z) = 1+0.51z^{-1}+0.7z^{-2}$

Figure 6.5 gives the spectrum and the autocorrelation function of the process. The accuracy in Table 6.1 is quite close to the Cramér-Rao bound for all sample sizes. That is remarkable for N equal to 10 and 20 because those sample sizes are less than the correlation length. This shows that the excellent performance for large samples is accompanied by very good model accuracy for finite and small samples always close to the Cramér-Rao bound. Therefore, the Burg algorithm is the preferred AR estimator.

This section treated only the estimation of AR models of AR(p) processes, with the true order considered as known. Later, the selection of the model order and the model type will be treated.

6.5 MA Estimation Methods

The convergence problems of maximum likelihood estimates, zeros on the unit circle, and selection problems stimulated the search for a robust algorithm to determine MA parameters. Osborn (1976) tried constrained optimisation of the likelihood function. Davidson (1981) described the inevitable problems, and Findley (1984) showed that ambiguities can be associated with the question, which is the best model. Priestley (1981) mentions many attempts to find approximate maximum likelihood estimators for MA models.

Durbin's (1959) method for MA estimation replaces a nonlinear estimation problem by two stages of linear estimation. Firstly, the parameters of a long AR model are estimated from the data. Afterward, MA parameters are computed by using the sequence of estimated AR parameters a_i as if they were an input signal x_n for the Yule-Walker algorithm. A derivation will be given here, based on ME manipulations.

The MA method of Durbin (1959) is based on the asymptotic theoretical equivalence of AR(∞) and MA(q) processes. Durbin's method for MA estimation gives outcomes with all zeros inside the unit circle and fulfills all requirements to be useful under all circumstances. Theoretically, a MA(q) model $B(z)$ is equivalent to an AR(∞) model $C(z)$ with $C(z)=1/B(z)$. Durbin's method uses the *estimated* parameters of a long AR model to approximate the MA model. Of course, the order of that long AR model has to be finite in estimation.

A new formulation of this old principle can be given with the model error and the approximate relations for interchanging the arguments in (5.61). Searching the best fitting MA(q) model for an AR(M) process is difficult. It requires a nonlinear solution that cannot be solved explicitly without a search procedure. Under the condition that both models are close or that the quotient of both models resembles white noise, an approximate solution is given by interchanging (5.61) as

$$\mathrm{ME}\left[\frac{\hat{B}_q(z)}{1},\frac{1}{C_M(z)}\right] \approx \mathrm{ME}\left[\frac{1}{\hat{B}_q(z)},\frac{C_M(z)}{1}\right] \qquad (6.23)$$

Then, (5. 62) gives the solution

$$\hat{B}_q(z) = \arg\min_{\tilde{B}_q(z)}\left\{\mathrm{ME}\left[\frac{1}{\tilde{B}_q(z)},\frac{C_M(z)}{1}\right]\right\} \qquad (6.24)$$

As the sequence of the operations in the ME had to be exchanged, it is clear that the resulting MA estimation method can be an approximation at best. The steps in this Durbin solution are

- estimate the autoregressive AR(M) model $C_M(z)$ of the data
- treat the parameters as a MA(M) process to find the autocorrelation with (4.11)
- use the first q autocorrelations to determine the parameters of an autoregressive model of order q with the Yule-Walker relations (5.24)
- those parameters are the parameters of the desired MA(q) model.

This method has good properties if a very high order AR model is used that is computed with the *true* MA(q) process. However, using *estimated* AR models generally gives disappointing results. Also using very high orders for the estimated intermediate AR model is not accurate. Mentz (1977) tried several modifications of Durbin's method, which are reported to be consistent but asymptotically inefficient.

Broersen (2000b) showed that the choice of order M of the AR model is very critical. It had been assumed without any proof that the best intermediate order would be the order of the best predicting AR model for the data. Simulations show that this order is not a good choice. A better choice is given by the AR order with the smallest mean square error of the parameters. Hocking (1976) showed that those two model orders for prediction and for parameter accuracy are different in ordinary regression theory. Simulation experiments will be required to determine

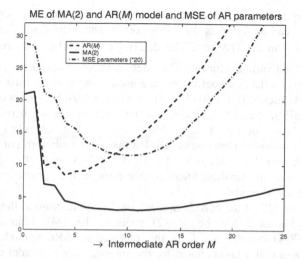

Figure 6.6. The average accuracy of the MA(2) model estimated with an intermediate AR(M) model, as a function of the AR order M in 1000 simulations with $N = 50$ of the MA(2) process with $B(z) = 1 - 0.24z^{-1} - 0.6z^{-2}$. Also the mean square error of the AR parameters is shown, multiplied by 20.

the best intermediate AR order for estimating a MA model. Also the accuracy of the MA model will be determined with the ME as a measure. That will be compared to the Cramér-Rao lower bound for the ME, which is equal to the number of estimated parameters. A comparison of models estimated with this MA algorithm and models estimated with other algorithms so far always had the result that the average ME of this method was smallest.

Figure 6.6 gives the average ME of MA(2) models as a function of the intermediate AR order M. The ME of that AR(M) model is also given. The ME of all AR models is greater than the ME of the MA models. Hence, it is also better to have MA candidates for accurate modeling of data of an unknown character. The best predicting AR model, if estimated from 50 observations, is the AR(4) model $C_4(z)$, both in theory and in simulations.

The theoretical best order for prediction follows the principle that every estimated parameter contributes one to the expected ME due to the estimation variance. If the bias in omitting a group of parameters is less than the omitted number of parameters, the expected bias of that group is less than the expected variance if they are included. They can better be left out as a group.

The mean square error of the parameters of an AR(M) model $C_M(z)$ is defined as

$$\text{MSE}\left[C_M(z)\right] = \sum_{i=0}^{M} \left(\gamma_i - \hat{c}_i\right)^2 + \sum_{i=M+1}^{\infty} \gamma_i^2 \tag{6.25}$$

where γ_i denote the true AR parameters of the AR(∞) model $\Gamma(z)$ which can be computed from the true MA process parameters with $\Gamma(z) = 1/B(z)$.

For ordinary regression, Hocking (1976) has shown that the best theoretical parameter order can be found as the lowest order where the deterministic residual variance is less than $(1+1/N\,)\sigma_\varepsilon^2$. The deterministic residual variance is defined here as the residual variance for biased models with fewer parameters than the true process if the true reflection coefficients are used without any estimation variance. That theoretical order is 10 for 50 observations of the MA(2) process of Figure 6.6. The smallest ME of the MA(2) model and the smallest mean square errors of the parameters are found for the estimated AR(10) model. The curves in Figure 6.6 are flat around their minima. It may be concluded that the simulations confirm the idea that the best intermediate AR order for Durbin's MA method is given by the AR(10) model with the smallest mean square error of the parameters, not by the best predicting AR(4) model.

The minimum ME of the MA(2) model was 3.1, quite close to the Cramér-Rao lower bound that is 2 for the MA(2) process. The ME obtained with the intermediate AR(4) model was 4.4, for the intermediate AR(25) model, it was 6.6. This demonstrates that a good choice for the intermediate AR order can give MA estimates close to the Cramér-Rao lower bound. The average ME of the best predicting AR(4) model was 8.4.

The fact that the ME of the MA(2) model was 3.1, which is rather close to the Cramér-Rao lower bound but still somewhat higher, is probably a finite-sample effect. The simulations for the same MA(2) process have been repeated with 250 observations. The results are given in Figure 6.7 with an enlarged scale for the parameter inaccuracy for visualizing the minimum. The best predicting order was 10 for 250 observations, in both theory and in the simulation result. The best order for the parameter accuracy was 18 in theory and in the simulations. Also the best

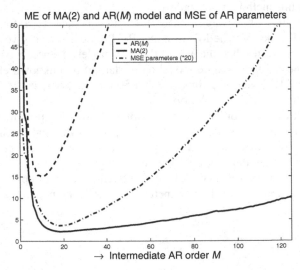

Figure 6.7. The average accuracy of the MA(2) model estimated with an intermediate AR(M) model, as a function of the AR order M in 1000 simulation runs of 250 observations of the MA(2) process with true $B(z) = 1 - 0.24z^{-1} - 0.6z^{-2}$. Also the mean square error of the AR parameters is shown, multiplied by 20.

MA(2) model was found with the intermediate AR(18) model. The best orders depend on sample size N. The minimum ME was 2.25, slightly above the Cramér-Rao bound, but that rather small difference can also be due to statistical uncertainty in the average of 1000 runs. All intermediate AR orders between 13 and 38 give a value below 3 as the ME of the MA(2) model, which is still rather accurate. The intermediate AR order can be chosen within a rather wide range without much loss in the accuracy of the final MA model. However, taking the best AR order for prediction as intermediate AR order would give ME = 3.9, and taking the intermediate AR order $N/2$ would give ME = 10.3 in the example of Figure 6.7. Too low or too high intermediate AR orders have a negative influence on accuracy.

The difference between the ME of the best AR model and the MA(2) model will increase with sample size. The minimum of the ME in Figure 6.7 is found at a higher order than in Figure 6.6, but it is also higher. All finite-order AR models are biased. Eventually, for N going to infinity, the bias contribution of the best AR model to the PE is only a small constant. However, multiplication by N in the ME gives an asymptotic infinite ME contribution in (5.40).

Many other examples and sample sizes have been tested in simulations. For MA processes, the best AR order for parameter accuracy is almost always higher than the best AR order for prediction in simulations, never lower. Approximately that highest order for the best parameter accuracy is also always the best intermediate AR order for MA estimation.

A MA(13) example has also been simulated that was sensitive for the triangular bias in Figure 3.4. The theory for the best predicting AR model gives AR order 60, and the simulations in Figure 6.8 also have the minimum at order 60. The best pa-

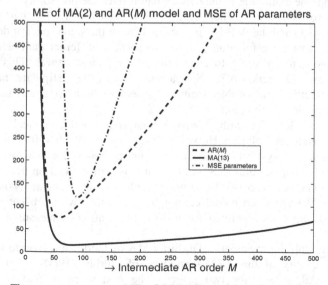

Figure 6.8. The average accuracy ME of MA(13) models estimated by an intermediate AR(M) model, as a function of the AR order M in 3000 simulation runs of 1000 observations of a true MA(13) process. Also, 5.10^{-5} times the mean square error of the parameters is shown. The MA(13) example is the same as in Figures 3.4 and 3.11.

rameter accuracy should be found at order 81 according to the theory, and the simulations had a minimum at order 89. The best MA(13) model was estimated for the intermediate AR order 81. The average ME value for the MA(13) model in the simulations was 14.9, close to the minimum achievable Cramér-Rao bound of 13.

This example is seen as a severe test. The ME due to triangular bias was 11,500 for 1000 observations of this process; see Table 5.2. If the true process order is known and if the best AR order is known, Durbin's MA estimation method with (6.22) is a very accurate method with accuracy close to the best achievable. The performance for data of an unknown character will be treated after the discussion of order selection. All three figures for MA estimation show that in those processes, the quality of estimated MA models can be much better than the quality of the best estimated AR model. This is a firm recommendation to try the class of MA models as candidates for practical data with unknown characteristics.

6.6 ARMA Estimation Methods

ARMA models are the most difficult to estimate. All problems of MA return, and also a new problem arises of splitting the dynamics in AR and MA parts. As in the MA estimation, maximum likelihood does not provide a reliable estimator.

Moreover, extra problems arise if ARMA($p+1,q+1$) models are estimated for a true ARMA(p,q) process. If the true order of a process is not known *a priori*, it will be necessary to estimate models of orders too high before it can be concluded that the lower orders give a better fit to the data. Estimation of one extra zero in the numerator and one extra pole in the denominator gives the possibility that the extra pole and zero are equal. They can cancel each other. In that case, the ARMA($p+1,q+1$) and the ARMA(p,q) models have the same spectral density and the same autocovariance functions, but they can have different parameters. If the extra pole and zero are close to each other, they almost cancel. This has been discussed already in Section 6.3.3. Stoica and Moses (1997) stipulate that there is yet no well-established algorithm, from both theoretical and practical standpoints, for ARMA parameter estimation.

A recent development with subspace analysis, called canonical correlation analysis, seemed a new solution. However, Bauer and de Waele (2003) have shown that this method always fails to produce accurate models for processes that are sensitive to the triangular bias (3.32) in the autocorrelation function. Hence, at least one complete class of processes has a poor result with subspace analysis. Because the quality of the estimated model cannot be guaranteed with subspace methods, those methods are not considered for routine application to measured stochastic data.

The successful improvement of Durbin's MA method is a reason to investigate the possible quality of the ARMA method of Durbin (1960). That also uses intermediate AR models for two stages. The first stage is finding an initial separation of the AR and the MA parts; the second stage uses the parameter estimates of the first stage as starting values for an improved ARMA estimate.

6.6.1   ARMA(p,q) Estimation, First-stage

The first ARMA method of Durbin uses both reconstructed residuals and previous observations as regressors for an explicit least-squares solution:

$$A_p(z)x_n = B_q(z)\,\varepsilon_n$$
$$C_M(z)\,x_n = \hat{\varepsilon}_n \qquad\qquad\qquad (6.26)$$

The residuals $\hat{\varepsilon}_n$ are reconstructed with a finite-order estimated long AR model $C_M(z)$

$$\hat{\varepsilon}_n = x_n + c_1 x_{n-1} + \cdots c_M x_{n-M} \qquad\qquad (6.27)$$

If the order of $C_M(z)$ were infinite and if $C_M(z)$ is exactly equal to the true $B_q(z)/A_p(z)$, the reconstruction of $\hat{\varepsilon}_n$ could be perfect, and the accuracy of the reconstruction is reasonable for finite length and estimated parameters. Parameters are estimated in Durbin's first method by minimising

$$\sum_{n=\max(p,q)+1}^{N} \left\{ x_n + \hat{a}_1 x_{n-1} + \cdots + \hat{a}_p x_{n-p} - \hat{\varepsilon}_n - \hat{b}_1\hat{\varepsilon}_{n-1} - \cdots - \hat{b}_q\hat{\varepsilon}_{n-q} \right\}^2 \qquad (6.28)$$

This ARMA(p,q) solution is not efficient and is not guaranteed to be stable and invertible. Therefore, the objective quality of (6.28) cannot be established or it is very poor because this estimate is not fully efficient, according to Durbin (1960). However, by computing the AR and the MA parameters at the same time, the dynamics are automatically divided into AR and MA parts. Many suggestions to improve efficiency have been made. Many of those improved methods are iterative, using the results of Durbin's first ARMA method as initial conditions. A second stage after the regression step in (6.28) is always necessary to obtain a satisfactory model.

If the estimated parameters are required only as an initial separation into an AR part with p parameters and a MA part with q parameters, several inefficient but simple alternative first-stage methods have been given by Broersen and de Waele (2002, 2005). They are called reduced-statistics estimators. Four different transformations of the long AR model have been used to find an initial splitting of the dynamics in the AR and the MA parts. Those methods use the estimated AR model $C_M(z)$ as a reduced-statistic input, and they do not use the data any further. The four methods are

- long AR
- long MA
- long covariance
- long inverse correlations

Klees *et al.* (2003) described a satellite signal processing application where the only available representation of the data is a given autocorrelation function. That autocorrelation can be transformed into a long AR model by the Levinson-Durbin recursion (5.24). Then, the only MA and ARMA estimators which are realisable must be derived entirely from the autocorrelation or from that long AR model.

6.6.2 ARMA(p,q) Estimation, First-stage Long AR

The first method, long AR, uses $C_M(z)$ itself as the basis for computations. The method was introduced by Graupe *et al.* (1975) as a single-stage ARMA method. The long AR method employs the relation

$$\frac{\hat{B}_q(z)}{\hat{A}_p(z)} \approx \frac{1}{C_M(z)}$$

or

$$\hat{B}_q(z)C_M(z) \approx \hat{A}_p(z) \qquad\qquad (6.29)$$

A serious problem is that those relations are approximations. They give exact results for exact polynomials, but in practice with estimated polynomials, they cannot be satisfied exactly. Standard estimation theory requires information about the statistical errors in these equations to formulate a good estimator. That information is not available here, and for that reason, it is not possible to derive a maximum likelihood estimator or any other efficient estimator from (6.29). By arbitrarily concentrating the inaccuracies in an unknown error signal ζ_m, the second representation in (6.29) can be written as

$$\sum_{i=0}^{q} \hat{b}_i c_{m-i} = \hat{a}_m + \zeta_m, \qquad m = 0, \cdots, M \qquad\qquad (6.30)$$

by equating the left- and right-hand side coefficients of z^{-m}. For the moment, it will be supposed that the orders p and q are known. Later, it will be shown how to select those orders for practical data.

   If an ARMA(p,q) model has to be estimated, it is known that the AR parameters have to be zero for orders $m > p$. With that knowledge, the q first-stage MA parameters can be estimated without knowing the first p AR parameters. MA parameter estimates can be found as a least-squares solution from the higher order equations in (6.30) as

$$\sum_{i=0}^{q} \hat{b}_i c_{m-i} = \zeta_m, \qquad m = p+1, \cdots, M \qquad\qquad (6.31)$$

by minimising the sum of ζ_m^2 for $p < m < M$. A solution exists for $M \geq p + q$. If $M = p + q$, there are just enough equations to estimate the q MA parameters with zero as the expectation for ζ_m. The equations are overdetermined for greater values of M and need a least-squares solution. This will be followed by calculation of the p AR parameters.

Having obtained these MA estimates, it seems attractive to determine an AR solution by solving the first p equations in (6.30). Substitution of the MA estimates from (6.31) yields $\hat{a}_m$ with $\zeta_m = 0$ for $m = 0, 1,..., p$. However, simulations have shown that better initial first-stage AR parameters for a given MA model can be found with the second-stage AR algorithm that will be derived later.

The informal derivation of this algorithm clarifies the idea that no statistical optimality can be claimed for this first-stage solution of the initial estimates. The same lack of statistical optimality applies to the next three estimators of initial estimates. It is hoped that at least one of the four estimators will be adequate for each different type of measured data. Simulations will be necessary to investigate the performance.

6.6.3 ARMA(p,q) Estimation, First-stage Long MA

The second reduced-statistics method, denoted long MA, is derived from an alternative evaluation of the first relation in (6.29). It uses an estimate for the infinite MA representation $G(z)$ of (4.63). This long MA model is calculated from the M parameters of $C_M(z)$. The length of the impulse response computed from $G(z)\, C_M(z) = 1$ can be chosen freely, much greater than M. Suppose that the impulse response practically died out at L.

$$\frac{\hat{B}_q(z)}{\hat{A}_p(z)} \approx \frac{1}{C_M(z)} = G_L(z)$$

$$\hat{A}_p(z)G_L(z) \approx \hat{B}_q(z) \tag{6.32}$$

Knowing that the MA parameters are zero for $m > q$, the p AR parameters for the initial $\hat{A}_p(z)$ follow as the least-squares solution from

$$\sum_{i=0}^{p} \hat{a}_i g_{m-i} = \zeta_m, \qquad m = q+1,\cdots,L \tag{6.33}$$

The different expressions for ζ_m in (6.30) and (6.33) indicate once more that the application of least squares to those approximate relations will not lead to theoretically optimal or efficient estimators. However, they can be useful as first-stage solutions.

6.6.4 ARMA(p,q) Estimation, First-stage Long COV

The third method for finding initial conditions for the AR part $\hat{A}_p(z)$ of the ARMA(p,q) model is denoted long COV because it uses the autocovariance function as reduced-statistics information. If the autocovariance or autocorrelation itself is given, it can be used immediately in this method. Otherwise, it has to be calculated from the long AR model. The formulas required to establish the relations between the AR parameters in the long AR model $C_M(z)$ and the autocovariances are based on the Yule-Walker equations, as in (5.5):

$$\sum_{i=0}^{M} c_i \hat{R}(m-i) = 0, \qquad m = 1, 2, \cdots, M$$

with $\qquad \hat{R}(-|k|) = \hat{R}(k), \qquad k > 0$ $\qquad\qquad$ (6.34)

Now the initial AR parameters $\hat{A}_p(z)$ are calculated with (4.69) and (4.70) by a least-squares solution:

$$\sum_{i=0}^{p} \hat{a}_i \hat{R}(m-i) = \varsigma_m, \qquad m = q+1, \cdots, M \qquad\qquad (6.35)$$

by minimising the sum of $\varsigma_m^{\,2}$ for $q + 1 < m < M$. It is assumed that M is greater than $p + q$.

6.6.5 ARMA(p,q) Estimation, First-stage Long Rinv

The fourth method, long Rinv, uses inverse correlations. Inverse correlations and spectra are defined by interchanging AR and MA polynomials (Priestley, 1981). The inverse correlation is found from the parameters of the intermediate AR model $C_M(z)$ with

$$R_{inv}(k) = \sum_{i=0}^{M-k} \hat{c}_i \hat{c}_{i+k}, \qquad k = 0, 1, \cdots, M \qquad\qquad (6.36)$$

Further, $R_{inv}(k)$ is zero for shifts greater than M, and it is symmetrical around zero shift. The initial estimates for the $r - 1$ MA parameters are found with the overdetermined equations:

$$\sum_{i=0}^{q} \hat{b}_i R_{inv}(m-i) \approx \varsigma_m, \qquad m = p+1, \cdots, M \qquad\qquad (6.37)$$

by minimizing the sum of $\varsigma_m^{\,2}$ for $p + 1 < m < M$. Afterward, the second-stage AR method is used to calculate the first-stage AR model from these estimated MA parameters, as in the long AR method before.

6.6.6 ARMA(p,q) Estimation, Second-stage

Five different estimators have been discussed as possibilities for the first stage. The first one uses the data and the residuals that are reconstructed with a long AR model in (6.28); the other four use only the long AR model $C_M(z)$ that was estimated from the data. Only the AR parameters $\hat{A}_p(z)$ of the first stage are required as initial estimates for the second stage. The estimated MA parameters of the first stage are not used in the second stage. The second-stage algorithm consists of two steps. The initial AR estimates $\hat{A}_p(z)$ are used first to compute the final MA model $\hat{B}_q(z)$. Then this final MA model is used to improve the first-stage AR estimate. The derivation is similar to the MA algorithm of (6.23) and (6.24). The MA parameters are estimated by

$$
\begin{aligned}
\hat{B}_q(z) &= \arg\min_{\tilde{B}_q(z)} \left\{ \text{ME}\left[\frac{\tilde{B}_q(z)}{\hat{A}_p(z)}, \frac{1}{C_M(z)} \right] \right\} \\
&\approx \arg\min_{\tilde{B}_q(z)} \left\{ \text{ME}\left[\frac{1}{\tilde{B}_q(z)}, \frac{C_M(z)}{\hat{A}_p(z)} \right] \right\}
\end{aligned}
\tag{6.38}
$$

In this derivation, use is made of the approximation due to interchanging the sequence of the elements of the ME, as in (5.61). After this final MA estimation, the intermediate AR model $C_M(z)$ is multiplied by this newly estimated MA model $\hat{B}_q(z)$. The solution for the improved AR parameters can be formulated as

$$
\hat{A}_p(z) = \arg\min_{\tilde{A}_p(z)} \left\{ \text{ME}\left[\frac{1}{\tilde{A}_p(z)}, \frac{1}{\hat{B}_q(z)C_M(z)} \right] \right\}
\tag{6.39}
$$

This yields the AR(p) model $\hat{A}_p(z)$ with the first p autocovariances equal to the first p autocovariances of the AR product model $\hat{B}_q(z)$ $C_M(z)$. The same computation is also used to calculate $\hat{A}_p(z)$ in the first stage for the long AR and the long R_{inv} methods where the overdetermined equations yielded an initial estimate for $\hat{B}_q(z)$.

The original method of Durbin (1960) used iterative updates of MA and AR parameters until the iterations converged. The two steps (6.38) and (6.39) in the second stage can be iterated if desired. Iteration of the second stage will give an improved model if poor or zero initial-stage AR estimates were used from the first stage. With zero parameters substituted in $\hat{A}_p(z)$, many second-stage iterations are required to converge to a somewhat acceptable estimated ARMA(p,q) model. In simulations with reduced-statistics, initial estimates up to 10 second-stage iterations did not improve the quality of the estimated models for the very best of the five types of initial estimates, at least in most simulation examples. Generally, second-stage iterations tend to converge for all first-stage methods to the same final value. That common value is slightly worse than the result that is obtained with the best type of the first stage after a single second-stage iteration. However, it will often improve other types of initial estimates that have poorer quality. In other words, poor second-stage estimates will mostly be improved with iterations in the second stage, but the best second-stage estimates will generally not improve and become worse by iterations.

The best choice for the order M of the long AR model has been studied by Broersen (2000b). It turns out to be the same principle as in MA estimation: the order of the *estimated* AR model with the smallest expected mean square of the AR parameters is the best choice.

6.6.7 ARMA(p,q) Estimation, Simulations

Four different examples have been chosen to demonstrate the necessity of different methods in reduced-statistics, first-stage solutions. The examples are

- ARMA(2,1)

$$x_n + 0.39x_{n-1} + 0.3x_{n-2} = \varepsilon_n - 0.9\varepsilon_{n-1}$$

- ARMA(3,2)

$$x_n - x_{n-1} + 0.88x_{n-2} - 0.5x_{n-3} = \varepsilon_n + 0.45\varepsilon_{n-1} - 0.5\varepsilon_{n-2}$$

- ARMA(4,3)

$$x_n - 3.36x_{n-1} + 5.09x_{n-2} - 3.42x_{n-3} + 0.92x_{n-4}$$
$$= \varepsilon_n - 1.8\varepsilon_{n-1} + 1.05\varepsilon_{n-2} - 0.12\varepsilon_{n-3}$$

- ARMA(5,4)

$$x_n + 3.9097x_{n-1} + 6.7395x_{n-2} + 6.4026x_{n-3} + 3.3521x_{n-4} + 0.7738x_{n-5}$$
$$= \varepsilon_n - 3.2795\varepsilon_{n-1} + 4.4493\varepsilon_{n-2} - 2.9597\varepsilon_{n-3} + 0.8145\varepsilon_{n-4}$$

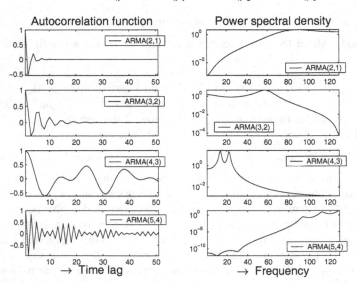

Figure 6.9. The autocorrelation function and the power spectral density of the four example processes. The ARMA(4,3) process has a long autocorrelation function, the ARMA(5,4) process has a large spectral range.

Figure 6.9 gives the autocorrelation functions and the spectral densities of the four examples. The first two processes have an autocorrelation function that is much shorter than 100, which will be the usual number of observations in the simulation results shown in the figures. The ARMA(4,3) process describes a spectrum with two low-order peaks. Therefore, this process has a very long autocorrelation function. The ARMA(5,4) process has all poles and zeros at the same radius of 0.95. This is realised by building parameters from reflection coefficients using (5.28) with $k_m = (0.95)^m$ for the AR parameters and $k_m = (-0.95)^m$ for the MA parameters. In this way, all poles of the generating process have the same radius of 0.95.

The last two examples have been selected especially as difficult processes, which had rather particular performance in simulations with numerous other ARMA processes. The first two examples are well-behaved processes.

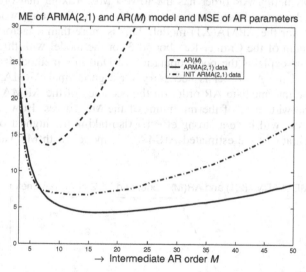

Figure 6.10. The average accuracy ME of the first- and second-stage ARMA(2,1) models estimated from the data as a function of the AR order M in 10,000 simulation runs of 100 observations of a true ARMA(2,1) process.

Figure 6.10 gives the results of the ARMA algorithm based on data, with (6.28) in the first stage as initial values. The first-stage estimates are denoted INIT; the full estimates completed with the second stage for the ARMA(2,1) example are denoted ARMA(2,1) data. The order of the intermediate AR model is varied from 3 to 50, to find the best choice for that model order. Both the initial-stage and the second-stage ARMA models have better accuracy than the AR model. This shows that it is a good idea to have ARMA models as candidates. The theoretical best AR order for prediction for the ARMA(2,1) example is 8 for $N = 100$, in agreement with the simulation result. The average ME of the AR(8) model was 13.6. The theoretically expected order with the smallest mean square error of the parameters was 15; the simulation result for the AR order with the best ARMA(2,1) in the second stage was 17. The smallest ME value was 4.17. The lowest ME value for the first-stage ARMA(2,1) model was 6.80. This first-stage ME also depends on the AR order because the reconstruction of the residuals with (6.27) is done with the AR model. It is remarkable that the AR order with the best result is somewhat different for the first- and the second-stage ARMA(2,1) models. The first stage has minimum 6.8 at order 12; the second stage has the minimum value 4.16 at order 17. Therefore, the best initial first-stage estimate does not belong to the best final estimate in the second stage. However, both curves are rather flat at the bottom.

For AR orders until seven, the initial stage ARMA model is better than the second stage. The second stage was no improvement, probably because the interchange of AR and MA parts in (6.38) and (6.39) is allowed only if the final ARMA model is accurate. The lower order models are least accurate. For higher intermediate AR orders, the second stage is certainly an improvement on the first-stage result.

The best predicting AR order has the lowest ME. Taking that order as inter-mediate, as is done or advised in most examples in the literature, would give the ME value 6.15 for the ARMA(2,1) model. This is more than a factor of 2 higher than the minimum of the Cramér-Rao bound 3 for the model with three estimated parameters. That explains the poor reputation of Durbin's method. It is clear that proper choice of the AR order is necessary for a good final ARMA model. The influence of the intermediate AR order on the accuracy of the ARMA estimates is moderate in the wide area of the minimum of the ME curves. However, taking a much lower order will have a strong effect. Also taking the highest order 50 will reduce the accuracy of the estimated ARMA(2,1) model, with ME values that are twice higher.

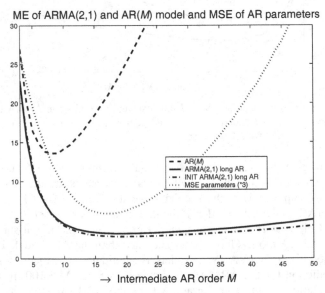

Figure 6.11. The average accuracy ME of the ARMA(2,1) long AR model estimated with an intermediate AR(M) model, as a function of the AR order M in 10,000 simulation runs of 100 observations of a true ARMA(2,1) process. Also the mean square error of the para-meters is shown, multiplied by 3.

Repeating the simulations of the same process with the long AR method of (6.31) in the first stage gives Figure 6.11. Here, the initial first-stage results are slightly better than the results with the second stage included, and iterations with the second stage cannot improve the result. In this example, a single-stage estimator would be sufficient and the best. However, even in this special example, the application of the second stage does not do much harm. The smallest MSE of the parameters was at order 17, and the smallest ME of the ARMA(2,1) model was found for the intermediate order M equal to 18. Also for ARMA processes, the best intermediate AR order is close to the theoretical order with the smallest parameter errors, as it was for MA models. The ME of the ARMA(2,1) model is 2.88 for the first stage only and 3.27 with the second stage included. Both values are close to the Cramér-Rao lower bound of 3.

Taking the order with the best parameter accuracy as the intermediate AR order is a good choice. However, the ME values for high intermediate AR orders increase much less for long AR than in Figure 6.10 for the reconstructed residuals as first stage.

All four reduced-statistics methods for the first stage have been tested in simulations to compare their results. Figure 6.12 gives the average ME after two stages; Figure 6.13 after only the first stage. For convenience, the theoretical orders for the best predicting AR model and the order for the best parameter accuracy have been indicated with arrows. The differences among the five first-stage methods are rather strong. The differences become smaller after the second stage, but they do not disappear.

The long AR method was the best reduced-statistics first-stage method for the ARMA(2,1) example for $N = 100$. This is seen in Figure 6.13. After the second stage in Figure 6.12, long AR and long Rinv are close competitors that almost

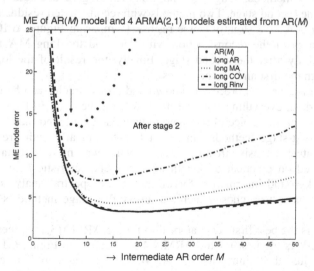

Figure 6.12. The average accuracy ME of the four second-stage ARMA(2,1) models estimated exclusively from an intermediate AR(M) model and of the AR(M) quality, as a function of the AR order M in 500 simulation runs of 100 observations of a true ARMA(2,1) process.

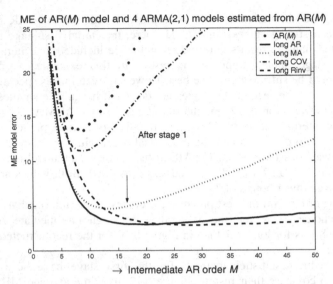

Figure 6.13. The average accuracy ME of the four first-stage ARMA(2,1) models estimated from an intermediate AR(M) model, as a function of the AR order M in 500 simulation runs of 100 observations of a true ARMA(2,1) process. The AR(M) quality is also given.

coincide over the whole range of intermediate AR orders. All methods give poor results for low intermediate AR orders. The best first-stage methods in this example have smooth dependence on the AR order. That dependence is much stronger for the long MA and for the long COV method.

At least two methods, long AR and long Rinv, give accurate ARMA(2,1) models after the second stage. Two methods with reconstructed residuals from data and with long MA give slightly worse results. Comparing Figures 6.10 and 6.12, the performances of the reconstruction with data and the long MA method are similar, especially after the second stage. Finally, the result of the long COV is poor, after both the first and second stages.

In simulations, the true process is known and all accuracies can be established. For practical data, everything must be derived from the data. In later chapters, it will be shown that the choice of the intermediate AR length and the choice among the different first-stage methods can be made automatically with order-selection criteria. The rather extensive treatment here explains why more first-stage methods have to be tried on practical data with unknown characteristics. Furthermore, it gives the background information for the automatic spectral analysis algorithm. The examples have been chosen such that every first-stage method is best somewhere.

Long MA is the best first-stage algorithm for the ARMA(3,2) process in Figure 6.14, long COV is the best for the ARMA(4,3) example in Figure 6.15, and long Rinv is only just the winner for the ARMA(5,4) process in Figure 6.16. The influence of the intermediate AR model order is moderate in all four examples, at least for the best first-stage method and as long as the order is above some critical value. The long COV method has peculiar behaviour in Figures 6.15 and 6.16.

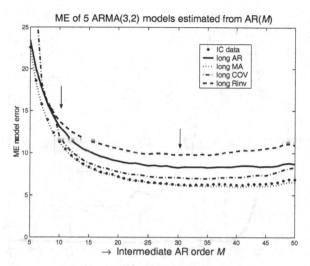

Figure 6.14. The average accuracy ME of the five second-stage ARMA(3,2) models estimated from an intermediate AR(M) model, as a function of the AR order M in 200 simulation runs of 100 observations of a true ARMA(3,2) process.

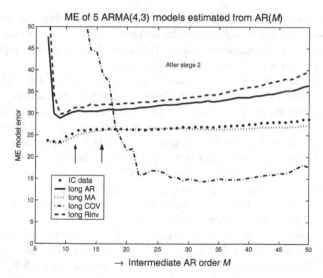

Figure 6.15. The average accuracy ME of the five second-stage ARMA(4,3) models estimated exclusively from an intermediate AR(M) model, as a function of the AR order M in 500 simulation runs of 100 observations of a true ARMA(4,3) process.

In the first two ARMA examples, the best of the four methods for the first stage gives a result that is close to the minimum achievable Cramér-Rao bound. The distance to that bound is somewhat greater in the last two examples. The difference from the Cramér-Rao bound was greatest for the ARMA(4,3) process, which has a true spectrum with two close and strong peaks. In this example, only the long COV

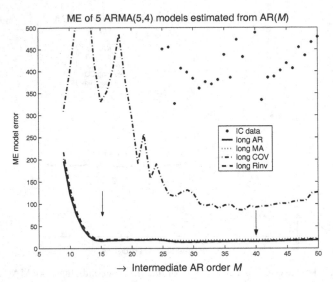

Figure 6.16. The average accuracy ME of the five second-stage ARMA(5,4) models estimated exclusively from an intermediate AR(M) model, as a function of the AR order M in 5000 simulation runs of 100 observations of a true ARMA(5,4) process

method gives more or less satisfactory results, if the intermediate AR order is high enough. These simulation results show the influence of the intermediate AR order. Moreover, they also demonstrate the well-known fact that no simple ARMA estimation algorithm is reliable in all circumstances.

In most examples, it is very important to take as an intermediate AR order, an order that is significantly higher than the best AR order for prediction. Taking that wrong and too low an intermediate order has been the main reason that the good accuracy of the algorithms developed here has not been recognized in the literature on ARMA estimation. For the best first-stage method for a particular signal, the choice of the intermediate AR order is not very critical, but the quality will always diminish if the maximum order were used, like $N/2$ in the examples.

In most examples, the performance of the ARMA algorithm based on the reconstruction of the residuals with (6.25) was close to the behaviour of the long MA method. That will be demonstrated once more in Figure 6.17 for the ARMA(4,3) example, with 1000 observations. This was the example with the worst quality for $N = 100$. What happens if more observations are available has been studied. The minimum ME of the long COV method was 7.9 for $N = 1000$, close to the minimum value of 7. All other methods give very inaccurate models. The first stages from data and from the long MA are quite close over the whole interval. The results of the first and of the second stage are very close for long COV; for the other methods, the second stage gave a significant improvement. The simulations have been repeated for $N = 10,000$. The global appearance was like Figure 6.17. The minimum ME of long COV was 7.0, the ME of long MA was 148, of long AR 1147, and of long Rinv 2104.

It turns out that the accuracy of the estimated ARMA(4,3) model for $N = 10,000$ is equal to the minimum obtainable value of 7 for the long COV method. Heavily

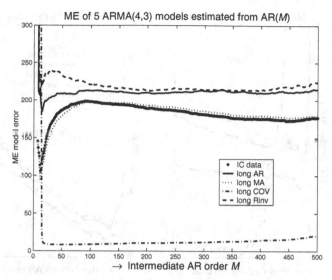

Figure 6.17. The average ME of the five second-stage ARMA(4,3) models, as a function of the AR order M in 100 simulation runs of 1000 observations of a true ARMA(4,3) process

biased models are found with the other four initial-stage methods. However, this bias disappears if the ARMA order is taken somewhat higher. The best accuracy for the initial stage with residuals reconstructed from the data was ME = 12 for the estimated ARMA(7,6) model. Sometimes, one or more initial-stage methods are biased if models of the true order are estimated. However, no examples have been found where all initial-stage methods were seriously biased. Furthermore, the bias always diminished or disappeared if the order was chosen somewhat higher than the true order, at the cost of an extra variance contribution to the ME for the additionally estimated parameters. Therefore, if models of a fixed *a priori* determined order are estimated, it is advisable to use all five initial-stage methods, followed by a second-stage estimation. It will turn out in the next chapter that an order-selection criterion can automatically select the best of the five solutions.

The results in Figure 6.17 and all previous figures in this section have been determined for a single iteration of the second-stage algorithm of (6.38) and (6.39). The question is still what improvement can be obtained with iterations of the second stage.

The most inaccurate ARMA(4,3) example has been used to test the result of 25 iterations of the second-stage algorithm for each of the five first-stage estimators. Figure 6.18 gives the results. The five methods converge more or less to the same ARMA models after iterations. The quality is much improved for the methods with poor quality in the first stage. However, the quality becomes worse for the best first-stage method, which was long COV for this example. After the iterations, long COV no longer gives the best result, but the long MA method does. Altogether, long COV after a single iteration gave the best result. The first stage with reconstructed residuals from the data remains close to the long MA first stage-line. It is curious that the long COV was the best before iterations and one of the worst after.

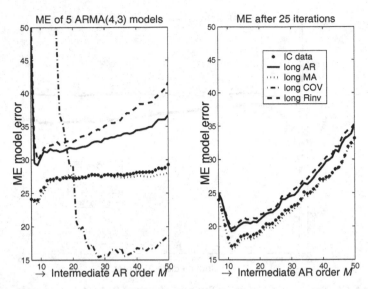

Figure 6.18. The average ME of the five second-stage ARMA(4,3) models, as a function of the AR order M in 500 simulation runs of 100 observations of a true ARMA(4,3) process, left is after one second-stage computation, and right gives the results of the same simulations after 25 second-stage iterations. Long COV is the best on the left-hand side. After iterations, it coincides with long AR, and long MA is the best.

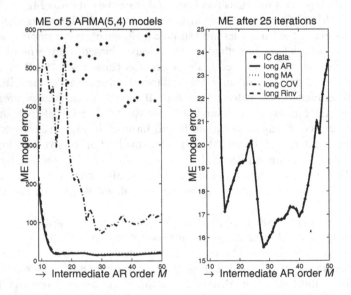

Figure 6.19. The average accuracy ME of the five second-stage ARMA(5,4) models, as a function of the AR order M in 500 simulation runs of 100 observations of a true ARMA(5,4) process, left is after one second-stage computation, and right gives the results of the same simulations after 25 second-stage iterations. All five lines coincide completely on the right-hand side. However, the minimum ME at AR order 28 is 1.0 lower on the left-hand side.

From this simulation and many others, it has been concluded that computationally demanding iterations do not improve the quality of the best result.

Figure 6.19 gives the result of iterations for the ARMA(5,4) example. Here, all five initial stages converge to exactly the same ME with the second-stage iterations. But the minimum ME after iterations is one greater than the best result after a single iteration of the second stage. A few simulation runs with very poor quality was the main cause for the poor average quality of the residuals reconstructed from the data after a single iteration. In most runs, the ME was about 100 but in some runs, much more than 1000.

At first sight, the result of the left-hand side of Figure 6.19 seems to indicate that the reconstruction of the residuals with (6.28) can sometimes deliver very poor models. However, this is only partially true. Examples can be found where the reconstruction of residuals as the first-stage method delivers a poor model of the fixed true process order. In those cases, a model of one or two orders higher is much better. Only if the algorithm is used to estimate models of predetermined orders, it may deliver a poor result.

So far, the results may seem disappointing. One or two of the five first-stage methods can give accurate results if the intermediate AR order is not chosen erroneously. It is not known in advance which first-stage method is the best. The intermediate AR order can be chosen quite well if the true process parameters are *a priori* known. It is the order of the best parameter accuracy then, where the residual variance computed with the true AR parameters of the process becomes less than $(1+1/N)\sigma_\varepsilon^2$. However, that information is not usually available in time series estimation. The automatic choice among the five first-stage methods and what should be the intermediate AR order will be treated after the next chapter on order selection.

6.7 Covariance Matrix of ARMA Parameters

6.7.1   The Covariance Matrix of Estimated AR Parameters

Kay (1988) reported large sample results for the covariance matrix for efficient estimates of AR parameters. They can also be used as good approximations for finite samples, as long as the AR order p is much less than N and the distance of the poles to the unit circle is not too small. The distance should be at least greater than $1/N$. The asymptotic covariance matrix for the estimated AR(p) parameters is given by

$$\mathrm{cov}(\hat{\alpha}_p) = E\left[(\hat{\alpha}_p - \alpha_p)(\hat{\alpha}_p - \alpha_p)^T\right]$$
$$= \sigma_\varepsilon^2 R_p^{-1} / N \qquad\qquad (6.40)$$

where R_p has been defined in (5.5). An extended covariance matrix includes the estimation of the residual variance, giving

$$\text{cov}(\hat{\alpha}_p, \hat{\sigma}_\varepsilon^2) = E\left\{ \begin{bmatrix} \hat{\alpha}_p - \alpha_p \\ \hat{\sigma}_\varepsilon^2 - \sigma_\varepsilon^2 \end{bmatrix} \begin{bmatrix} \hat{\alpha}_p - \alpha_p \\ \hat{\sigma}_\varepsilon^2 - \sigma_\varepsilon^2 \end{bmatrix}^T \right\} = \begin{bmatrix} \dfrac{\sigma_\varepsilon^2}{N} R_p^{-1} & 0 \\ 0 & \dfrac{2\sigma_\varepsilon^4}{N} \end{bmatrix} \tag{6.41}$$

This is the equation which one would expect by treating the lagged x_n values as ordinary regressors in a linear regression problem. The estimates of the parameters are not correlated with the estimate of the residual variance, which follows from the zero in the off-diagonal position in (6.41). Furthermore, for an AR(1) process, it follows

$$\text{var}(\hat{a}_1) = \frac{1 - a_1^2}{N} \tag{6.42}$$

For an AR(2) process, the variances of both parameters are equal and become

$$\text{var}(\hat{a}_1) = \text{var}(\hat{a}_2) = \frac{1 - a_2^2}{N} \tag{6.43}$$

More generally, as the matrix R_p is symmetrical about both the diagonal and the anti-diagonal, its inverse will have the same property. Therefore, the variance of estimated AR(p) parameters is symmetrical which means

$$\text{var}(a_1) = \text{var}(a_p),$$

$$\text{var}(a_2) = \text{var}(a_{p-1})$$

and so on.

In all AR, MA and ARMA processes, the covariance between estimated parameters and the estimated innovation variance is equal to zero. The lower bound for the estimated variance of the innovation variance for unbiased models is given by

$$\text{var}(\hat{\sigma}_\varepsilon^2) > \frac{2\sigma_\varepsilon^4}{N} \tag{6.44}$$

Kay and Makhoul (1983) derived formulas for the variance of estimated reflection coefficients by treating them as transformations of the parameters and using (2.32).

6.7.2 The Covariance Matrix of Estimated MA Parameters

The covariance matrix for efficiently estimated MA(q) parameters becomes

$$
\operatorname{cov}(\hat{\beta}_q, \hat{\sigma}_\varepsilon^2) = E\left\{ \begin{bmatrix} \hat{\beta}_q - \beta_q \\ \hat{\sigma}_\varepsilon^2 - \sigma_\varepsilon^2 \end{bmatrix} \begin{bmatrix} \hat{\beta}_q - \beta_q \\ \hat{\sigma}_\varepsilon^2 - \sigma_\varepsilon^2 \end{bmatrix}^T \right\} = \begin{bmatrix} \dfrac{\sigma_\varepsilon^2}{N} R_{zz}^{-1} & 0 \\ 0 & \dfrac{2\sigma_\varepsilon^4}{N} \end{bmatrix},
\tag{6.45}
$$

where R_{zz} denotes the $q \times q$ autocovariance matrix of the signal z_n that is created with an AR(q) model with the parameters of the MA(q) process and the same innovation variance

$$
z_n + b_1 z_{n-1} + \cdots + b_q z_{n-q} = \varepsilon_n
\tag{6.46}
$$

R_{zz} looks like R_p in (5.5). In principle, maximum likelihood estimates will approach the efficiency of this covariance matrix asymptotically. In finite samples, the variance found with (6.46) is a fair approximation of the variance of Durbin's method with the best intermediate AR order. It is not necessary to make a detailed study of finite sample properties because the highest order that can be estimated efficiently for MA models must necessarily be much lower than the order of the intermediate AR order. It cannot be as high as $0.5N$, but only $0.1N$ or $0.2N$. For low orders, the asymptotic expressions are reasonably accurate, and no finite-sample formulas have been developed.

6.7.3 The Covariance Matrix of Estimated ARMA Parameters

The covariance matrix of ARMA parameters is given in Kay (1988). The most important property is that the parameter variance becomes infinite if both the AR and the MA order are greater than the true order. That is easily explained because multiplying a quotient of a canceling pole-zero pair $(1-\alpha z^{-1})/(1-\alpha z^{-1})$ by an arbitrary ARMA(p,q) model does not change the quotient $B(z)/A(z)$, the autocorrelation function, or the spectrum. However, many different parameter values in the ARMA($p+1,q+1$) model yield exactly the same second-order moments, and the true values of the parameters cannot be found, at least not with the given estimation methods. The estimation of the parameters of the ARMA($p+1,q+1$) model will not converge to the true ARMA(p,q)ARMA(1,1) process parameters for $N \to \infty$, although the estimated autocorrelation and the estimated spectrum will have a vanishing error with respect to the ARMA(p,q) process, at least if numerical problems are avoided in the estimation procedure.

   The covariance matrix for efficiently estimated ARMA(p,q) parameters becomes the inverse of the Fisher information matrix (Kay, 1988) and is given by

$$\text{cov}(\hat{\alpha}_p, \hat{\beta}_q, \hat{\sigma}_\varepsilon^2) = E\left\{ \begin{bmatrix} \hat{\alpha}_p - \alpha_p \\ \hat{\beta}_q - \beta_q \\ \hat{\sigma}_\varepsilon^2 - \sigma_\varepsilon^2 \end{bmatrix} \begin{bmatrix} \hat{\alpha}_p - \alpha_p \\ \hat{\beta}_q - \beta_q \\ \hat{\sigma}_\varepsilon^2 - \sigma_\varepsilon^2 \end{bmatrix}^T \right\} = \frac{\sigma_\varepsilon^2}{N} \begin{bmatrix} R_{yy} & -R_{yz} & 0 \\ -R_{yz}^T & R_{zz} & 0 \\ 0 & 0 & 1/(2\sigma_\varepsilon^2) \end{bmatrix}^{-1} \tag{6.47}$$

The result for the estimated variance given by (6.44) is independent of the parameters and also equal to the accuracy of estimating the variance of a white noise process. The process y_n is an AR(p) process with $A_p(z)$ as parameters, the process z_n is an AR process with $B_q(z)$ as parameters, and the matrix R_{yz} has elements

$$[R_{yz}]_{i,j} = E\left[y(n)z(n + i - j)\right]. \tag{6.48}$$

The elements of those matrices are easily found with a variant of Yule-Walker equations that are used directly for the elements of the symmetrical matrices R_{yy} and R_{zz}:

$$y_n + a_1 y_{n-1} + \cdots + a_p y_{n-p} = \varepsilon_n$$
$$z_n + b_1 z_{n-1} + \cdots + b_q z_{n-q} = \varepsilon_n . \tag{6.49}$$

Multiplying the first equation by $z_n, z_{n-1}, \ldots, z_{n-p}$, the second by $y_{n-1}, \ldots, y_{n-p}$, and taking expectations yields

$$E[y_n z_n] + a_1 E[y_{n-1} z_n] + \cdots + a_p E[y_{n-p} z_n] = E[\varepsilon_n z_n]$$
$$E[y_n z_{n-1}] + a_1 E[y_{n-1} z_{n-1}] + \cdots + a_p E[y_{n-p} z_{n-1}] = E[\varepsilon_n z_{n-1}]$$
$$\vdots$$
$$E[y_n z_{n-p}] + a_1 E[y_{n-1} z_{n-p}] + \cdots + a_p E[y_{n-p} z_{n-p}] = E[\varepsilon_n z_{n-p}]$$

and

$$E[z_n y_{n-1}] + E[b_1 z_{n-1} y_{n-1}] + \cdots + E[b_q z_{n-q} y_{n-1}] = E[\varepsilon_n y_{n-1}]$$
$$\vdots$$
$$E[z_n y_{n-p}] + E[b_1 z_{n-1} y_{n-p}] + \cdots + E[b_q z_{n-q} y_{n-p}] = E[\varepsilon_n y_{n-q}]$$

This can be rewritten with the general property $R_{yz}(k) = R_{zy}(-k)$ and with the property that y_n and z_n can be written as a linear combination of ε_n and only previous values of ε:

$$r_{yz}(0) + a_1 r_{yz}(1) + \cdots + a_p r_{yz}(p) = \sigma_\varepsilon^2$$
$$r_{yz}(-1) + a_1 r_{yz}(0) + \cdots + a_p r_{yz}(p-1) = 0$$
$$\vdots$$
$$r_{yz}(-p) + a_1 r_{yz}(-p+1) + \cdots + a_p r_{yz}(0) = 0$$

and

$$r_{yz}(1) + b_1 r_{yz}(0) + \cdots + b_q r_{yz}(-q+1) = 0$$
$$\vdots$$
$$r_{yz}(p) + b_1 r_{yz}(p-1) + \cdots + b_q r_{yz}(p-q) = 0.$$

Together this gives $2p + 1$ equations for $2p + 1$ unknowns, which can be solved with

$$
\begin{bmatrix}
a_p & a_{p-1} & \cdots & a_1 & 1 & 0 & 0 & 0 & 0 \\
0 & a_p & & & & 1 & 0 & & \\
0 & 0 & a_p & \ddots & & & 1 & 0 & \\
0 & & 0 & a_p & \ddots & & & 1 & 0 \\
0 & & & 0 & a_p & & & a_1 & 1 \\
0 & & & 0 & 1 & b_1 & & b_q & 0 & 0 \\
0 & 0 & & 1 & & \ddots & b_q & 0 & & 0 \\
0 & 1 & & \ddots & & b_q & 0 & & & 0 \\
1 & b_1 & \cdots & b_q & 0 & 0 & 0 & 0 & 0
\end{bmatrix}
\begin{bmatrix}
r_{yz}(p) \\
r_{yz}(p-1) \\
\vdots \\
r_{yz}(1) \\
r_{yz}(0) \\
r_{yz}(-1) \\
\vdots \\
\vdots \\
r_{yz}(-p)
\end{bmatrix}
=
\begin{bmatrix}
\upsilon_\varepsilon^2 \\
0 \\
\vdots \\
0 \\
0 \\
0 \\
\vdots \\
\vdots \\
0
\end{bmatrix}.
$$

This completes the computation of terms required for the covariance matrix for estimated parameters of an ARMA(p,q) process. The $p \times q$ matrix R_{yz} becomes

$$
R_{yz} =
\begin{bmatrix}
r_{yz}(0) & r_{yz}(-1) & & & r_{yz}(-q+2) & r_{yz}(-q+1) \\
r_{yz}(1) & r_{yz}(0) & r_{yz}(-1) & \ddots & & \vdots \\
& r_{yz}(1) & r_{yz}(0) & & & \vdots \\
\vdots & & & \ddots & & r_{yz}(1) \\
& & & & & r_{yz}(0) \\
r_{yz}(p-2) & & & & & \vdots \\
r_{yz}(p-1) & r_{yz}(p-2) & & & & r_{yz}(p-q)
\end{bmatrix}
\tag{6.50}
$$

with $r_{yz}(k) = E[y_n z_{n+k}]$.

The accuracy of estimated autocorrelations and spectra is found by considering them as functions of the estimated parameters. Friedlander and Porat (1984) give some results for spectral estimates as well as for integrated spectra. That latter result can be applied to the minimum obtainable value for the ME, which is equal to the number of estimated parameters. With the Taylor expansion result of (2.32) and the covariance matrix (6.47) of the estimated ARMA parameters, formulas for the variance can be derived. Ninness (2003) gives an improved mathematical analysis and Ninness (2004) included also the effect of not knowing the true process order in the accuracy analysis.

6.8 Estimated Autocovariance and Spectrum

6.8.1 Estimators for the Mean and the Variance

The estimator for the mean value has been defined in (3.19) as

$$\hat{\mu}_x = \frac{1}{N}\sum_{i=1}^{N} x_i \tag{6.51}$$

In all further results, it is assumed that the mean of the data is subtracted before the signal processing starts. Not subtracting the mean value may have strange effects in the estimation of time series models.

The estimator for the variance is given in (3.20) as

$$\hat{\sigma}_x^2 = \frac{1}{N-1}\sum_{i=1}^{N}\left(x_i - \hat{\mu}_x\right)^2 \tag{6.52}$$

Subtracting the measured average gives a loss of one degree of freedom for an unbiased estimate. The variance $\hat{\sigma}_x^2$ of the process is generally estimated directly from the data without a model for the data.

Porat (1994) proved that only the first $p - q$ estimates of autocovariances are efficient for an ARMA(p,q) process. That shows that (3.20) is an efficient estimator for the variance of an AR(p) processes and also for ARMA(p,q) processes with $p > q$. According to the theory, the variance estimator (6.52) is not efficient for ARMA(p,q) with $q \geq p$ and for all MA(q) processes. Asymptotically, the maximum likelihood principle should provide an efficient estimator for the variance in those circumstances. However, the finite-sample properties of maximum likelihood estimation are poor and those estimates have not been considered. No existing estimator claims better accuracy. Therefore, (6.52) is the estimator for the variance of a signal that is used in all circumstances.

6.8.2 Estimation of the Autocorrelation Function

Estimation of an autocorrelation *function* requires knowledge of or the estimation of the parameters $\hat{A}(z)$ and $\hat{B}(z)$ of the time series model of the data. The lagged products estimator (3.31) provides only biased estimates for individual lags, which can hardly be interpreted as estimating a function. Furthermore, the lagged product bias has two components in (3.33), the triangular bias and a contribution arising from the fact that the autocorrelation is the quotient of two estimated stochastic variables.

Porat (1994) derived the important result that only the first p lagged product estimates are efficient for AR(p) processes and only $p - q$ are efficient for an ARMA(p,q) process with $p > q$. Efficiency is a special statistical property. Priestley (1981) defined relative efficiency as the quotient of the variances of two different unbiased estimators. An estimator is efficient if it has the smallest possible variance of all unbiased estimators. That lowest variance bound is called the Cramér-Rao lower bound.

Lagged product estimators are not efficient for ARMA(p,q) with $q \geq p$ and for all MA(q) processes (Porat, 1994). The length of the lagged product estimator with nonzero estimates is in principle equal to $N - 1$. Taking more observations gives a longer nonzero estimate for the autocorrelation function. The reason that this lagged product estimator has been preferred in the past was probably that no other estimator was available.

The time series estimators for the autocorrelation function use the theoretical relations between true parameters and their correlations by substituting the estimated parameters. The general solution for the autocorrelation function of ARMA processes is given by (4.62)

$$\hat{r}(k) = \sum_{m=-q}^{q} \left[\hat{r}_v \left(k + m \right) \hat{r}_{MA}(m) \right], \qquad \forall k$$

$$\hat{\rho}(k) = \hat{r}(k) / \hat{r}(0) \tag{6.53}$$

For AR models, the estimated parameters $\hat{A}_p(z)$ are transformed into reflection coefficients with (5.30), and those give the autocorrelation function $\hat{r}_v(k)$ with (5.31). This solution can also be written as the solution of the Yule-Walker equations (5.4) with estimated parameters.

$$\hat{\rho}(k) + \hat{a}_1 \hat{\rho}(k-1) + \cdots + \hat{a}_p \hat{\rho}(k-p) = 0 \ , k = 1, \cdots, p \tag{6.54}$$

It is possible to start the solution arbitrarily with $\sigma^2 = 1$ or any other value and to use division by $\hat{r}(0)$ after convolution with the MA autocovariance to obtain the autocorrelation function starting with one.

The estimated moving average autocovariance uses (4.61) with the estimated parameters

$$\hat{r}_{MA}(k) = \sum_{i=0}^{q} \hat{b}_i \hat{b}_{i+|k|} \ , \ -q \leq k \leq q \tag{6.55}$$

Some care is required with the scaling because $\hat{r}_{MA}(0)$ is not normalized.

The autocovariance function follows always from the autocorrelation function by multiplying by $\hat{\sigma}_x^2$ with $\hat{R}(0) = \hat{\sigma}_x^2$. The estimated autocovariance function is always symmetrical.

The autocorrelation is found by substituting estimated parameters in the true relations. This is valid for all ARMA processes. For AR(p) processes, the reverse is also true: p parameters follow from p autocorrelations. This has been used in (6.24) to derive the MA estimator and also for the first- and the second-stage ARMA estimators. However, the reverse is not valid for estimated MA and ARMA models. Of course, $p + q$ true autocorrelations are enough to find the exact parameters of an ARMA(p,q) process with $p + q$ equations and without measurement errors. However, estimated MA or ARMA parameters have estimation errors and they require more than $p + q$ estimated autocorrelations. The simulations of the best intermediate AR order have demonstrated this. It is clear that high intermediate AR orders give much better estimates than the AR(2) model in Figures 6.6 and 6.7. The AR(2) model defines efficient estimates for the first two lags with the smallest possible estimation errors. But that is not good enough to estimate two MA parameters efficiently. MA parameters become better if more estimated AR parameters are used.

6.8.3 The Residual Variance

It is not often necessary to have an explicit estimate for the variance of the innovations ε_n. Due to the differences between the residual variance and the prediction error, this might become a confusing quantity in estimation practice. Nevertheless, if some explicit expression is required, it is advisable to use (5.47) with true values replaced by estimates:

$$\hat{\sigma}_\varepsilon^2 = \hat{\sigma}_x^2 / P_g\left[\hat{A}(z), \hat{B}(z)\right] \tag{6.56}$$

P_g is defined as the power gain or ratio of output and input power of an ARMA model. This choice of $\hat{\sigma}_\varepsilon^2$ guarantees that $\hat{\sigma}_\varepsilon^2$, $\hat{\sigma}_x^2$ and the ARMA process with $\hat{A}(z)$, $\hat{B}(z)$ together are a consistent description of the process. This has the elegant advantage that if new data are generated with innovation variance $\hat{\sigma}_\varepsilon^2$ and parameters $\hat{A}(z)$, $\hat{B}(z)$, then the output variance will have $\hat{\sigma}_x^2$ as an expectation.

Generally, the variance $\hat{\sigma}_x^2$ of the process will be an unbiased estimate. This means that $\hat{\sigma}_\varepsilon^2$ will be a biased estimate of σ_ε^2, using the estimated parameters for this computation of the innovation variance. The reason is that the square of unbiased estimated parameters is always greater than the square of the true parameters due to the estimation variance.

6.8.4 The Power Spectral Density

The estimated spectral density can be scaled with a constant such that the integral becomes equal to $\hat{\sigma}_x^2$. Any other scaling is also allowed. The spectral density shows how the power is divided over the frequency range. For the total power of the data, (6.52) can be used.

Using the estimated parameters gives a unique estimated spectral density as

$$\hat{h}(\omega) = \frac{\hat{\sigma}_\varepsilon^2}{2\pi} \frac{\left|\hat{B}(e^{j\omega})\right|^2}{\left|\hat{A}(e^{j\omega})\right|^2} \tag{6.57}$$

The method of Burg for AR and the improved methods of Durbin that have been treated for MA and ARMA models give estimated models that can be used with this formula to compute an estimated spectrum. This shows that the parameters of a time series model, together with the variance of the exciting white noise, determine the spectral density.

The normalized spectral density is given by

$$\hat{\phi}(\omega) = \frac{1}{2\pi} \frac{1}{P_g\left[\hat{A}(z), \hat{B}(z)\right]} \frac{\left|\hat{B}(e^{j\omega})\right|^2}{\left|\hat{A}(e^{j\omega})\right|^2} \qquad (6.58)$$

The actual computation of (6.58) for a finite number of frequencies is carried out with the fast Fourier transform (FFT), adding zero parameters until the required length of the Fourier transform carries out the FFT of a short parameter vector. An AR(1) process is exactly the same as an AR(M) process with a_1, completed with $M - 1$ AR parameters equal to zero. In this way, the required number of frequencies for evaluating (6.57) and (6.58) with the FFT is obtained.

The program that is used for spectral computations in the ARMASA toolbox of Broersen (2002) does not use any information about the process variance. If K equidistant points are demanded, the program calculates the spectrum at K frequencies given by

$$0, \frac{0.5}{K-1}, \frac{1}{K-1}, \cdots, \frac{0.5(K-2)}{K-1}, \frac{0.5(K-1)}{K-1}$$

if the sampling time is one. Otherwise, the frequencies are multiplied by $1/T_s$, where T_s denotes the sampling time. The spectrum is then multiplied by T_s. The normalization is such that the sum of all spectral estimates for all $2K - 2$ frequencies in the frequency domain equals $(2K - 2)T_s$. Those frequencies are

$$0, \frac{1}{(2K-2)T_s}, \frac{2}{(2K-2)T_s}, \cdots, \frac{1}{2T_s}, \cdots, \frac{2K-3}{(2K-2)T_s}$$

Only K frequencies are given; the last $K - 2$ spectral values have not been given because they are the mirrored results of $K - 2$ frequencies between 0 and $1/2T_s$.

Generally, the autocovariance function (6.53) and the spectrum (6.58) are not related by a finite FFT transformation, which is often used in digital computations. The FFT relates N points in the time domain to N points in the frequency domain. The true autocovariance function of ARMA processes is generally infinitely long and the spectrum is a continuous function for $-\pi \leq \omega \leq \pi$. Both do not fulfill the requirements for a practical FFT.

6.9 Exercises

6.1   Find one important reason why the estimator

$$\hat{a}_1 = -\frac{1}{N-1} \sum_{n=2}^{N} \frac{x_n}{x_{n-1}}$$

for the parameter of an AR(1) process is not treated in this chapter.

6.2   Let x_n be a normally distributed zero mean AR(1) process

$$x_n + a_1 x_{n-1} = \varepsilon_n$$
$$\mu_\varepsilon = 0$$
$$\sigma_\varepsilon^2 = 1.$$

The estimator of the Yule–Walker algorithm from N observations would be given by $\hat{a}_{1,\text{YW}} = -\hat{r}(1)/\hat{r}(0)$, where the estimated autocovariances are the usual biased lagged product estimates.
Derive an asymptotic expression for the variance of this estimator as the quotient of two stochastic variables. The variance expressions for the lagged product estimator for the autocorrelation are given in (3.34) to (3.40). The asymptotic expression has terms with order of magnitude $1/N$; terms with order of magnitude $1/N^2$ and still higher negative powers of N are neglected.

6.3   A new estimator for the process of Exercise 6.2 is defined as

$$\hat{a}_{1,\text{NEW}} = \frac{-2\hat{r}(1)}{1 + \sqrt{1 + 4\,\hat{r}(1)^2}}.$$

What is the asymptotic expectation of this estimator? All terms with order of magnitude $1/N$ may be neglected here.

6.4   Prove that the variance of this new estimator of Exercise 6.3 is

$$\text{var}\left(\hat{a}_{1,\text{NEW}}\right) = \frac{1}{N} \frac{\left(1 - a^2\right)\left(1 + 4a^2 - a^4\right)}{\left(1 + a^2\right)^2}.$$

Would you prefer the Yule-Walker estimator or the new one?

6.5   Prove the asymptotic equivalence

$$\prod_{i=1}^{K}(1 - v_i) = 1 - \frac{K}{N}.$$

6.6   What is the asymptotic covariance matrix of the two MA parameters estimated from N observations of a MA(2) process with the true parameter vector $[1, b_1, b_2]$.

6.7   Prove the asymptotic relation

$$\text{var}[\hat{a}_1] = \text{var}\left[\hat{a}_p\right]$$

for the estimation of p parameters from N observations of an AR(p) process.

6.8   What is the relation between Exercise 5.16 and the Yule-Walker method of AR parameter estimation?

6.9   Give a reason why the Burg method is often preferred to the Yule-Walker method for the estimation of the parameters of an AR model from stochastic data with unknown characteristics.

6.10   Give a reason why the Burg method is mostly preferred to the least-squares methods for the estimation of the parameters of an AR model from stochastic data.

6.11   Give a reason why the Burg method is mostly preferred to the maximum likelihood method for the estimation of the parameters of an AR model from stochastic data.

6.12   Given an ARMA(1,1) process with AR parameter α and MA parameter β. Prove

$$\text{cov}\left[\hat{a},\hat{b}\right] = \frac{1}{N}\frac{1-\alpha\beta}{(\alpha-\beta)^2}\begin{bmatrix}(1-\alpha^2)(1-\alpha\beta) & (1-\alpha^2)(1-\beta^2) \\ (1-\alpha^2)(1-\beta^2) & (1-\beta^2)(1-\alpha\beta)\end{bmatrix}.$$

AR Order Selection

7.1 Overview of Order Selection

In many problems, models of different types and orders can be estimated, and the best is not always known *a priori*. An old problem is the fitting of a polynomial to measured data. It is well known that an estimated polynomial of order $M - 1$ fits precisely to M given data points, but it is equally known that the same polynomial will have a poor fit to new data. The question is to select the best order of a polynomial that is estimated from the current data and that will fit well to new data of the same type. Every extra order of the polynomial will give a better fit to the current data, but only statistically significant orders improve the fit to new data.

A related problem is subset selection. Which regressors are important for a measured independent variable and which are not? This latter question can also be applied to polynomial fitting by letting the sequence of the polynomials free. Hierarchical ordering of the polynomials prescribes the sequence, generally X^0, X^1, X^2, ..., where X is the measured regressor vector. A model with p parameters or regressors included will contain all X^k, $k < p$. If the sequence is free, every power X^k can be the next regressor in a model. Subset selection is much more difficult and less reliable than hierarchical order selection because any hierarchically ordered model has only one neighbouring model with one parameter more and one with one parameter less. Limiting the number of candidate models makes selection easier. Mallows (1973) described C_p, a selection criterion from regression theory. Hocking (1976) showed the relations of C_p with several statistical tests. He also proved that the order with the smallest error of the parameters is not always the same as the model with the best fit to the data.

In autoregressive order selection, Akaike (1970, 1974, 1978) introduced several criteria. Many others followed with similar criteria. Ulrych and Bishop (1975) reported the practical problem that the order selected depended on the highest candidate order. The consistent criteria that were introduced by Rissanen (1978, 1986), Akaike (1978), and Hannan and Quinn (1979) could under some conditions converge to the true process order if the number of observations approaches infinity. However, in practice, only a finite number of observations is available, and the theoretical properties for infinite sample size are not relevant for finite-sample practice. Shibata (1976, 1984) derived mathematically how much selection quality suffered from overfit with the explicit assumption that underfit was impossible. Overfit denotes the situation where the model has too many parameters; underfit happens if not all important orders are included. Overfit models give extra inaccuracy due to estimation variance errors, and underfit models suffer from bias errors.

Broersen (1985) described the finite-sample character of order selection if the candidate model orders are not very small in comparison with sample size. He showed that finite-sample properties depend heavily on the estimation algorithm that is used; see Figures 6.1 - 6.4 for AR estimates with the algorithm of Burg. For AR estimates, the Yule-Walker method, the Burg method, and the least-squares method require different finite-sample adaptations (Wensink, 1996). Also Hurvich and Tsai (1989) and Hurvich *et al.* (1990) introduced a small sample improvement to Akaike's AIC criterion. Surprisingly, their improved AIC_C criterion is not dependent on the method of estimation. It gives good results only in small sample AR estimation when applied to Burg estimates. Broersen and Wensink (1996, 1998) introduced several improvements to finite-sample selection criteria for autoregressive order selection. Finally, finite-sample peculiarities and the desired asymptotic behaviour have been combined in a single criterion by Broersen (2000a).

Recent books on order selection have been written by Choi (1992), Burnham and Anderson (1998), and Miller (1990). Miller's book is not concerned with time series. However, it describes the different types of bias that are also relevant for order selection in time series. They are caused by the double use of the observations, first to estimate parameters and afterward to select a model. It is clear that a parameter that occasionally has a higher estimate than its expectation will also have a greater probability of being selected. Shibata (1976, 1984) studied this effect for time series. Those aspects are often neglected in the theory of order selection. Kuhlback (1959) has defined a discrepancy measure to express the difference between probability density functions. An order-selection criterion based on these statistical measures for probability density functions will resemble closely a criterion that searches for the smallest prediction error.

It is rather difficult to take all peculiar effects into account and derive a theoretical preference for an order-selection criterion along strict mathematical lines. Bias, consistency, finite-sample effects, the number of observations and the estimation method may have their influence. The approach here will be to present several selection criteria without trying to be complete. Many equivalent criteria have been presented in the past, and it is not necessary to explain all of them. The most important criteria will be compared in simulation studies. Which type of simulation is decisive will be discussed.

Unfortunately, many simulation studies that have been carried out lead to dubious conclusions. It can be explained easily how wrongly conducted simulation studies have led to disappointing results. Akaike's AIC criterion is known to be sensitive to overfit by selecting too many parameters and almost completely free of underfit by including too few parameters. On the other hand, consistent criteria are safe against overfit at the cost of a high probability of underfit. Many simulation studies compare only criteria from those two categories. Their results depend on whether the examples used are more prone to underfit or to overfit. This chapter will show that the best criterion is generally a compromise between overfit and underfit.

Often, selection criteria are developed only for AR processes and applied afterward to AR(p), MA(q), and ARMA(p,q) processes. Only some special finite-sample criteria of Broersen (2000a) are exclusively applicable to AR order selection, and most other criteria can be applied to all AR, MA, and ARMA processes. The number of estimated parameters is denoted K. For an ARMA(p,q) process with subtracted mean, K will be $p + q + 1$. The subtraction of the mean itself will generally not lead to the selection of another order of the time series model but the estimated model parameters may be different.

In linear regression, the observations are modeled as the true response values plus additive noise. Without noise, the estimated parameters would become equal to the true values, for all sample sizes N. This is different from the estimation of time series models, where the white input innovations are necessary to obtain a time series. Without input innovations ε_n in (4.53), no output x_n would be found. The estimated linear regression parameters converge to the true values for smaller noise and greater sample size. Their accuracy depends on the noise level. The accuracy of estimated time series parameters, however, is independent of the level of the innovations in (6.42) and (6.43). Therefore, innovations in time series and noise in regression have some mathematical correspondence but also a number of differences. As the expectations of parameters in regression are simply the values obtained by substituting zero for the additive noise, the treatment and the interpretation of order selection are somewhat simpler in linear regression. For that reason, it is presented here.

7.2 Order Selection in Linear Regression

This section about linear regression has the same notation as Section 2.5 that is specific for this section only, except for the variance of the noise that will be denoted σ_ε^2 here. The regression Equation (2.39) has K regressors. In matrix notation,

$$Y = X\beta + \varepsilon \tag{7.1}$$

The expectation of RSS$_K$, the residual sum of squares for the model with all K regressors included, can be derived by simple matrix calculations. With given values of the regressors, the best prediction that can be made for the output Y is $X\beta$. Estimates $\hat{b}$ of (2.41) are found by minimising the sum of squared errors

$$\text{RSS}_K = [Y - X\hat{b}]^T [Y - X\hat{b}] \tag{7.2}$$

RSS$_K$ can be calculated as

$$
\begin{aligned}
\mathrm{RSS}_K &= [Y - X\hat{b}]^T [Y - X\hat{b}] \\
&= [Y - X(X^T X)^{-1} X^T Y]^T [Y - X(X^T X)^{-1} X^T Y] \\
&= Y^T [I_N - X(X^T X)^{-1} X^T]^T [I_N - X(X^T X)^{-1} X^T] Y \\
&= Y^T [I_N - X(X^T X)^{-1} X^T] Y \\
&= Y^T Y - Y^T X\hat{b} ,
\end{aligned}
\tag{7.3}
$$

where $\hat{b}$ from (2.41) has been substituted. I_N denotes the unit matrix of dimension N. The matrix M_K is introduced as a shorthand notation for the projection matrix on all K regressors,

$$
M_K = X(X^T X)^{-1} X^T
\tag{7.4}
$$

Further,

$$
E[Y] = X\beta
$$

$$
E\left[Y^T Y\right] = E\{trace[YY^T]\} = N\sigma_\varepsilon^2 + \beta^T X^T X \beta
\tag{7.5}
$$

where the equality trace (BA) = trace (AB) has been used. Applying this rule to the next to last line of (7.3), the expectation of RSS_K becomes

$$
\begin{aligned}
E(\mathrm{RSS}_K) &= E\{ trace([I_N - M_K] YY^T) \} \\
&= E\left\{ trace\left[(I_N - M_K)\left(\sigma_\varepsilon^2 I_N + X\beta\beta^T X^T \right) \right] \right\} \\
&= \sigma_\varepsilon^2 \, trace(I_N - M_K) \\
&= \sigma_\varepsilon^2 \, trace(I_N) - \sigma_\varepsilon^2 \, trace(M_K) \\
&= \sigma_\varepsilon^2 \, trace(I_N) - \sigma_\varepsilon^2 \, trace(I_K) \\
&= \sigma_\varepsilon^2 (N - K)
\end{aligned}
\tag{7.6}
$$

What happens if some of the true parameters in the regression equation are very small will be investigated. Then it would be possible that the standard deviation in estimated parameters would become greater than their true values. Suppose for a moment that the true value of a parameter is zero, which indicates that one of the regressors has no influence at all. Suppose that the true relation is a straight line $y(n)=x(n)+\varepsilon$. Then, using $x(n)$ and $x^2(n)$ as possible regressors, the true parameter for $x^2(n)$ would be zero. The influence of such a superfluous regressor can also be studied by introducing additional uncorrelated rows of N random numbers, which are considered extra regressors. The expectation of the estimated parameter values for those q extra or nonsense regressors would be zero, but with (7.6), the residual sum of squares RSS_{K+q} would become

$$
E(\mathrm{RSS}_{K+q}) = \sigma_\varepsilon^2 (N - K - q)
\tag{7.7}
$$

This demonstrates that the expectation of the residual variance reduces with σ_ε^2 for all parameters, whether they belong to important or to nonsense regressors. Important regressors have an extra reduction of the residual variance that is generally much greater than σ_ε^2. Excluding such regressors would give an extra bias term in RSS_{K-r} in addition to the variance contributions. Superfluous uncorrelated regressors have no influence on the bias in RSS_{K-r}, only on the reduction due to the estimation variance.

By taking less than K regressors, it is possible that a bias component will also be present in the residual sum of squares RSS_{K-r}. In order selection or in subset selection, models are considered where only a part of the regressors has been included in the model. Order selection is the expression for selection in hierarchically nested models, e.g., with successive orthogonal polynomials as candidate models. Subset selection is the term used for the selection of some arbitrary set of regressors, without any sequence.

A lower order model has p parameters, whereas r parameters are left out, with $p + r = K$. Without loss of generality, the p parameters can be attributed to the first p regressors because the individual numbering of the regressors can be changed without mathematical consequences. Therefore, the equations describe order selection as well as subset selection. The $N \times K$ matrix X is partitioned into the $(N \times p \mid N \times r)$ matrices $(X_p \mid X_r)$. The $K \times 1$ vector β is partitioned in the $p \times 1$ vector β_p and the $r \times 1$ vector β_r. The parameter vector $\hat{b}_p$ for the first p regressors is estimated by minimising the residual sum of squares for a subset RSS_p

$$\text{RSS}_p = [Y - X_p \hat{b}_p]^T [Y - X_p \hat{b}_p] \tag{7.8}$$

The solution for the p parameters is given by

$$\hat{b}_p = (X_p^T X_p)^{-1} X_p^T Y \tag{7.9}$$

which is a biased estimate for β_p if the true values β_r for the omitted regressors are not equal to zero. It follows elementarily that

$$
\begin{aligned}
E\{\hat{b}_p\} &= E\left[(X_p^T X_p)^{-1} X_p^T Y \right] \\
&= E\{(X_p^T X_p)^{-1} X_p^T [X\beta + \varepsilon]\} \\
&= (X_p^T X_p)^{-1} [X_p^T X_p \beta_p + X_p^T X_r \beta_r] \\
&= \beta_p + (X_p^T X_p)^{-1} X_p^T X_r \beta_r
\end{aligned}
\tag{7.10}
$$

This gives no bias if β_r equals 0 or if X_p and X_r are orthogonal, meaning that $X_p^T X_r = 0$.

The residual sum of squares RSS_p has been minimised to find $\hat{b}_p$. The expectation of RSS_p becomes

$$E\left[\mathrm{RSS}_p\right] = E\{[Y - X_p \hat{b}_p]^T [Y - X_p \hat{b}_p]\}$$

$$= E\{[Y - X_p (X_p^T X_p)^{-1} X_p^T Y]^T [Y - X_p (X_p^T X_p)^{-1} X_p^T Y]\}$$

$$= E\{Y^T [I_N - X_p (X_p^T X_p)^{-1} X_p^T]^T [I_N - X_p (X_p^T X_p)^{-1} X_p^T]Y\}$$

$$= E\{Y^T [I_N - X_p (X_p^T X_p)^{-1} X_p^T]Y\} \qquad (7.11)$$

With the shorthand notation

$$M_p = X_p (X_p^T X_p)^{-1} X_p^T \qquad (7.12)$$

$$E\left[\mathrm{RSS}_p\right] = E\{Y^T (I_N - M_p)Y\}$$

$$= E\left\{trace\left[(I_N - M_p)(\sigma_\varepsilon^2 I_N + X \beta \beta^T X^T)\right]\right\}$$

$$= (N - p)\sigma_\varepsilon^2 + \beta^T X^T (I_N - M_p) X \beta \qquad (7.13)$$

The right-hand side of this expression can be simplified by explicitly writing out the submatrices

$$\beta^T X^T \left[I_N - M_p\right] X \beta = \beta^T \begin{bmatrix} X_p^T \\ X_r^T \end{bmatrix} \left[1 - X_p (X_p^T X_p)^{-1} X_p^T\right] \begin{bmatrix} X_p & X_r \end{bmatrix} \beta$$

$$= \begin{bmatrix} \beta_p^T & \beta_r^T \end{bmatrix} \begin{bmatrix} 0 & 0 \\ 0 & X_r^T X_r - X_r^T X_p (X_p^T X_p)^{-1} X_p^T X_r \end{bmatrix} \begin{bmatrix} \beta_p \\ \beta_r \end{bmatrix}$$

$$= \beta_r^T \left(X_r^T X_r - X_r^T M_p X_r\right) \beta_r$$

$$= \beta_r^T X_r^T \left(I_N - M_p\right) X_r \beta_r$$

The residual result can be related to the partitioned covariance matrix of the complete $K \times 1$ parameter vector $\hat{b}$, whose expectation is the true vector β and is given by

$$\mathrm{cov}(\hat{b}, \hat{b}) = E\left\{[\hat{b} - \beta][\hat{b} - \beta]^T\right\}$$

$$= E\left\{\left[(X^T X)^{-1} X^T (X \beta + \varepsilon) - \beta\right]\left[(\varepsilon^T + \beta^T X^T) X (X^T X)^{-1} - \beta^T\right]\right\}$$

$$= E\left\{(X^T X)^{-1} X^T \varepsilon \varepsilon^T X (X^T X)^{-1}\right\}$$

$$= (X^T X)^{-1} X^T E(\varepsilon \varepsilon^T) X (X^T X)^{-1} = \sigma_\varepsilon^2 (X^T X)^{-1} X^T I_N X (X^T X)^{-1}$$

$$= \sigma_\varepsilon^2 (X^T X)^{-1} \qquad (7.14)$$

The covariance matrix can be partitioned to give

$$\text{cov}\left(\hat{b},\hat{b}\right)=\text{cov}\begin{bmatrix}\hat{b}_p\hat{b}_p^T & \hat{b}_p\hat{b}_r^T \\ \hat{b}_r\hat{b}_p^T & \hat{b}_r\hat{b}_r^T\end{bmatrix}=\sigma_\varepsilon^2\left(X^TX\right)^{-1}$$

$$=\sigma_\varepsilon^2\begin{bmatrix}X_p^TX_p & X_p^TX_r \\ X_r^TX_p & X_r^TX_r\end{bmatrix}^{-1}=\begin{bmatrix}\text{cov}_{pp} & \text{cov}_{pr} \\ \text{cov}_{rp} & \text{cov}_{rr}\end{bmatrix} \qquad (7.15)$$

A standard partitioned matrix result in Searle (1982) gives for symmetrical A and D and all inverses existing

$$\begin{bmatrix}A & B \\ B^T & D\end{bmatrix}^{-1}=\begin{bmatrix}A^{-1}+FE^{-1}F^T & -FE^{-1} \\ -E^{-1}F^T & E^{-1}\end{bmatrix}$$

$$E=D-B^TA^{-1}B$$

$$F=A^{-1}B \qquad (7.16)$$

which yields for the lower right rectangle of the parameter covariance matrix

$$\text{cov}_{rr}=\sigma_\varepsilon^2\left[X_r^TX_r-X_r^TX_p\left(X_p^TX_p\right)^{-1}X_p^TX_r\right]^{-1}$$

$$=\sigma_\varepsilon^2\left[X_r^T\left(I_N-M_p\right)X_r\right]^{-1} \qquad (7.17)$$

This can be substituted in the residual sum of squares in (7.13):

$$E\{\text{RSS}_p\}=\left(N-p\right)\sigma_\varepsilon^2+\beta^TX^T\left[I_N-M_p\right]X\beta$$

$$=\left(N-p\right)\sigma_\varepsilon^2+\beta_r^TX_r^T[I_N-M_p]X_r\beta_r$$

$$=\sigma_\varepsilon^2\left(N-p+\beta_r^T\,\text{cov}_{rr}^{-1}\,\beta_r\right) \qquad (7.18)$$

This shows that every regressor in the model gives a decrease in the *expectation* of the residual sum of squares RSS_p that is at least equal to the variance of the noise.

Furthermore, regressors with truly nonzero parameters have a bias contribution to RSS_p that depends on the value of the true parameter and on the covariance matrix of the parameters that are left out. Suppose that $r=1$ or $p=K-1$ and only the last regressor is excluded from the estimated model. Then, the 1×1 covariance matrix cov_{rr} is just the variance of that last parameter. If it is difficult to estimate that last parameter accurately, that means that the variance of that parameter is great. The inverse of the variance is small and the bias contribution to the expectation of RSS_p is small.

The real purpose of the estimation is not to make the residuals as small as possible but to make the description $X_p\hat{b}_p$ as close as possible to the true relation without noise, which is given by $X\beta$. Hence, a quality measure can be introduced, the *scaled subset model error* J_p, which is defined as

$$J_p = \left(X_p \hat{b}_p - X\beta \right)^T \left(X_p \hat{b}_p - X\beta \right) / \sigma_\varepsilon^2 \qquad (7.19)$$

This measure requires knowledge of the true parameters, which is not available in practice. As such, it has the same meaning as the model error (ME), introduced in (5.40) as an accuracy measure for time series models. J_p can be used in Monte Carlo computer simulations with artificially generated data with a random generator. It is also useful in theoretical derivations where the true regression equation $X\beta$ is supposed to be known. It is a natural demand to look for the particular subset $p*$ with the smallest J_{p*} of all possible subsets of the K available regressors. In words

Which estimated subset is closest to the response that would have been obtained if noise-free measurements could have been made?

This is not necessarily the unbiased subset with all nonzero parameters included. It is possible that omitting a small parameter and its regressor from the regression equation gives an additional bias component in J_{K-1} that is smaller than the increased variance that is involved in the estimation of the complete model with all K regressors included. If the bias in leaving out a regressor is smaller than the standard deviation in including a regressor with an estimated parameter, an estimated model without that regressor is better.

With some manipulation, the expectation of the subset model error J_p can be written as

$$E\left(J_p \right) = p + \beta_r^T \, \mathrm{cov}_{rr}^{-1} \, \beta_r \qquad (7.20)$$

There is a strong relation between this subset model error J_p and the expectation of the residual sum of squares RSS_p in (7.18) of the subset model. Substitution gives

$$E\left(J_p \right) = E(\mathrm{RSS}_p) / \sigma_\varepsilon^2 - N + 2p \qquad (7.21)$$

It is possible to define a selection criterion that is based on quantities that are known in the practice of estimation and has the same expectation as J_p. This is the order-selection-criterion C_p of Mallows (1973), defined as

$$C_p = \mathrm{RSS}_p / s^2 - N + 2p \qquad (7.22)$$

In this criterion, s^2 is defined with the residual sum of squares of the complete model with all K regressors included in the estimation. This is an unbiased estimate if all true regressors are between those K, given by

$$s^2 = \frac{\mathrm{RSS}_K}{N - K} \qquad (7.23)$$

The expectation of s^2 is the true variance σ_ε^2. With that expectation result substituted, it turns out that the expectation of J_p and C_p are the same. C_p is calculated for all subset models with $p = 1,\ldots,K$, and the model with the smallest value of C_p is selected.

In linear regression, the number of regressors is often denoted as the size of a subset. In polynomial regression, subsets generally consist of nested hierarchical models of increasing order, the regressors of which may be made orthogonal. The formula for C_n shows that the decrease in RSS_p when including one extra parameter, should at least be $2s^2$ to make the extra parameter significant. Introducing r extra parameters gives the minimally required increase of $2r$ in C_p. Hence, the new residual variance RSS_{p+r} should be at least $2rs^2$ less than RSS_p to give a smaller value of C_p. The idea is to calculate C_p for all possible 2^K subsets if K regressors are available as candidates. Subset selection is looking for the subset with the smallest value of C_p. That subset is selected. In hierarchical models with a fixed sequence of the regressors, the total number of candidate subsets is only K. It is clear why this reduced number of candidates improves selection quality. The probability that the single next hierarchical regressor has an estimate that is significantly greater than its expectation is much smaller than the probability that one of all remaining regressors has that property.

An interpretation as a weak parameter criterion is that a group of r parameters is weak and can better be omitted from the selected model if it causes a residual reduction less than $2rs^2$. For a single parameter,

$$\hat{b}_i^2 > 2 \, \text{var}\left(\hat{b}_i\right) \tag{7.24}$$

or a parameter estimate has to be greater than $\sqrt{2}$ times its standard deviation to be statistically significant. For the last r parameters, this can be written as

$$\hat{b}_r^T \, \text{cov}_{rr}^{-1} \hat{b}_r > 2r \tag{7.25}$$

to make sure that $C_p > K$, where K is the criterion value for the complete model C_K due to the specific choice (7.23) for s^2.

Hocking (1976) gives an extensive survey of the properties of C_p. The criterion C_p has a relation with the t - statistic for comparing two subsets with only one additional regressor in the larger one:

$$C_{K-1} - C_K = \hat{b}_K^2 / \text{vâr}\left(\hat{b}_K\right) - 2 \;, \tag{7.26}$$

which has as distribution $t^2 - 2$. For larger subsets,

$$C_p - C_K = \hat{b}_r^T \, \text{côv}_{rr}^{-1} \hat{b}_r - 2r = r(F_{r,N-K} - 2) \tag{7.27}$$

where the estimated covariance equals the true covariance with s^2 instead of the true variance σ_ε^2 and $F_{r,N-K}$ is the central F-distribution with $N - K$ degrees of freedom. Central means that it is based on the hypothesis that $\hat{b}_r$ equals zero. There is a strong relation between looking for the minimum of C_p and testing of F-statistics with significance level 2. On the one hand, the introduction of C_p was based on the relationship between the expectation of C_p and J_p. This leads to the single level 2 for the F-test, whereas also other levels of significance are quite common in using F-tests. This principle can and will be used in defining other coefficients for the penalty factor of additional parameters in an order-selection criterion. Cavanaugh (2004) treats the correspondence between order-selection methods in regression and in time series.

With the choice of J_p, the quality criterion is chosen as the fit of the estimated model to precisely the interval that is covered by the regressors in (2.39). One consequence is that extrapolation of regression equations is generally very inaccurate. It might be reasonable if the selected regression equation occasionally represents a true physical relationship, but extrapolation is never supported by statistical properties. A different purpose of order selection might be that the best fit is sought for only an interval of the regressors. Hocking (1976) showed that such a purpose would require a dedicated order-selection criterion. If the accuracy of the estimated parameters were the purpose, as in (6.25), the best model is characterized by

$$\beta_r^T \operatorname{cov}_{rr}^{-1} \beta_r < 1 \tag{7.28}$$

according to Hocking (1976). That gives the result that the parameter accuracy becomes better if those last r parameters are excluded when they give a residual reduction less than s^2 if their expectations are substituted. This order cannot be selected properly from practical data because each individual parameter contributes the same amount s^2 to the residual reduction due to variance. That order plays a role in the intermediate order of the AR model used for MA and ARMA estimation.

As a comparison, the best prediction order with minimum J_p is found with (7.25)

$$\beta_r^T \operatorname{cov}_{rr}^{-1} \beta_r < r \tag{7.29}$$

In this case, the expectations of the residual reductions due to the bias and due to the variance of omitted parameters would be exactly equal. The requirements for the best parameter accuracy are stronger, and the order with the best parameter accuracy will generally be higher than the order with the best prediction. Therefore, intermediate AR orders for MA and ARMA estimation need special treatment.

7.3 Asymptotic Order-selection Criteria

The order selection criterion FPE, the final prediction error, has been described by Akaike (1970) as the first order-selection criterion for AR processes. The FPE uses the relation between the unbiased expectations of the residual variance s_K^2 in (6.17) and the prediction error PE(K) in (6.18):

$$\text{FPE}(K) = s_K^2 \, \frac{N+K}{N-K} \qquad\qquad (7.30)$$

The application of this order-selection criterion is simple: Compute the residual variance for all candidate models, compute the FPE of all candidates, and select the model order with the smallest FPE. For models with orders lower than the true process orders, the residual variance will have significant contributions due to true parameter values. Omitting those low order parameters introduces a bias in the prediction error. Hence, the value of the FPE for biased models will be greater than that without significant bias contributions. The FPE is the transform of the residual variance in the expected prediction error. Therefore, the expectation of the FPE is PE(K), and it is obvious that selecting the model with the smallest FPE is a sensible choice. Figure 7.1 illustrates the principle. The averages of PE and FPE are almost the same, and PE and FPE virtually coincide in the figure. For the parameters below the true order 5, which are statistically significant, the decrease in both the empirical residual variance RES and the PE is determined mainly by the true parameter values. The small difference is given with (7.30) by a factor $2m/N$ for model order m. The lines really start to diverge visually at the true order 5, where the PE increases with the model order and the residual variance keeps decreasing. Finite-sample deviations are small in Figure 7.1 because the highest candidate model order is small in comparison to the sample size. The average looks very regular; the result of individual runs has more variation. One peculiar individual result is given in Figure 7.2. That example has been selected because the minimum

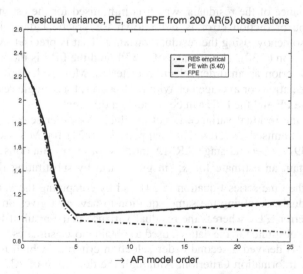

Figure 7.1. Average of the residual variance, the prediction error, and the order-selection criterion FPE from Burg estimates in 200 simulation runs of 200 AR(5) observations, as a function of the AR model order

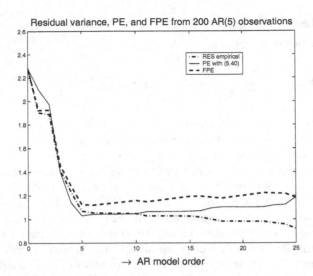

Figure 7.2. The residual variance, the prediction error, and the order-selection criterion FPE from Burg estimates in a single simulation run of 200 AR(5) observations, as a function of the AR model order

of the FPE was again decreasing at the highest orders in the specific simulation run, whereas the true PE increases. Generally, the difference between the PE and the FPE is much smaller, and individual plots look like Figure 7.1.

One important issue in Figure 7.1 is that knowledge of the true system parameters has been used to compute the PE with (5.40) or (5.46). That is possible in simulations, but in real experiments, only the residual variance s_K^2 is known.

That is the variance of the residuals, which is minimised for the estimation of the parameters. The prediction error is not known for practical data; it has to be approximated somehow using the residual variance. That is precisely what is done in the FPE criterion (7.30). The average of the PE and the FPE is the motivation to use the FPE criterion as an order-selection criterion. Most order-selection theory deals with expectations or average behaviour. However, for a single realisation, the behaviour of the PE and the FPE can be somewhat different.

Sometimes, the residual variance is not available for order selection because it has not been minimised directly. That happens in (6.24) for MA models and in (6.38) and (6.39) for second-stage ARMA models. In those situations, it is always possible to obtain an estimate for s_K^2 in given data by substituting the estimated parameters in the time series Equation (5.34) and by computing the variance of the remaining residuals. Applying the same equation to new data gives an estimate for the prediction error (PE), whereas the residual variance is the result if the estimated parameters are substituted in the data x_n, used to obtain those estimates.

Akaike (1974) derived a second order selection criterion, which is the famous AIC, Akaike's Information Criterion. Although the definition of AIC is based on the maximum likelihood estimate of the residual variance, it has become a common practice to apply it to any estimate of the residual variance s_K^2, obtained

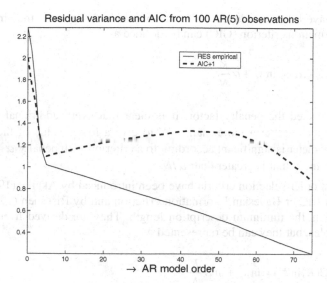

Figure 7.3. Average of the residual variance and AIC from Burg estimates in 2000 simulation runs of 100 AR(5) observations, as a function of the AR model order. The global minimum of AIC is found at very high orders if candidates with orders above $N/2$ are allowed. The global minimum of the average of AIC is at order 5 if the highest candidate order is not taken too high.

by arbitrary methods of parameters estimation. With this extension, AIC is defined as

$$\mathrm{AIC}(K) = \ln s_K^2 + 2\frac{K}{N} \tag{7.31}$$

Figure 7.3 shows the residual variance for the AR Burg algorithm and the AIC criterion of (7.31) that is derived from the logarithm with an additional penalty that depends on the number of estimated parameters. AIC has a local minimum at order 5. This would also be the global minimum if the highest candidate order is not too high. However, high AR orders are often necessary as candidates for selection, e.g., as intermediate orders for MA and ARMA estimation. The behaviour of the residual variance is regular, as in Figure 6.2. However, the order-selection criterion AIC has a undesirable effect at higher model orders. The same local high order minimum would be found with the FPE of Figure 7.1 if the highest candidate orders were as high as $N/2$.

Both the FPE and the AIC criterion could give dubious results when applied to high-order AR candidate models. Ulrych and Bishop (1975) reported the preference for too high orders and also the selection of the highest candidate order. Therefore, order selection depended on the available candidates instead of exclusively on the characteristics of the data. Furthermore, Shibata (1976) showed that AIC had a tendency for overfit. Even asymptotically, the minimum of AIC does not converge to the true AR order.

Criteria have been introduced under various names. To show the similarity, a general information criterion (GIC) can be defined as

$$GIC(K, \alpha) = \ln s_K^2 + \alpha \frac{K}{N} \qquad (7.32)$$

where α is called the penalty factor. It is clear that every additional parameter must reduce the first term more than α/N to give a lower value of the criterion (7.32). A parameter is significant according to a criterion with penalty α if it causes a decrease in $\ln s_K^2$ that is greater than α/N.

Consistent order-selection criteria have been introduced by Akaike (1978) under the acronym BIC or Bayesian Information criterion and by Rissanen (1978) as the MDL criterion, the minimum description length. They are derived from different basic principles, but they can be represented as

$$GIC(K, \ln N) = \ln s_K^2 + \ln N \frac{K}{N} \qquad (7.33)$$

For the usual consistent criteria, the penalty factor α is given as $\ln N$.

Hannan and Quinn (1979) also introduced a criterion with the smallest possible penalty factor that leads to consistency:

$$GIC(K, 2\ln \ln N) = \ln s_K^2 + 2\ln \ln N \frac{K}{N} \cdot \qquad (7.34)$$

So far, all criteria have been derived with asymptotic theory. Hurvich and Tsai (1989) suggested a small sample correction to AIC, called AIC$_C$ as

$$AIC_C(K) = \ln s_K^2 + \frac{1 + \frac{K}{N}}{1 - \frac{K+2}{N}} \qquad (7.35)$$

The small-sample correction in this criterion is an asymptotic correction to AIC. Later, dedicated small-sample criteria for AR order selection will be presented.

7.4 Relations for Order-selection Criteria

For large N, or rather for large N and small K, the FPE of (7.30) is closely approximated with the first order Taylor approximation:

$$FPE(K) \approx s_K^2 \left(1 + \frac{2K}{N}\right) \qquad (7.36)$$

This approximate way to describe the FPE is used to compare the candidate orders FPE($p-1$) and FPE(p) for N observations of an AR(p) process. The order p is used here to indicate that the calculations apply only to AR models. The residual reduction between the AR(p-1) model and the AR(p) model is given by the square of the additionally estimated parameter, as in the Levinson-Durbin recursion of (5.24):

$$s_p^2 = s_{p-1}^2 \left(1 - \hat{k}_p^2\right) \tag{7.37}$$

It follows by the same approximation of (7.36) that

$$\text{FPE}(p-1) \approx s_{p-1}^2 \left[1 + 2(p-1)/N\right] \tag{7.38}$$

and combining the above equations yields

$$
\begin{aligned}
\text{FPE}(p) &\approx s_p^2 \{1 + 2p/N\} \\
&\approx s_{p-1}^2 \left(1 - \hat{k}_p^2\right)\{1 + 2p/N\} \\
&\approx s_{p-1}^2 \{1 + 2p/N - \hat{k}_p^2\}
\end{aligned} \tag{7.39}
$$

The approximation in the last line supposes that the square of the parameter has a magnitude of about $1/N$ and terms with $1/N^2$ are neglected. That is a property of all asymptotic approximations to neglect terms that are a factor of N smaller; those terms vanish asymptotically. A comparison of (7.38) and (7.39) shows that the square of the last parameter must be greater than $2/N$ to be statistically significant and to give a reduction in FPE(p) with respect to FPE(p-1).

The variance of the last parameter or of the last the reflection coefficient is given by (6.16) as

$$\text{var}\left[\hat{k}_p\right] = \frac{1 - k_p^2}{N} \tag{7.40}$$

For final parameters with a magnitude of about $1/\sqrt{N}$ or less, the variance (7.40) can be approximated asymptotically by $1/N$. This gives a very simple expression for the last parameter to be statistically significant, meaning that including the parameter will give a lower value for the FPE:

> **A parameter is statistically significant**
> **if its squared estimate is greater than**
> **twice its estimation variance.**

If the estimate for the last parameter is greater than $\sqrt{2}$ times its standard deviation, it should be included.

> *An extra parameter is statistically significant and should*
> *be selected if the expected increase of the residual*
> *variance due to the bias, if the parameter is omitted, is*
> *greater than the expected decrease of the residual*
> *variance due to the estimation variance if it is included.*

A very rough derivation of Akaike's order-selection criterion AIC is found by taking the natural logarithm of FPE, using $\ln(1+\delta) \approx \delta$,

$$
\begin{aligned}
\ln\left[\text{FPE}(p)\right] &\approx \ln\left[s_p^2\left(1+2p/N\right)\right] \\
&= \ln\left[s_p^2\right] + \ln\left[1+2p/N\right] \\
&\approx \ln\left[s_p^2\right] + 2p/N \\
&= \text{AIC}(p) .
\end{aligned}
\tag{7.41}
$$

This clarifies why the criteria FPE(p) and AIC(p) will almost always select the same order. However, this sloppy derivation does not give justice to the much deeper implications, which are more visible in a derivation that approximates the Kullback-Leibler discrepancy. An elegant derivation from basic statistical theory is the basis for AIC and that will be given later.

All order selection criteria, which have been discussed, have a strong relation. It requires some manipulation to compare C_p and AIC(p). Dividing C_p of (7.22) by N to transform from a residual sum of N squares to the residual variance gives

$$
\begin{aligned}
C_p/N &= \text{RSS}_p\big/Ns_K^2 - 1 + 2p/N \\
&= s_p^2/s_K^2 - 1 + 2p/N
\end{aligned}
\tag{7.42}
$$

By assuming that the residual variance of the models of order K and of order p are close to each other, with a quotient close to one, it follows again by using $\ln(1+\delta) \approx \delta$, that

$$
C_p/N \approx \ln\left(s_p^2\right) - \ln\left(s_K^2\right) + 2p/N
\tag{7.43}
$$

As s_K^2 is a constant for all different model orders, the minimum of C_p is found as

$$
\begin{aligned}
\min\left(C_p\right) &\approx \min\left\{\ln\left[s_p^2\right] - \ln\left[s_K^2\right] + 2p/N \right\} \\
&= \min\left\{\ln\left[s_p^2\right] + 2p/N \right\} = \min\left\{\text{AIC}(p)\right\} .
\end{aligned}
\tag{7.44}
$$

This shows that the application of a logarithm in selection criteria gives the possibility of selecting the best order without having or using knowledge of the true value of the variance of the innovations. This unknown value is the same for all different candidate models and can be omitted in looking for the minimum of selection criteria. This normalizing value was required in C_p but is not necessary in the GIC criteria (7.32).

7.5 Finite-sample Order-selection Criteria

Finite-sample theory for AR processes and the result of simulations have shown that the asymptotic expression for the variance $1/N$ can, for AR estimation with Burg's method, better be replaced by the inverse of the true number of degrees of freedom, which is is $1/(N + 1 - i)$ for order i. Broersen (1990), Broersen and Wensink (1993), and Wensink (1996) have also investigated the small sample performance of other AR estimation methods. Empirical approximations for the variance of reflection coefficients with the true value zero are given by

$$v_{i,YW} = \frac{N-i}{N(N+2)}, \qquad i = 1, \cdots, N-1$$

$$v_{i,B} = \frac{1}{N+1-i}, \qquad i = 1, \cdots, N-1$$

$$v_{i,LS} = \frac{1}{N+2-2i}, \qquad i = 1, \cdots, \frac{N}{2}$$

$$v_{i,LSFB} = \frac{1}{N+1.5-1.5i}, \qquad i = 1, \cdots, 0.6N$$

$$v_0 = 1/N \tag{7.45}$$

Those are the results for the Yule-Walker method, Burg's method, unilateral AR least squares, and the combined forward-backward AR least-squares algorithms, respectively. If the mean is subtracted from the signal before the parameters are estimated, this can be included in the finite-sample variance coefficient by taking $v_0 = 1/N$, the same for all estimation methods. The highest AR order for which parameters can be estimated depends on the estimation method. The method is indicated with the acronym of the estimation method in the index: $v_{i,YW}$, $v_{i,B}$, $v_{i,LS}$, $v_{i,LSFB}$. In the following, all v_i without further indication will be those of the Burg method and general formulas will use $v_{i,.}$ with a dot denoting a method to be specified.

The variance coefficients for the Yule-Walker are smaller than the asymptotic value $1/N$, as given in (6.16). That is possible because they are biased estimates. The other coefficients are greater than $1/N$. Least squares can estimate K parameters from $N - K$ residuals, leaving $N - 2K$ degrees of freedom, which explains the factor 2 in the denominator of the coefficient $v_{i,LS}$. Likewise, the Burg algorithm estimates one reflection coefficient of order K from $N - K$ residuals and $v_{K,B}$ is almost $1/(N - K)$, apart from the constant one. The constants in the denominators of (7.45) have been introduced to obtain a still better approximation. This is an empirical result, found by estimating parameters in a white noise signal. The average variance of the ith reflection coefficient in a large number of repeated

Monte Carlo simulations was closely approximated by the finite-sample variance coefficient v_i. If the model order is less than $0.1/N$, the difference between the finite sample variance v_i and the asymptotic variance $1/N$ is often negligible for practical purposes. For higher orders, however, the difference may become important. The v_i are used to adapt order-selection criteria to the method of estimation.

For all estimation methods, the finite-sample expectations of the prediction error and the residual variance are much better described by finite-sample expressions than by asymptotic theory. The expressions for white noise signals have been given in (6.17) and (6.18). For AR(p) processes, a correction should be given for the first p orders. That correction uses the deterministic relation (5.25) for all intermediate orders and keeps the statistical formula that has been obtained for white noise. Without a formal derivation, the approximate finite-sample expressions for the residual variance and the prediction error of intermediate orders p for general AR(K) processes have been given by Broersen (1990) as

$$E_{FS}\left[s_p^2\right] = \sigma_x^2 \prod_{i=1}^{p}\left[\left(1-k_i^2\right)\left(1-v_{i..}\right)\right] \tag{7.46}$$

and

$$E_{FS}\left[\text{PE}(p)\right] = \sigma_x^2 \prod_{i=1}^{p}\left[\left(1-k_i^2\right)\left(1+v_{i..}\right)\right] \tag{7.47}$$

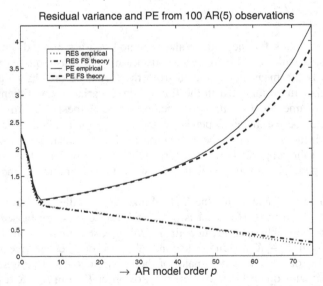

Figure 7.4. Average of the residual variance and the prediction error from Burg estimates in 500 simulation runs of 100 AR(5) observations, as a function of the AR model order. The theoretical finite sample FS expressions (7.46) and (7.46) are also plotted for the Burg method.

The results are reasonably good approximations for all p and for all estimation methods. Furthermore, they are quite accurate if p is equal to or greater than the true order. The limited accuracy is not considered important here because the statistical magnitude of the $v_{i,}$ will be much smaller than the deterministic influence of statistically significant parameters.

Those empirical relations belong much more to the derivation of relations in physics than to mathematics. In physics, it is usual to look for relations for observed phenomena, as in Figure 7.4; in mathematics, axioms are used to derive exact relations. The empirical lines in Figure 7.4 are given for rather high AR orders; the same true AR process has been used in Figure 7.1 with a lower maximum candidate order and with more observations. The theoretical formulas are good approximations for the empirical curves. The finite-sample residual variance is equal to the asymptotic result for the Burg method; the finite PE differs significantly from the asymptotic approximation. It might be possible to derive still more accurate theoretical approximations for the residual variance and the PE for AR Burg estimates for model orders greater than $N/2$. For the purpose in mind, finite sample order-selection criteria, the description is considered accurate enough.

The problem in time series analysis is that the validity of the axioms can be proved only for the asymptotic case with finite orders and an infinitely growing sample size N. However, the sample size N in practice is always limited, and the true AR representation of most processes would be of infinite order because many practical processes have some noise. All analog signals that are represented on a digital computer have at least quantisation noise, and often more noise sources are present. An AR(p) process with additive white noise is theoretically a finite-order ARMA(p,p) process, which is equivalent to an AR(∞) process. Therefore, the two requirements to apply asymptotic theory, both finite model orders and infinite sample sizes, are not met in practical time series data. The finite-sample description is the first attempt to describe the empirical practical relations. Mathematicians may not appreciate the loose mathematical style that has to be used in finite-sample theory, despite the fact that the outcomes are closer to reality than the more elegant asymptotic formulas. It may be expected or considered a challenge that a more strict and elegant mathematical basis of finite-sample theory will be developed in the future.

The finite-sample criterion (FSC) is the equivalent of the FPE criterion (7.30) for order selection in finite sample theory. It is also the method for deriving an expectation for the prediction error PE(p) from the observed residual variance that is known and minimised in the estimation; see Broersen and Wensink (1993). FPE has been introduced as the quotient of the prediction error in (6.18) and the residual variance in (6.19). Likewise, the quotient of the white noise results (6.21) and (6.22) defines the finite-sample equivalent:

$$\text{FSC}(p) = s_p^2 \frac{\prod_{i=0}^{p}(1+v_i)}{\prod_{i=0}^{p}(1-v_i)} \tag{7.48}$$

The expectations in (7.46) and (7.47) have shown that the deterministic contributions to the prediction error and the residual variance are the same and they disappear in the quotient. That is a reason to apply FSC also to real-life data. This indicates that the FSC is a good way to transform the residual variance into the prediction error.

The difference between FPE and FSC is very small if the order p is less than, say, $N/10$. This means that application of FPE and FSC will almost always select the same model order. If higher model orders are considered candidates for selection, FSC outperforms FPE.

The two criteria, FPE and FSC, are conceptually simple. That is the main reason to derive them. However, the criteria to be used in practice are variants of Akaike's famous AIC criterion. The selection results of AIC and FPE are almost always the same, the differences are marginal. For only low orders as candidate models, FSC would also select the same order as AIC.

In finite-sample theory, the residual variance decreases with the product of $1-v_i$., where asymptotic theory gives $1-1/N$. Taking the logarithm of $FSC(p)$, Broersen and Wensink (1993) derived the finite-sample equivalent of AIC for the selection of the order of AR processes:

$$\ln[FSC(p)] = \ln\left[s_p^2 \frac{\prod_{i=0}^{p}(1+v_i)}{\prod_{i=0}^{p}(1-v_i)} \right]$$

$$= \ln\left[s_p^2 \right] + \ln\left[\prod_{i=0}^{p}(1+v_i) \right] - \ln\left[\prod_{i=0}^{p}(1-v_i) \right]$$

$$\approx \ln\left[s_p^2 \right] + 2\sum_{i=0}^{p} v_i \qquad (7.49)$$

With this result, the finite-sample equivalent of AIC(p), the finite information criterion, FIC(p), is defined as

$$FIC(p,2) = \ln\{s_p^2\} + 2\sum_{i=0}^{p} v_i , \qquad (7.50)$$

where 2 in FIC(p,2) denotes the penalty factor 2 before the summation in the definition.

Figure 7.5 shows that the global performance of FIC(p,2) as an order-selection criterion is more desirable than that of AIC. FIC has only one minimum. At orders around 25, the difference FIC(p,2) − FIC(p − 1,2) is also greater than the difference AIC(p) − AIC(p − 1). That would become very important if the true order were around 25. For orders lower than $N/10$, the difference between AIC and FIC is negligible. Finite-sample corrections are required if the maximum candidate AR order is greater than about $N/10$.

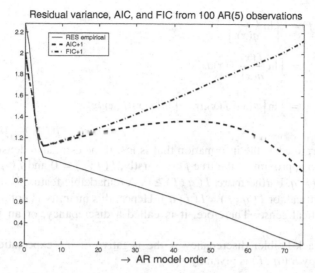

Figure 7.5. Average of the residual variance, AIC, and FIC(p,2) from Burg estimates in 2000 simulation runs of 100 AR(5) observations, as a function of the AR model order. The global minimum of AIC is found at high orders if candidates with orders above $N/2$ are allowed. The global minimum of the average of FIC(p,2) is at order five, independent of the highest candidate order.

The more general finite sample criterion with penalty α becomes

$$\mathrm{FIC}(p,\alpha) = \ln\left(s_p^2\right) + \alpha \sum_{i=0}^{p} v_i \qquad (7.51)$$

Historically, most attention in the time series literature has been devoted to AR processes. Therefore, most derivations have been given exclusively for AR.

So far, order-selection criteria are based on a transform of the residual variance in the prediction error. Order selection can be interpreted as looking for the best predicting model. However, a very strong additional argument comes from the statistical theory of Kullback (1959). That relation was already mentioned by Akaike (1974), who also pointed at the relation with the regression theory of Mallows (1973).

7.6 Kullback-Leibler Discrepancy

In the statistical theory, a measure has been defined that describes the difference between two probability density functions. It can be used for evaluating the difference between true and estimated probability density functions. It is the Kullback-Leibler information quantity $I(q;f)$. Kullback (1975) defined this as the expectation of $\ln[f(x)/q(x)]$, where $f(x)$ is the true probability density function and $q(x)$ is the probability described by the estimated model:

$$I(q;f) = E\left[\ln\frac{f(x)}{q(x)}\right]$$

$$= \int_{-\infty}^{\infty}\ln\frac{f(x)}{q(x)}f(x)dx$$

$$= \int_{-\infty}^{\infty}\ln[f(x)]f(x)dx - \int_{-\infty}^{\infty}\ln[q(x)]f(x)dx \qquad (7.52)$$

It can be interpreted as the information that is lost if the estimated density function $q(x)$ is used to approximate the true $f(x)$. Firstly, $I(f;f) = 0$ and $I(q;f)$ can be zero only if $f = q$. Furthermore, $I(q;f) \geq 0$. A remarkable feature is that $I(q;f)$ is not symmetrical, or $I(q;f) \neq I(f;q)$. Hence, this quantity is not a distance in the mathematical sense. Therefore, it is called a discrepancy, or an information quantity.

The Kullback-Leibler discrepancy is the negative of the expectation of Bolzmann's entropy $B(q;f)$ is given by

$$B(q;f) = \ln\frac{q(x)}{f(x)} = -\ln\frac{f(x)}{q(x)} \qquad (7.53)$$

Thus, minimising the Kullback-Leibler discrepancy is equivalent to maximising Bolzmann's entropy. However, the minimisation is subject to a constraint, which is the model of the information in the data. A good model contains all information, leaving only noise. Maximising the expectation of this uncertainty leaves the information that is justified or supported by the model. It should be stressed that x denotes the integration variable in (7.53); in other words, the data are not directly present in the definition of the Kullback-Leibler information but are felt only by their influence on the estimated or approximating model distribution $q(x)$.

It is interesting how this information measure can be applied to order selection in time series models. Remember that a time series model represents the autocovariance function, which is all there is to know for a vector of multivariable normally distributed variables. Therefore, the search for the best estimated autocorrelation function is the same as that for the model with the minimum of the Kullback-Leibler discrepancy. The first term in $I(q;f)$ of (7.52) contains only the true probability density $f(x)$, which is a constant for all possible models and need not be taken into account in looking for the minimum. The function $\ln[q(x)]$ has a strong relation with the likelihood function that has been maximised to estimate the parameters by the maximum likelihood principle.

The Kullback-Leibler index has been defined as

$$\Delta(q;f) = -2\int_{-\infty}^{\infty}\ln[q(x)]f(x)dx = -2E\ln[q(x)] \qquad (7.54)$$

With this new definition, the Kullback-Leibler discrepancy or information is given by

$$I(q; f) = 1/2 * [\Delta(q; f) - \Delta(f; f)] \tag{7.55}$$

This quantity cannot be determined in practice because the true $f(x)$ will not be known. It is only known in simulation experiments. Using this discrepancy as a quality measure, the second term is not important. It is identical for all different probability densities $q(x)$ and has no influence on which estimated density $q(x)$ will produce the smallest value for the Kullback-Leibler discrepancy. Therefore, an order-selection criterion will be derived with only the first term in (7.55).

Now suppose that all probability density functions are Gaussian. In time series, the true probability density function $f(X)$ represents the joint probability of N observations $X = x_1, ..., x_N$ that could be generated with the true $A(z)$, $B(z)$ and σ_ε^2. The model density $q_X(x)$ is completely characterized by $\mu_X{}^*$ and by $R_X{}^*$, which can be computed with the estimated $A*(z)$, $B*(z)$ and σ_ε^{*2}. The general expression involving the likelihood function of the data is

$$\Delta(q_X; f) = -2E \ln[q_X(X)]$$

$$= -2E \ln \left\{ \frac{1}{2\pi^{N/2} |R_X^*|^{1/2}} \exp\left[-\frac{1}{2}(X - \mu_X^*)^T R_X^{*-1}(X - \mu_X^*)\right] \right\} \tag{7.56}$$

The expectation denotes the N dimensional integral with the true $f(X)$ of N observations. The likelihood of time series observations is often approximated by the likelihood of the innovations. The density function q_x represents the density function of η_n that belongs to the estimated model with parameters $A*(z)$ and $B*(z)$, whereas the true parameters are $A(z)$ and $B(z)$. This simple relation is often used as an asymptotic approximation for the statistics of the data. Apart from values at the beginning of the interval, the observations can be related directly to the innovations that generated those observations. For an AR(1) process with parameter a and σ_ε^2, an approximation can be given that has no special treatment of the initial terms:

$$q_X(x | x_0) = \frac{1}{\sigma_\varepsilon^N \sqrt{2\pi}^N} \exp - \frac{\sum\limits_{n=1}^{N}(x_n + ax_{n-1})^2}{2\sigma_\varepsilon^2}$$

$$q_\varepsilon(\varepsilon) = \frac{1}{\sigma_\varepsilon^N \sqrt{2\pi}^N} \exp - \frac{\sum\limits_{n=1}^{N} \varepsilon_n^2}{2\sigma_\varepsilon^2} \tag{7.57}$$

In practice, the true process quantities are unknown. Therefore, only estimated values for mean and variance and an assumed type of the probability density function can be used to approximate Kullback-Leibler index Δ. The variables η_n belong to a model with $A*(z)$, $B*(z)$, and σ_ε^{*2} for which the Kullback-Leibler index has to be computed.

They can be derived from the variables X, with the observations $x_1, x_2, ..., x_N$, with density $f_X(x)$ as

$$\eta_n = \frac{A^*(z)}{B^*(z)} x_n \qquad (7.58)$$

and are supposed to have the same shape of the probability density function as ε_n. Therefore, the joint density of η_n becomes

$$q_\varepsilon(\eta) = \frac{1}{\sigma_\varepsilon^{*N} \sqrt{2\pi}^N} \exp - \frac{\sum_{n=1}^{N} \left[\frac{A^*(z)}{B^*(z)} x_n \right]^2}{2\sigma_\varepsilon^{*2}} . \qquad (7.59)$$

An asymptotic approximation for the Kullback-Leibler index, for N approaching infinity and with omission of conditioning of the initial observations x_n for $n < 1$, can be evaluated as

$$\Delta(q_\varepsilon; f) \approx -2E \ln \left\{ q_\varepsilon \left[\frac{A^*(z)}{B^*(z)} (x_1, \cdots, x_n) \right] \right\}$$

$$= -2E \ln \left[q_\varepsilon(\eta_1, \cdots, \eta_n) \right]$$

$$= -2E \ln \frac{1}{\sigma_\varepsilon^{*N} \sqrt{2\pi}^N} \exp - \frac{\sum_{n=1}^{N} \eta_n^2}{2\sigma_\varepsilon^{*2}}$$

which gives

$$\Delta(q_\varepsilon; f) \approx N E \left(\ln \sigma_\varepsilon^{*2} \right) + N \ln 2\pi + E \left(\frac{N \eta_n^2}{\sigma_\varepsilon^{*2}} \right) \qquad (7.60)$$

$$= N \ln \sigma_\varepsilon^{*2} + N \ln 2\pi + \frac{N}{\sigma_\varepsilon^{*2}} E \left(\eta_n^2 \right)$$

Only η_n is a function of X and requires integration with $f(X)$ to determine the expectation with the true density given by $A(z)$, $B(z)$, and σ_ε^2. The other quantities $A^*(z)$, $B^*(z)$, and σ_ε^{*2} are given as the model characterization of the probability density for which the Kullback-Leibler index has to be computed.

The constant 2π is not important in looking for a minimum of $\Delta(q_\varepsilon; f)$. The first term is given by the estimate for the residual variance s_K^2, obtained when estimating a time series model with K parameters. The third term can be approximated with the same result used in deriving the FPE criterion (7.30), where the estimate for the prediction error has been used that could be based on the residual variance. This gives

$$\Delta(q_{\varepsilon};f) - N \ln 2\pi - N \approx N \ln\left(s_K^2\right) + N\frac{1+K/N}{1-K/N} - N$$
$$\approx N \ln\{s_K^2\} + 2K \tag{7.61}$$

This equation yields the well-known selection criterion of Akaike as an estimator for the Kullback-Leibler index:

$$\text{AIC}(K) = \ln\{s_K^2\} + 2K/N \tag{7.62}$$

This derivation shows that the background of AIC has a sound statistical basis. The previous derivation of AIC as a Taylor approximation of the logarithm of FPE incidentally gave the same result.

Cavanaugh (1999) introduced the symmetrical Kullback divergence measure

$$2J(q;f) = 2I(q;f) - 2I(f;q)$$
$$= \Delta(q;f) - \Delta(f;f) + \Delta(f;q) - \Delta(q;q)$$

As in (7.55), the second term is a constant for all possible $q(x)$. Furthermore, Cavanaugh (1999, 2004) proved that the last two terms together have as expectation K, the number of estimated parameters that is involved. He divided $J(q;f)$ in the contribution $-2E \ln [q(x)]$ plus three remaining terms that have the expectation K. With this result, a second order-selection criterion, based on the symmetrical Kullback divergence has been introduced as

$$\text{KIC}(K) = \ln\{s_K^2\} + 3K/N \cdot \tag{7.63}$$

A comparison with AIC(K) of (7.62) shows that only the penalty term for the parameters has been changed from 2 to 3.

Those asymptotic approximations are poor for AR models if the order K is greater than say $N/10$. Then finite-sample theory should be used, which gives

$$\Delta(q_{\varepsilon};f)/N - \ln 2\pi - 1 \approx \ln\{s_K^2\} + \frac{\displaystyle\prod_{i=0}^{K} 1 + v_i}{\displaystyle\prod_{i=0}^{K} 1 - v_i} - 1$$

This yields the finite-sample information criterion FSIC(K), derived by Broersen and Wensink (1998) as an alternative for the asymptotic criterion AIC(K):

$$\text{FSIC}(K) \approx \ln\left(s_K^2\right) + \frac{\displaystyle\prod_{i=0}^{K} 1 + v_i}{\displaystyle\prod_{i=0}^{K} 1 - v_i} - 1 \tag{7.64}$$

This criterion turns out to give good protection against the increased variance of $\ln(s_K^2)$ for higher values of K.

Finally, three different derivations of an order-selection principle give the same penalty factor 2 for additional parameters

- C_p of Mallows (1973) that minimises the difference from noiseless regression data

- FPE of Akaike (1970) that minimises the prediction error in time series

- AIC of Akaike (1974) and FSIC that minimise the Kullback-Leibler index or maximise the expectation of entropy

Furthermore, differences in C_p have also been expressed as F-statistics with the hypothesis that the additional parameters are zero. This latter point of view allows an investigation of the influence of taking levels of significance in F-tests other than the fixed penalty factor 2 that appears in the other derivations.

7.7 The Penalty Factor

The discussion is first restricted to the penalty α for AR(p) models. Every additional reflection coefficient above the true order p has the expectation zero and the asymptotic approximation for the variance is $1/N$. The probability density function is normal, and the probability density function of N times the squared reflection coefficient has χ^2 density with one degree of freedom. The asymptotic criterion for order $p + 1$ for a small estimated reflection coefficient k_{p+1} can be written as:

$$
\begin{aligned}
\mathrm{GIC}(p+1,\alpha) &= \ln\left[s_p^2\left(1-\hat{k}_{p+1}^2\right)\right]+\alpha\frac{p+1}{N} \\
&\approx \mathrm{GIC}(p,\alpha)-\hat{k}_{p+1}^2+\frac{\alpha}{N}
\end{aligned}
\tag{7.65}
$$

It follows that order $p + 1$ has a smaller criterion value if

$$
\hat{k}_{p+1}^2 > \frac{\alpha}{N}
\tag{7.66}
$$

The probability that order $p + 1$ will be selected for a true AR(p) process instead of p is asymptotically given by the probability $\chi^2 > \alpha$, if the possibility of underfit is excluded. For FPE, AIC, and similar asymptotic criteria with penalty 2, the probability that $\chi^2 > 2$ is 15.7%. Using a higher penalty factor $\alpha = 3$ for the additional parameter would give a probability of overfit given by $\chi^2 > 3$ which is 8.3%, and $\chi^2 > 10$ gives 0.0016%. The higher the penalty factor, the smaller the probability of overfit.

This simple reasoning can be applied to the choice between the candidate orders p and $p + 1$. For more overfit possibilities with higher orders, the combinatorial probabilities for the selection of a specific overfit order become somewhat more complex. Suppose for the moment that underfit is still not possible. Therefore, the selected order is always at least p or higher.

Order $p + 2$ is selected if

- the residual reduction of the two additional orders together is at least $2\alpha/N$
- the residual reduction of the second order separately is at least α/N
- the residual reduction of all orders $p + 2 + k$ in comparison with order $p + 2$ is less than $\alpha k /N$, for all k.

Shibata (1976) has theoretically described those probabilities. Moreover, every order overfit gives asymptotically an expectation of the increase of the normalized prediction error PE $/\sigma_\varepsilon^2$ with a magnitude of $1/N$. Multiplying the probability of overfitting exactly k orders by k and summing all possibilities gives the cost of overfit $C_{over}(\alpha, p)$, measured on the same scale as the model error (ME). The index p is the true order of the AR process. That cost has been given in a very elegant result by Shibata [1984], using the mathematical properties of χ^2 as

$$C_{over}(\alpha, p) = \sum_{i=1}^{\infty} pr(\chi_{i+2}^2 > \alpha i) \qquad (7.67)$$

The value of $C_{over}(\alpha, p)$ is 2.56 for $\alpha = 2$ and 0.85 for $\alpha = 3$. Those numbers have been calculated with the assumption that infinitely many overfit orders are available, but they are already quite accurate for 10 possible overfit orders, at least for $\alpha > 2$. The cost of overfit is computed with the explicit assumption that also the true order p was a candidate; selecting the true order gives no contribution to the cost in (7.67).

Figure 7.6 shows that the costs of overfit will increase sharply if α is less than 2. Therefore, penalty factors less than 2 have not been considered in practice. The asymptotic cost of overfit does not depend on the true order p. The asymptotic expectation of the ME equals p if the AR(p) model is estimated. If the model is order selected with GIC(p,α) and if the lowest candidate for selection is order p, the expectation for the ME of the selected order is given by

$$E[\text{ME}(\text{selected with } GIC(p,\alpha))] = p + C_{over}(\alpha, p) \qquad (7.68)$$

Considering only the possibility of overfit is reasonable in practice if all model orders are candidates and the true reflection coefficients in the AR(p) process are much greater than $1/\sqrt{N}$. Overfit is always a danger in order selection, unless the highest candidate order is the best to select. Hence, the above given expected loss in accuracy in the normalized ME or PE$/\sigma_\varepsilon^2$ will always be present.

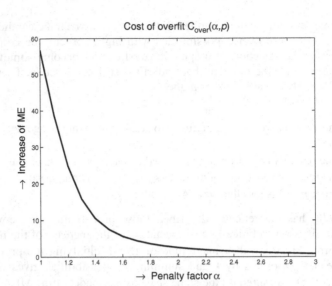

Figure 7.6. Cost of overfit as a function of the penalty factor, measured as an increase in the expected model error (ME) of the selected model due to the selection. Only the true order and higher orders are taken as candidates for selection.

How much of the inaccuracy of the estimated and selected model is due to overfit remains unknown in individual real-life experiments. That would require *a priori* knowledge of the order. Overfit is purely a statistical matter.

However, if the true final reflection coefficient is less than about $3/\sqrt{N}$, the probability of selecting an order less than p should also be considered. That might happen in the individual estimation of a time series model when the actual estimate is one or two standard deviations smaller than its expectation due to the estimation variance.

If the true final parameters are small, it is possible that they are not selected with a selection criterion. The probability of underfit depends strongly on the true parameters of each specific example. Therefore, it is largely deterministic. Often, all parameters until order p give a very significant relative reduction of the residual variance, much greater than $1/N$. As a reduction of α/N is minimally required to include that order, it is usual that the first p parameters are included for all sensible values of α, say, α less than 5 or 6. In this situation, the risk or cost of underfit is negligible. Without the possibility of underfit, the highest values of α give the best accuracy for the selected model. The reason is that the cost of overfit decreases for greater values of α in Figure 7.6. On the other hand, the true representation of a MA process has an infinite AR order, with ever-smaller true AR parameter values at higher orders. In such cases, all finite AR order models are underfitted, but the bias contribution of the higher orders will be very small. The standard deviation caused by the variance of high-order estimated parameters contributes much more to inaccuracy than the bias of parameters if they are omitted.

The best compromise for α remains a matter of taste. The cost of underfit is present in those specific examples only where the magnitude of the reduction of the scaled residual variance by including one more order is about α/N. That is the value that gives criteria equal to (7.66) for orders $p - 1$ and p:

$$E[\text{PE}(p-1)] = \sigma_x^2 \prod_{i=1}^{p-1} (1-k_i^2) \left(1 + \frac{p-1}{N}\right)$$

$$\approx E[\text{PE}(p)] \frac{1}{(1+1/N)(1-k_p^2)}$$

$$\approx E[\text{PE}(p)] \left(1 + k_p^2 - \frac{1}{N}\right) \tag{7.69}$$

This derivation uses the asymptotic approximation where products with terms $1/N^2$ are neglected if terms $1/N$ are present.

Furthermore, the expectation of a squared estimated reflection coefficient is biased by the estimation variance, which is equal for the first and the last parameters in (6.39). For an AR(p) process, it follows for small values of k_p that

$$E\left[\hat{k}_p^2\right] \approx k_p^2 + \frac{1-k_p^2}{N} \approx k_p^2 + \frac{1}{N} \tag{7.70}$$

The other way around, for a measured value $\hat{k}_p^2 = \alpha/N$, the expected value of k_p^2 is given by $(\alpha-1)/N$. With (7.65), it follows that if GIC(p,α) = GIC($p-1,\alpha$), then

$$\frac{E[\text{PE}(p-1)]}{\sigma_\varepsilon^2} \approx \frac{E[\text{PE}(p)]}{\sigma_\varepsilon^2} (1 + \frac{\alpha-1}{N} - \frac{1}{N})$$

$$\approx (1 + \frac{p}{N})(1 + \frac{\alpha-2}{N}) \approx (1 + \frac{p+\alpha-2}{N}) \tag{7.71}$$

The expectations of the prediction errors for the orders $p - 1$ and p are equal for $\alpha = 2$. The always present cost of overfit has been quantified in (7.67) and in Figure 7.6: 2.56/N for $\alpha = 2$ and 0.85/N for $\alpha = 3$. It has been shown that no cost of underfit was expected for $\alpha = 2$. For arbitrary values of α, if greater than 2, the expected cost of underfit caused by selecting the AR($p - 1$) model can be defined. It is the increase of the ME value, caused by excluding the order p with the critical parameter from the model selected. The critical value is defined by GIC(p,α) = GIC($p - 1,\alpha$) in (7.65)

$$E\{\text{ME}[\text{GIC}(p-1,\alpha)]\} = p + C_{\text{under}}(\alpha, p) \tag{7.72}$$

With (7.71), the underfit cost becomes

$$C_{\text{under}}(\alpha, p) = \alpha - 2 . \tag{7.73}$$

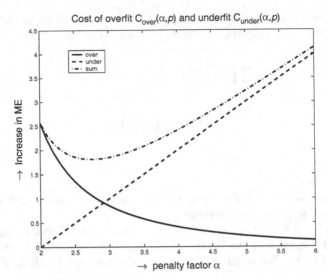

Figure 7.7. Costs of overfit, underfit, and the sum of both as a function of the penalty factor, measured as an increase in the expected model error (ME)

The cost of overfit is always applicable, the cost of underfit is applicable only if the last parameter is less than $\sqrt{(\alpha/N)}$.

Figure 7.7 gives both the cost of overfit and of underfit, as well as the sum of both. The cost of overfit decreases gradually for higher values of the penalty factor. It will eventually become zero for very high penalties. That is the property that is used in the consistent order-selection criteria of (7.33) and (7.34). By letting the penalty grow with N, the cost of overfit vanishes. That is a nice property if the cost of underfit may be neglected. Figure 7.6 shows that the cost of underfit increases linearly with the penalty factor. If some lower limit of the true final parameter values were known, the highest penalty that gives no risk of underfit would be the best choice for the penalty factor. Otherwise, the best choice for the penalty factor is a compromise.

It is remarkable that neither the true order p nor the number of observations N plays a role in Figure 7.7. Obviously, the best compromise for the penalty is independent of p and N. Penalty 2 gives only costs of overfit, no cost of underfit. Hence, the AIC criterion will never leave out statistically significant parts at the cost of often including insignificant details in the selected model. High penalties will have small overfit costs and give mainly costs of underfit. Broersen (1996) proposed a compromise and takes the value where overfit and underfit are equal as the best penalty, and de Waele (2003) used the formula

$$\alpha_{opt} = \arg \min_{\alpha} \max \left[C_{under}(\alpha, p), C_{over}(\alpha, p) \right] \tag{7.74}$$

to find the same compromise for α, which is 2.915. Another compromise would follow as the value for which the sum is minimal. That value for α is 2.72. The compromise is made between two heterogeneous types of cost. All finite-sample arguments have been left out. Therefore, it would be presumptuous to use such a precise number, and the result of this discussion is summarized by taking as a penalty

$$\alpha_{opt} = 3 \tag{7.75}$$

In practice, a model cannot have overfit and underfit costs at the same time. By taking the penalty 3, it can be concluded that the expected selection cost has a maximum of one.

It is quite remarkable that two completely different derivations have found the same penalty factor 3. The first was by Cavanaugh (1999), who introduced a symmetrical Kullback divergence, instead of the Kullback-Leibler discrepancy that is given in (7.52). The second used a compromise between underfit and overfit costs. Simulations will be used to illustrate the differences between the factors 2 and 3 in the practice of order selection.

The expected selection cost with maximum one should be compared with the cost of estimating an extra parameter, which also makes an asymptotic minimal contribution of one to the ME for each estimated parameter in (5.41). In other words, selection among many candidates has the same cost in model quality as the estimation of one extra parameter. An alternative formulation of this principle is that avoiding order selection by taking a fixed model order gives better model quality in asymptotic theory only if precisely the best order is taken. Taking the fixed order one too high gives the same expected loss as taking a selected order. Taking it too low gives underfit errors. Guessing a fixed order that is two or more orders wrong gives a disadvantage in comparison with order selection with penalty 3. Finite-sample results of selection will be obtained later in Monte Carlo simulations. The selection results will be compared with the accuracy of fixed model orders. Examples will be used with possible costs of overfit and underfit.

For AR models, a discussion of the cost of underfit underpins the choice of the penalty factor. For small values of $\hat{k}_p$, the standard deviation can with (6.16) be approximated by $1/N$. The unilateral probability to be further from the expectation k_p than 1.96 times the standard deviation equals 2.5% for a normally distributed estimate $\hat{k}_p$. In other words, if the true last parameter k_p equals about $3/\sqrt{N}$, there is still a nonvanishing probability that the actual estimate $\hat{k}_p$ is twice the standard deviation smaller than the true value and the estimated $\hat{k}_p^2$ will become $\approx 1/N$. This gives a residual reduction that is too small to be selected with the usual values of 2 or 3 for α in GIC(p, α). On the other hand, if the true parameter k_p equals zero, there is almost a 5% probability that the absolute value of $\hat{k}_p$ exceeds $2/\sqrt{N}$, which will lead to selection by the criterion GIC(p, α) if the penalty factor α is less than 4. Final true parameters must be greater than four or five times their standard deviation $1/\sqrt{N}$ to make it practically sure that in no single simulation run too small an estimate is found that is not selected. A final true AR(p) parameter value of 0.01 would require more than 200,000 observations to be almost never missed in the practice of order selection or in a proper *conditional* theory that includes the estimation uncertainty. However, the *unconditional* expectation for the statistical significance of the final AR parameter is $1/\sqrt{N}$ and 10,000 observations are sufficient to let a final parameter of 0.01 be significant in unconditional theory, based on the expectation of the parameter. The reasoning for the last reflection coefficient k_p can be extended to processes where both k_p and k_{p-1} have true values near $\sqrt{\alpha}/\sqrt{N}$. In those critical cases, a large estimate for the last parameter $\hat{k}_p$ can compensate for a smaller estimate for $\hat{k}_{p-1}$. This *conditional* theory is much more complicated than the study of Shibata (1976), which treated only the probability of overfit. A combined theoretical study of overfit and underfit may be feasible, but the bias results depend on the particular AR(p) process, especially on the true values of the last parameters. Therefore, simulations will be used to demonstrate the importance of underfit.

For all purposes where Akaike's AIC criterion with penalty 2 has been a good choice in the past, a better choice will probably be a similar criterion GIC(p, α) of (7.32), the generalized information criterion, with the value 3 for the penalty α. This criterion can also be recommended for MA and ARMA models, where the asymptotic theory gives a satisfactory description of the behaviour of the residual variance.

A particular and well-known choice of α leads to problems with the underfit discussion here. That is the choice $\alpha = \ln N$ in (7.33) or $\alpha = 2\ln\ln N$ in (7.34) that is sometimes used in so-called consistent criteria. This dependence of α on N can lead to serious underfit problems for increasing sample sizes N, because α can grow without bound for $N \rightarrow \infty$. Moreover, in the discussion above, it is clear that an increasing α is a weapon against overfit at the cost of underfit. No reason can be given to make this overfit/underfit compromise a function of sample size. Consistent criteria are popular for mathematicians if they want to know with certainty that they describe the truth. Those criteria are not at all attractive for practical use where the best fitting model is the purpose. The literature has given much attention to consistent criteria because they will select the true order for $N \rightarrow \infty$. A mathematical condition for the applicability of consistent criteria is that the true AR order must be finite and that the final AR parameters are greater than some fixed small number. Under those conditions, consistent criteria have excellent theoretical properties because they can reveal the truth if the sample size increases more and more. However, the costs of underfit are neglected, and the costs of overfit go to zero. If the truth has an infinite order, like AR models of MA or ARMA processes, consistent criteria will have poor performance due to the underfit costs. These follow from Figure 7.7 and are equal to $\alpha - 2$. Therefore, examples can be constructed for the final parameter with a value somewhat smaller than $\sqrt{(\ln N / N)}$ that would not be selected with penalty $\ln N$ and that would give an unbounded underfit error of $\ln N - 2$.

Numerous simulation results in the literature support the use of consistent criteria. However, the favourable conclusions for consistent criteria have always been obtained from examples where no underfit costs were possible or probable, by choosing the last parameter of the simulated processes much greater than $\sqrt{(\ln N / N)}$. Taking smaller values would lead to the conclusion that the performance of consistent criteria is poor. Moreover, the usual application is to select the best model for a given and fixed number N of observations. Then, $\ln N$ is only a constant, dependent on N. It is only important what happens for that particular value of N. There may be a good reason to take a value for the penalty α different from 2, but there are hardly any reasons to let the penalty depend on N. Therefore, consistent criteria have no practical importance. That will be demonstrated in simulations.

For AR models, finite-sample theory remains necessary. It is easy to combine a penalty α with finite-sample results using the finite-sample variance coefficients v_i, of (7.45), giving the finite information criterion of (7.51).

A recent development for automatic model selection used subspace techniques. Bauer and de Waele (2003) describe an automatic procedure for ARMA estimation and selection. They found one important disadvantage, directly related to the fundament of subspace modeling. Subspace autocorrelations are subject to the triangular bias that also plays a role in the lagged product autocovariance function (3.30). It means that examples can be found that are sensitive to triangular bias and subspace methods are very inaccurate for those examples. Furthermore, order selection in subspace methods has some peculiar properties. Bauer (2001) used estimated singular values and a penalty function that depends on sample size. Singular values have the property that the dividing line between significant and insignificant values depends on the actual true process parameters. It is not possible to establish critical values for singular values without knowing the process or without looking at the estimated singular values. In other words, order-selection criteria cannot rely on the white noise behaviour that is found in the PE in Figure 7.1 for all orders above the true order. No such characteristic exists for singular values. Bauer and de Waele (2003) used a consistent penalty factor for selection. However, it is not possible to make an objective compromise between overfit and underfit, as in Figure 7.7, if the critical level depends on true singular values. Bauer and de Waele (2003) conclude that the influence of spectral leakage is considerable in examples that are sensitive to triangular bias. As long as a method is not reliable for all types of measured data, it is less useful than methods without such flaws.

The compromise for penalty 3 in the FIC criterion of (7.51) has been derived by balancing overfit costs with the cost of underfit with only one critical parameter value at the end. It is also possible to derive a compromise for more critical values at the highest AR orders. Suppose that the critical value of the last two parameters of an AR(p) process is determined by

$$\text{GIC}(p-2,\alpha) = \text{GIC}(p-1,\alpha) = \text{GIC}(p,\alpha) \tag{7.76}$$

Then the underfit cost would become $2(\alpha - 2)$. That would make the line of the underfit costs two times steeper in Figure (7.7) and the line of the overfit costs would stay the same. The minimum of the sum is found at $\alpha = 2.3$ now. More generally, by considering the possibility of more underfit orders, the penalty of the best would become very close to 2. Penalty 3 is a good compromise to protect against one order of overfit. However, if two models are compared where the number of parameters differs by more than 3 or 4, the best penalty becomes close to 2.

7.8 Finite-sample AR Criterion CIC

The asymptotic problem has been resolved sufficiently. The Kullback-Leibler derivation yields AIC(p) with penalty 2. The compromise between bias and variance errors gives the value 3 for the penalty factor in GIC if one order of underfit is considered. This same value 3 follows with a symmetrical Kullback divergence by Cavanaugh (1999). The situation has some extra complications for finite-sample theory. Asymptotic theory is always applicable to low-order models with less than $N/10$ parameters. That gives the finite-sample criterion FIC(p,α) of (7.51) with $\alpha = 3$ as the best compromise for low-order finite samples.

The Kullback-Leibler derivation yields the criterion FSIC(p) in (7.64) with remarkably good finite-sample performance; see Broersen (2000). Even if the maximum model order is as high as $N - 10$ or higher, the selection results of FSIC(p) are not seriously influenced by the curious finite sample-effects that other criteria display, including the FIC(p,α) criterion with $\alpha = 3$ that was the favourite based on asymptotic arguments. However, using a Taylor approximation, the low-order penalty factor in FSIC(p) turns out to be 2. FSIC(p) with a smaller penalty for low-order models mostly selects more parameters in AR simulations than FIC($p,3$), whereas the accuracy of the latter model is better if only low-order candidates are considered. On the other hand, FIC($p,3$) can sometimes select orders above $N/2$ where low orders are much better. It would be desirable to combine the good performance of FIC($p,3$) of (7.51) for low orders with the good performance of FSIC(p) of (7.64) for much higher orders.

Broersen (2000) showed that the increasing uncertainty of the estimated residual variance s_p^2 can give problems at very high model orders, say, about $0.5N$. Due to the increased standard deviation of the estimated residual variance, it occurs more often that the sample residual variance is very much smaller than its expectation. FSIC(p) is a criterion with an order-dependent penalty. A compromise between the best low-order penalty 3 and the desirable extra protection against very high candidate orders is found in the combined information criterion CIC(p), that is defined by Broersen (2000) as

$$\text{CIC}(p) = \ln\left(s_p^2\right) + \max\left[\frac{\displaystyle\prod_{i=0}^{p}1+v_i}{\displaystyle\prod_{i=0}^{p}1-v_i}-1 \, , \, 3\sum_{i=0}^{p}v_i\right] \qquad (7.77)$$

where v_i is $1/(N+1-i)$ for AR estimates by Burg's method. For other estimation methods, the v_i coefficients of (7.45) can be substituted in CIC(p). The criterion CIC takes the highest of two penalty factors. Almost the same selection results as with CIC can be obtained with a number of similar criteria combining the asymptotic penalty factor of 3 at lower orders with the good performance of FSIC(p) of (7.64) for higher model orders.

The candidate order for selection with CIC(p) can be chosen as high as $N - 1$ without numerical complications, but generally $p = N/2$ is taken as an upper boundary for accurate estimation. For large data sets, the maximum candidate order can be restricted to 250, 500, or 1000 to limit computing time. However, especially if an order were selected close to that chosen maximum candidate order, it is advisable to verify whether a relaxation of the experimenter's restrictions on the maximum candidate order influences the order selected.

Asymptotic AR order-selection criteria can give wrong orders if the candidate orders are higher than $0.1N$. Using higher penalties or consistent criteria cannot cure that problem. For autoregressive order selection, three categories of criteria have been discerned. They take into account

- the asymptotic expectation of s_p^2

    FPE, AIC, GIC, AIC, AIC_C, C_p

- the finite-sample expectation of s_p^2

    FSC, FIC

- the finite-sample expectation and the finite-sample variance of s_p^2

    FSIC, CIC

Figure 7.8 gives the effective penalties of the finite-sample criterion FSIC(p) (7.64) with the finite-sample variance coefficients of (7.45) substituted for the different estimation methods. The order-dependent part of the criteria after $\ln(s_p^2)$ is shown. As a comparison, the penalties of the AIC (7.31) and the improved AIC_C criterion (7.35) are also given. It is notable that AIC and FSIC for the Yule-Walker method are very similar. That means that Yule-Walker estimation and AIC are a

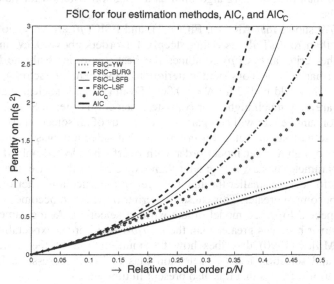

Figure 7.8. Penalty of FSIC for four AR estimation methods, as a function of model order. The penalty factor of AIC and AIC_C is divided by N.

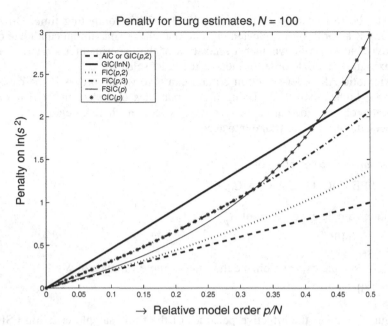

Figure 7.9. Penalty of order-selection criteria for AR Burg estimates as a function of model order. $N = 100$ determines the penalty ln N of the consistent criterion.

perfect finite-sample match. To a lesser degree, FSIC for Burg and AIC_C are also similar for model orders below $0.25N$. That is one reason and also an explanation why those combinations are often used in practice. It also explains why AIC_C is less suitable than the AR Burg algorithm as a selection criterion for other estimation methods.

Figure 7.9 shows that AIC(p), FIC(p,2), and FSIC(p) are very close for AR orders less than $N/10$. FSIC is rising steeply for orders above $0.3N$, much more than the other criteria. CIC(p) combines the best low-order with the best high-order performance. The consistent criterion has the penalty factor 4.6 for 100 observations; it would be 9.2 for $N = 10,000$. From (7.73), it is clear that the costs of underfit are very much higher for consistent selection criteria. Simulations with a variation of sample size will compare the behaviour of all selection criteria.

A final remark about selection criteria is that selection bias is present after selection. Too high a order model order with overfit is selected only if that extra estimated parameter has a value greater than expected. In that case, the prediction error seems to become smaller through the eyes of the selection criterion, although it is really becoming greater. This causes the criterion value to become smaller than *a priori* expected for fixed model orders without selection. At the same time, the prediction error becomes greater than the *a priori* fixed order expectation without selection. Miller (1990) describes how the multiple use of the same data for estimation and selection leads to selection bias in linear regression. Experience from simulations teaches that it is also present in time series.

7.9 Order-selection Simulations

The following procedure has been used in many simulations to construct different interesting examples easily. The true parameters of a generating process of an arbitrary order p are built from reflection coefficients using (5.28) with $k_m = \beta^m$. In this way, all poles of the generating process have the same radius β. The example used here is slightly different and has an interesting looking autocorrelation function. This example can be done by taking $k_m = \beta^m$ and afterward taking a different sign for k_1 that becomes $-\beta$. Figure 7.10 gives the poles of this example for $\beta = 0.7$, and Figure 7.11 shows the true autocorrelation function and the power spectral density of the true AR(11) process.

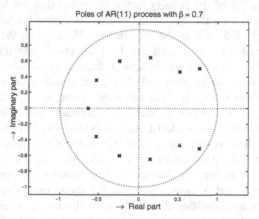

Figure 7.10. Poles of the AR(11) example with $k_1 = -0.7$ and $k_i = 0.70^i$, $i = 2, \ldots, 11$

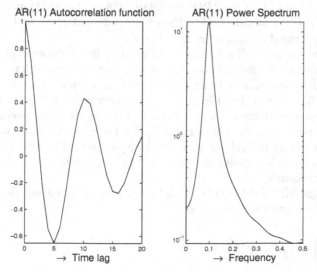

Figure 7.11. Autocorrelation function and power spectral density of the AR(11) example of Figure 7.10 that is used for the simulations for AR order-selection criteria

This example process has been used to compare order-selection criteria. It is applied to sample sizes varying from 20 to 5000. At size 20, the sample size is shorter than the correlation length in Figure 7.11 and only a few parameters are statistically significant. For 5000 and more observations, all 11 parameters are significant. The criterion FPE (7.30) gives the same results as AIC (7.31) in almost all circumstances, and FSC (7.48) is the same as FIC(p,2) (7.50). Therefore, the results of FPE and FSC are not reported.

The best fixed order model is not always the true AR(11) model but depends on the sample size in Table 7.1. Hence, knowledge of the true process order is not sufficient to know the best order for estimation. Table 7.1 shows that for every sample size, a fixed order can be established for which the average model error (ME) is smaller than the average ME that is obtained from selected models. The analysis in Chapter 7.5 of the penalty factor suggested that the loss in quality due to selection, measured on the ME scale, would be approximately one. Table 7.1 gives approximately 2.5 or 3 as the loss in the ME between the best fixed order model and the model selected with the best selection criterion. The reason is twofold. In the first place, finite-sample effects play a role, giving a higher variability to estimated parameters and also to the residual variance, which is the basis for order-selection criteria. That can cause some wrong order to be selected occasionally. The second reason is that the theoretical derivation of the penalty factor supposed that only one underfit order was possible. In practice in Table 7.1, several underfit orders are candidates, the best order is still an underfit order, and also orders higher than the best order are still lower than the true order. This demonstrates that the asymptotic theoretical derivation for the best penalty factor 3 is not completely applicable. However, Table 7.1 shows that penalty 3 in FIC(p, α) is always better than penalties 2 and 4, for all sample sizes. Therefore, the theoretical analysis that led to penalty 3 is strongly supported by the simulations.

The sum over the columns of the ME values in Table 7.1 for the fixed model orders ranges from 118 for the AR(8) model to 5988 for the AR(1) model. The sum for the true AR(11) model is 151. Therefore, order selection will improve the model quality, even if the true order were known.

The absolute value of all reflection coefficients is 0.7 in the example. It follows from (7.36) that the prediction error will be the same for two orders if $k^2 = 1/N$. This would give the values of Table 7.2 as asymptotic critical values for N.

The sample size must be at least N_{crit} to let order p be significant. Comparing this to the best fixed order in Table 7.1 shows that the agreement would be perfect if the AR(4) model were the best for $N = 20$. That single deviation is due to finite-sample effects where critical values are given by v_i instead of $1/N$. Therefore, the best fixed order can be determined easily for a process with uniformly decreasing reflection coefficients.

The average ME of the AR(11) model is close to the finite-sample expectation of the ME that is given with (7.47) as

$$E_{\text{FS}}[\text{ME}(p)] = N\left[\prod_{i=1}^{p}(1+v_i)-1\right]$$

(7.78)

Table 7.1. Average ME of Burg estimates of the AR(11) process of Figure 7.11 with parameters 1., – 0.714, 0.2116, 0.2004, 0.1594, 0.1178, 0.0838, 0.0578, 0.0386, 0.0243, 0.0141, and 0.0198. The average is over 10,000 simulation runs for models selected with 13 different criteria. The maximum candidate AR order is $N/2$, or 11 for $N = 15$ and 20. The fixed order ME and the finite-sample expectation of the ME of the AR(11) model are also given.

$N \rightarrow$	15	20	50	100	200	500	1000	2000	5000
GIC(p,2)=AIC	35.6	20.3	26.4	27.8	21.4	13.2	13.4	13.7	14.1
FIC(p,2)	23.1	15.2	14.1	12.5	12.2	12.7	13.3	13.7	14.1
GIC(p,3)=KIC	28.2	15.7	13.1	11.3	*11.4*	*12.2*	*12.8*	*13.3*	*13.4*
FIC(p,3)	14.7	12.1	11.4	11.1	*11.4*	*12.2*	*12.8*	*13.3*	*13.4*
GIC(p,4)	21.3	13.8	12.5	11.9	12.1	12.9	13.6	14.0	14.2
FIC(p,4)	12.4	12.3	12.6	12.0	12.2	12.9	13.6	14.0	14.2
GIC(p,2lnln$N$)	36.6	19.1	14.6	11.3	11.6	12.6	13.4	14.1	14.4
FIC(p,2lnln$N$)	23.2	14.2	11.4	11.1	11.6	12.6	13.4	14.1	14.4
GIC(p,ln$N$)	30.7	15.7	12.4	12.6	13.6	15.4	16.9	18.2	19.5
FIC(p,ln$N$)	16.4	12.1	12.4	12.8	13.8	15.4	16.9	18.2	19.5
AIC$_C$(p)	10.2	10.6	10.8	11.0	11.6	12.5	13.1	13.6	14.1
FSIC(p)	*9.9*	*10.3*	*10.6*	*10.7*	*11.4*	12.4	13.0	13.6	14.1
CIC(p)	10.4	11.0	11.4	11.1	*11.4*	*12.2*	*12.8*	*13.3*	*13.4*
AR(1)	11.4	14.7	34.7	68.2	135.4	337.3	673.8	1347	3366
AR(2)	7.9	9.2	17.4	31.0	58.1	139.6	275.3	546.7	1361
AR(3)	*6.8*	*7.2*	10.3	16.2	28.2	64.7	125.5	247.3	612.7
AR(4)	7.6	7.5	8.2	10.7	16.2	33.2	61.7	118.8	290.2
AR(5)	8.77	8.3	*7.5*	8.4	10.9	18.9	32.5	59.7	141.5
AR(6)	11.2	10.0	8.1	*8.1*	9.1	12.8	19.3	32.3	71.4
AR(7)	14.2	11.8	8.9	8.4	*8.6*	10.3	13.2	19.4	37.9
AR(8)	18.7	14.4	10.2	9.3	9.1	*9.7*	11.0	13.7	22.2
AR(9)	24.0	17.5	11.5	10.3	9.8	9.9	*10.3*	11.4	15.0
AR(10)	31.3	21.6	13.0	11.5	10.8	10.5	10.6	*10.8*	12.0
AR(11)	41.0	26.6	14.5	12.6	11.8	11.3	11.2	11.1	*11.0*
E_{FS}[ME(11)]	33.0	22.0	13.8	12.2	11.6	11.2	11.1	11.1	11.0

Table 7.2. Critical sample size N_{crit} as a function of the best fixed order p of the AR model

p	1	2	3	4	5	6	7	8	9	10	11
N_{crit}	3	5	9	18	36	73	148	301	615	1254	2558

The results for N equal to 15 and 20 are less accurate because the data length is shorter than the correlation length there. That is always a reason for poor accuracy of the theoretical relations with either asymptotic or finite-sample approximations. However, it is still possible to select the order; it is only difficult to determine the accuracy without simulations.

FIC(p,α) and GIC(p,α) are identical for greater N, except for $\alpha = 2$. It is better to use finite-sample equations if the highest candidate order for selection is greater than $N/10$. If the highest candidate order is smaller than $N/10$, FIC(p,3), GIC(p,3) and CIC(p) will almost always select the same AR order. Therefore, the use of finite-sample equations is advisable for all sample sizes.

   None of the criteria is best in all circumstances. The average of the ME or the sum of the ME values can be taken as a method of formulating a preference for one of the criteria because the ME is more or less independent of sample size. The sequence for the descending sum of the average model errors ME over the different sample sizes in Table 7.1 is

Best group with sum of ME is about 107, in descending quality:

   FSIC(p), CIC(p), AIC$_C$(p), FIC(p,3)

Middle group where the sum of ME is about 115:

   FIC(p,4), FIC(p,2lnln$N$), GIC(p,4), FIC(p,2), GIC(p,3)

Final group with sum ME from 137 to 180:

   FIC(p,ln$N$), GIC(p,2lnln$N$), GIC(p,ln$N$), AIC(p)

The best is FSIC(p) where the sum is 1.3 less than for AIC$_C$(p). Those two criteria behave similarly in the AR Burg estimates, although they have been derived from quite different points of view. The difference between the sum of FSIC(p) and CIC(p) is 0.9 and CIC would have the smallest sum if one more column with many observations were added. It is clear that penalty 3 is best for higher N, better than 2 and 4. Eventually, for $N \rightarrow \infty$, the risk of underfit will disappear completely in the true AR(11) process. Then, higher penalty factors will become better in true finite-order AR simulations. But the performance of the consistent order-selection criteria with penalty ln N in Table 7.1 shows that this desired behaviour requires many more than 5000 observations. Because of the asymptotic quality, CIC(p) is preferred. However, all criteria from the best group are quite acceptable criteria in small or finite samples.

   It is remarkable that the two criteria that are most widely known and discussed in the literature, AIC(p) of Akaike and the consistent criterion GIC(p,ln$N$), perform poorly in this simulation, as well as in many other simulations. Generally, simulations with a small probability of overfit tend to find AIC(p) preferable, whereas simulations with a small probability of underfit have a preference for the penalty ln N, or the highest of the penalties that is tested. It seems to be a good idea to perform simulations with both the possibilities of underfit and of overfit present. That is always an unprejudiced way to treat measured data.

   To verify the expected performance for very large sample sizes, the order-selection simulation with the same example of Figure 7.11 has been repeated with $N = 5,000,000$. Finite sample and asymptotic criteria give the same results then. Therefore, only one ME value is given for each penalty in Table 7.3. As could be expected in this situation without any danger or even possibility of underfit, the highest penalty gives the best result. In this case, that is the consistent order-

Table 7.3. Average ME of selected models from 5,000,000 observations of the AR(11) process of Figures 7.8 and 7.11. The average is over 200 simulation runs for models selected with 15 different criteria. Maximum candidate AR order is 1000.

Penalty	ME
2	13.16
3	11.73
4	11.40
$2\ln\ln N = 5.4$	11.29
$\ln N = 15.5$	11.03

Table 7.4. Average ME of AR models estimated from data of an MA(11) process with parameters 1., -0.714, 0.2116, 0.2004, 0.1594, 0.1178, 0.0838, 0.0578, 0.0386, 0.0243, 0.0141, and 0.0198. The average is over 100 simulation runs for models selected with different criteria. Maximum candidate AR order is $N/2$, or 11 for $N = 15, 20$.

N	15	20	50	100	200	500	1000	2000	5000
GIC(p,2)=AIC	44.9	20.7	31.8	46.6	34.9	30.5	31.5	36.0	41.7
FIC(p,3)	10.9	13.3	17.7	21.5	24.9	30.0	33.5	36.6	43.8
FIC(p,4)	9.3	11.9	18.5	23.8	26.2	31.9	36.8	40.6	48.0
GIC(p,ln$N$)	33.6	15.4	20.1	26.0	29.8	41.6	45.3	57.1	70.2
AIC$_C$(p)	9.0	_11.4_	18.1	20.8	23.7	29.7	_31.3_	36.0	_41.7_
FSIC(p)	_8.8_	11.5	_16.9_	_20.4_	_23.5_	_29.5_	31.5	_35.6_	41.7
CIC(p)	9.3	11.8	17.8	21.5	24.9	30.0	33.5	36.6	43.8

selection criterion with penalty ln$N$. However, a fixed penalty factor of 25 would still be better. The value of the penalty where overfit becomes probable is about 44.

Table 7.4 presents selection results that have been obtained with a MA(11) process. The zeros of the MA process are the locations of the poles in Figure 7.10. As the true AR order of MA processes is ∞, all estimated AR models are necessarily underfitted and biased. The best fixed order model will always increase with sample size.

The sequence of decreasing sum of the ME over all sample sizes is:

FSIC(p), AIC$_C$(p), CIC(p), FIC(p,3), FIC(p,4), GIC(p,2)=AIC and GIC(p,ln$N$).

Also in this case, the finite-sample criteria FSIC(p) and CIC(p) and also AIC$_C$(p) give the best result. In this example with only biased AR models, the ME will keep increasing for greater sample sizes. The best AR model is found with penalty 2 for greater values of N because the costs of underfit are most important in this example. However, in practical signal processing applications, MA models also will be used as candidates and then it is quite certain that a MA model will be chosen for greater sample sizes. Therefore, the choice for the preferred selection criterion should be based on Table 7.1, where CIC(p) is preferred to the other good criteria because of its better performance with increasing N.

7.10 Subset Selection

Subsets of parameters or of reflection coefficients do not have the hierarchical structure of the usual AR estimation algorithms. Therefore, selection becomes more difficult and less accurate. Broersen (1990b) studied subsets of AR parameters. Under rather strict conditions, subset models of AR parameters could be found. The most important condition was that the highest AR order was limited in the search for a subset. The main problem of subsets is that even in white noise, many parameters are accepted in a subset because they are greater than $\sqrt{2/N}$ or $\sqrt{3/N}$, as penalty 2 or 3 is used in the selection criterion $GIC(K, \alpha)$ of (7.32) for subset selection.

Subsets of reflection coefficients have been examined by Broersen (1986). If the order of the reflection coefficients in the true subset is known, it is nevertheless possible that the estimation of only those truly nonzero reflection coefficients gives a subset model with very poor quality. This happens if reflection coefficients are close to one in absolute value. This disadvantageous effect is much stronger than the small profits that can be found in other examples. Also replacing a number of very small estimated reflection coefficients, say, less than $\sqrt{1/N}$, by zero sometimes has an undesirable effect. Subsets of parameters as well as subsets of reflection coefficients have so few advantages and so many disadvantages that they are not treated any further here.

7.11 Exercises

7.1   Derive the expectation of J_p, as given in Equation (7.20).

7.2   Find an example AR process and sample size where the performance of the order-selection criterion $GIC(p,10)$ will be better than the performance of $CIC(p)$. Show it in a simulation example.

7.3   Find an example AR process, sample size, and maximum candidate order where the performance of the order-selection criterion $GIC(p,1)$ will be better than the performance of $CIC(p)$. Show it in a simulation example.

7.4   Find an example AR process and sample size where the expectation of the order-selection criterion $GIC(p,3)$ is approximately the same for orders 0, 1, 2, and 3. Find out in a simulation experiment which order is selected. Compare the ME of the selected order with the ME of the fixed order models of the orders 0, 1, 2, and 3.

7.5   Given a MA(2) process with parameters [1, 0.8, 0.6]. Find the theoretically expected order of the best predicting AR model for $N = 10$, 100, 1000, and 10,000.

7.6   Under which conditions is it possible to estimate models with an accuracy that is lower than the Cramér-Rao lower bound.

8

MA and ARMA Order Selection

8.1 Introduction

The best estimation methods for MA and ARMA models use an intermediate AR model discussed in Chapter 6. The theoretical best order for an intermediate AR model is ∞ if the true parameters are known. A MA(q) process with polynomial $B_q(z)$ is equivalent to an AR model of infinite order with the parameters of $1/B_q(z)$. For estimated AR models to be used as intermediates for MA and ARMA estimation, this order ∞ has been replaced in the past by the *best predicting AR order*. This was an important reason that those methods had a poor reputation. In simulations, it turned out that the best order for an estimated intermediate AR model is finite. It was equal to the order of the AR model, which has the smallest mean square error (6.25) of the parameters. Taking a higher intermediate order had a negative influence on the accuracy of the MA and ARMA models. Taking a lower intermediate AR order was often still more detrimental. The problem is that the mean square errors of the parameters cannot be computed for given data. Therefore, the intermediate order has to be chosen or selected from the measured data.

A second order that has to be selected is the MA order q in MA estimation or the two ARMA orders p and q. ARMA models present a new phenomenon, not found in AR and in MA models. In pure AR and pure MA models, each model of order p has one single neighbouring model with one parameter less and a single neighbour with one parameter more. That makes order selection a hierarchical problem with nested models as candidates. Each higher order model comprises the parameters of all lower order models, unless the possibility of subsets of parameters is considered.

The number of ARMA(p',q') candidate models with $p' \leq L$ and $q' \leq L$ is L^2. If all ARMA(p',q') models were available for order selection, the true order model would have many close competitors and it may be expected that the performance of order-selection criteria would deteriorate. It would be related more to subset selection with arbitrary subsets than to selection in a hierarchically nested class of candidate models, where each higher order model contains all parameters of lower order models. Nested selection has L candidate models if L is the highest order considered, whereas there are 2^L possible subset models. In subset selection, each possible subset model is a next candidate for selection, and each estimated parameter that seems statistically significant individually is included, even if many previous model orders were not significant. Theoretically, the selected subset size will increase with the number of candidate models, whereas the selected hierarchical model order is independent of L. Solo (2001) investigated the costs of subset selection instead of hierarchical search: the choice between all possible subsets selects many more parameters but leads to a poor model in time series. The explanation is simple. If many closely competing candidates with similar unconditional accuracy are available for selection, generally one of those candidates is an estimated model that seems to fit much better than the others and that one will be selected. The performance of L^2 ARMA(p',q') candidate models will be somewhere between L hierarchical models and 2^L subset models.

In contrast with the hierarchical models, the ARMA(p,q) class of models will have several ARMA($p+k,q-k$) models with the same number of parameters and many more with one parameter more or with one parameter less. ARMA models can be made hierarchical by considering only ARMA($r,r-1$) or only ARMA(r,r) models. At the cost of probably missing the true process orders among the possible models, the order-selection problem is made much simpler. This problem will be discussed, and it will be shown that the loss caused by not having the proper ARMA(p,q) model as a candidate for order selection is largely compensated for by the fact that selection from a small number of candidates is more accurate. Broersen and de Waele (2004) have shown that a limitation is hardly of influence on the average quality of selected ARMA models.

8.2 Intermediate AR Orders for MA and ARMA Estimation

Theoretically, different model orders can be the best for various purposes. Two important rather general purposes are

- prediction with the model
- accuracy of the model parameters.

Those orders are often the same. As an example, consider an AR(p) process where all p reflection coefficients are much greater than $1/\sqrt{N}$. Therefore, all parameters will be statistically significant, and the estimated model of order p will be the best for all purposes.

However, both MA and ARMA processes are represented exactly by an AR(∞) process. No matter how large N becomes, the reflection coefficients at very high orders will always tend to zero and become small in comparison with $1/\sqrt{N}$. In those cases, the best order for prediction with an AR model and the AR model with the closest approximation of the parameters will generally not be the same. The parameter accuracy is the infinitely long sum of squared errors between estimated and true parameters. Hocking (1976) has given theoretical arguments that the AR order q is the best for the parameter accuracy if the sum of the true values of a_i^2 for $i = q + 1$ until ∞ is less than $1/N$; see also (7.28) which gives the same result if applied to the true AR parameters of orders greater than q. This is quite different from the best model order for prediction which is found as the order q' for which all individual a_i^2 for $i > q'$ are less than $1/N$, or more precisely the order q' for which each sum of r squared values of a_i^2 for $q' < i < q'+r$ is less than r/N, for each value of r; see (7.29) for the linear regression case. For AR selection, that order can be found by using the order-selection criterion GIC(q',α) of (7.32), with penalty α equal to 1, applied to the intermediate residual variances of models with the true parameters, as found with the Levinson-Durbin recursion (5.24). Penalty one can be made plausible by considering that the usual penalty 2 gives the same weight to bias and variance contributions. Substitution of the true parameters in lower order models removes the variance contribution and leaves only the bias.

In simulation experiments, the best intermediate theoretical AR order for MA and ARMA estimation turned out to be the order for which the parameters are most accurate, not the model order for the best prediction. All order-selection criteria that have been described in Chapter 7 select the best model order for prediction. It is not possible with known criteria to select the best model order for parameter accuracy from measured data. It would require a penalty α of about $1 + 1/N$. Figure 7.6 gives the cost of overfit as a function of the penalty. Values of α close to 1 make the probability of overfit almost 100%. No practical useful model orders for any purpose would be selected with those penalties. The estimation variance of parameters for given data would completely determine which order is selected. That means that the intermediate AR order that turns out to be the best in simulations with known true MA parameters cannot be selected from data in practice with any known selection criterion.

Many simulations have been done to find a good practical choice for the AR order; see Broersen (2000b). That order is used in the ARMASA program, where it is selected automatically. It is based on the AR order K that is selected with the CIC criterion (7.77) for the best predicting AR model and on the order q of the MA(q) model that is computed.

Table 8.1. Average of 100 simulation runs of the ME of MA(4) models estimated from different intermediate AR orders from 500 observations of the MA(4) process, as a function of the radius β of the zeros of the MA polynomial

AR order	$\beta = -0.3$	$\beta = 0.6$	$\beta = -0.9$
q	3.00	3.38	180.01
$2q$	3.49	4.05	20.36
K	*3.00*	3.47	7.21
$2K$	3.49	4.05	6.11
$K+q$	3.49	*4.03*	6.91
$2K+q$	3.64	4.15	*5.02*
$3K+q$	3.81	4.19	5.30
$N/2$	30.18	31.08	21.45
Best order	3.00	4.05	4.34

Table 8.1 shows that too low an AR order gives very poor results if the zeros are close to the unit circle. The best average result is found for the intermediate order $2K + q$. This order has been called a sliding window order, because it depends on the number of MA parameters that is estimated. The choice of the AR order turns out to be particularly important if zeros are close to the unit circle. Both orders too low and too high would give a poor result then. The final row gives the ME for the AR order that is found by using (6.23) and the known process parameters. That order is unknown in practice but using the practical order $2K + q$ gives almost the same results for MA(4) quality given in Table 8.1. They are close to the Cramér-Rao lower bound which is four for the MA(4) process.

Table 8.2. Average of 100 simulation runs of the ME of MA(q) models estimated from different intermediate AR orders from 100 observations of the MA(q) process, as a function of the order q of the MA polynomial, with radius $\beta = 0.8$ of the zeros

AR order	$q = 1$	$q = 2$	$q = 3$	$q = 4$
q	20.38	17.55	4.77	*4.07*
$2q$	9.33	4.20	4.69	6.17
K	3.42	5.01	4.59	4.47
$2K$	1.87	3.83	4.25	6.11
$K+q$	2.43	4.60	4.61	6.02
$2K+q$	1.74	*3.06*	*3.68*	5.57
$3K+q$	*1.66*	3.29	4.14	6.13
$N/2$	4.33	8.27	15.74	19.37
Best order	1.38	2.98	3.35	5.26

Table 8.2 shows that for different MA orders, the sliding window choice $2K+q$ always gives also a MA quality close to what can be found by using the theoretical best intermediate order. Therefore, to estimate the MA(q) parameters with the method of (6.24), the AR($2K + q$) model is used as the intermediate AR(L_{MA}) model, with

$$L_{MA} = 2K + q \qquad (8.1)$$

Broersen (2000b) has chosen that order as a good practical compromise. Generally, the upper limit for L_{MA} is taken as $N/2$ or about 1000 if N is greater than 2000 and (8.1) would give a higher result.

In ARMA estimation with small parameter values, the intermediate order is not very critical. But if poles and zeros approach the unit circle, it becomes very important to have an intermediate AR model of sufficiently high order. Table 8.3 gives some results. Based on previous results of Broersen (2000b) and also on the results in Table 8.3 for ARMA(p,q) estimation, the sliding window AR($3K + p + q$) model is used for automatic spectral analysis with ARMASA. Using higher AR orders can have a negative influence on the accuracy of the final MA or ARMA model. However, lower AR orders might often lead to poor accuracy of the MA or ARMA models. The choice

$$L_{ARMA} = 3K + p + q \qquad\qquad (8.2)$$

is also given by Broersen (2002b). If the roots are not too close to the unit circle, the ME values are close to the Cramér-Rao lower bound 5 for the ARMA(3,2) process.

Table 8.3. Average of 100 simulation runs of the ME of ARMA(3,2) models estimated from different intermediate AR orders from 1000 observations of the ARMA(3,2) process, as a function of the common radius β of the poles and zeros

AR order	β = − 0.5	β = 0.8	β = − 0.95	β = − 0.99
$p+q$	5.25	98.8	876.64	1247.1
$2(p+q)$	5.30	8.99	102.68	389.7
K	*5.15*	5.72	14.46	39.29
$2K$	5.32	5.34	6.23	19.86
$K+p+q$	5.29	*5.27*	12.6	38.28
$2K+p+q$	5.76	5.28	*5.95*	19.58
$3K+p+q$	5.89	5.32	5.99	*11.84*
Best order	4.74	5.12	5.75	11.84

8.3 Reduction of the Number of ARMA Candidate Models

It has been discussed that using hierarchical ARMA models will reduce the required computation time and improve the quality of order selection. The reason is that ARMA(p,q) models may have several ARMA($p+k,q-k$) neighbours with the same number of parameters which may be of comparable quality. The choice between a large number of good models is much more difficult than the choice between only a few serious candidates. If the estimate of a parameter with a true value on the boundary of statistical significance is greater than its true value, it will be selected and if its estimate is smaller, it will not be selected. If there is only one candidate for overfit, the selection problem is concentrated on the statistical properties of the estimated AR parameters of order p and order $p+1$. If there are more competing candidate models, the model that occasionally has the largest parameter estimates will probably be selected. If there are many serious candidates, there will often be one among them that occasionally has higher estimates for the given data, and that particular candidate model will be selected for those data.

In the ARMAsel program, the standard is that only hierarchical ARMA($r,r-1$) models are considered as candidates. This particular choice is inspired by the fact that models of those orders are good discrete time approximations for many continuous time processes (Priestley, 1981, p. 382). The first higher order candidate has two parameters more, the closest lower order candidate has two parameters less. This nesting also gives an important reduction in the computation time required. Only L instead of L^2 models have to be evaluated.

The closest biased model has two parameters less. Only if both the last MA and the last AR parameter of the ARMA($r,r-1$) are small, will underfit be probable.

Asymptotically, the probability of one order overfit for ARMA(r,r–1) models is given by the probability that the chi-squared distribution with two degrees of freedom exceeds 2α, where α is the penalty factor. This is smaller than the probability that the chi-squared distribution with one degree of freedom exceeds α, which applies to AR or to MA models where only one parameter more is involved. Therefore, the probability of overfit of ARMA order selection with the same penalty factor is much less than that of AR or MA selection. The price to be paid is that the true ARMA(p,q) model may not always be among the candidates for selection and some additional parameters with zero expectation have to be included in the model selected. For increasing sample size, all biased underfit models will have very large ME values. In that case, the best choice for an ARMA(p,q) process will have both $r \geq p$ and $r - 1 \geq q$. This requires the estimation of a maximum of $r - 1$ extra parameters with the true values of zero if the true process were ARMA($1,r$–1). The asymptotic increase in the Cramér-Rao lower boundary will be from r to $2r - 1$ in this worst case, which is considered acceptable for data with an unknown character.

It turns out that the loss in accuracy is still much less than described above in many finite-sample simulations because many ARMA($r+k,r$–1–k) models have about the same ME. Unfortunately, theoretical evaluation of the accuracy of underfit models is rather difficult and would depend on the true values of all parameters. It can be expected that estimated parameters for those different models with the same total number of parameters are strongly correlated if they are estimated from the same data. Conditional expectations are required for theoretical results for the costs of selection, and it is not attractive to make such a study, even if it were possible.

Therefore, simulations have been used to investigate the behaviour of neighbouring ARMA models. The simulations have been made with an ARMA(7,2) processes with AR parameters given by the reflection coefficients

$$1, -\beta, (-\beta)^2, (-\beta)^3, (-\beta)^4, (-\beta)^5, (-\beta)^6, (-\beta)^7 \qquad (8.3)$$

and with MA parameters given by reflection coefficients

$$1, \beta, \beta^2 \qquad (8.4)$$

The AR and the MA parameters are computed with (5.28) that relates AR parameters to reflection coefficients. This choice for generating reflection coefficients gives an ARMA process with all poles and all zeros at the same radius $|\beta|$. It gives the convenient possibility of generating different levels of significance for the parameters of the true process and creating examples where biased underfit models are the best, as well as processes where only unbiased models are attractive candidates for selection. The ARMA(7,2) process is chosen such that all competitive ARMA models are in the class ARMA($r, < r$) if β is large enough. In this way, the conclusions can be extrapolated to the full class of all ARMA($\leq r, \leq r$)

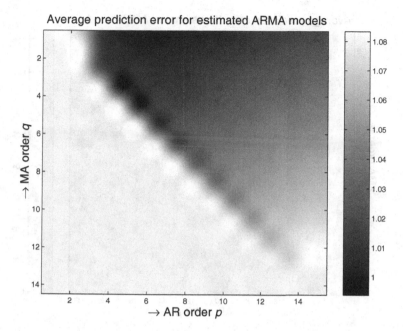

Figure 8.1. Average of 4000 simulation runs of the PE of ARMA(p,q) models estimated from 250 observations of the ARMA(7,2) process with $\beta = 0.7$ as a function of the model orders

models. The variation in β gives an opportunity for studying the performance of the penalty factor in the most difficult selection examples where underfitted, biased models are expected to be the best candidates among estimated models.

The average of many closely related ARMA models is shown in Figure 8.1, where the average prediction error of a number of ARMA(p,q) models is given for $q < p$. The PE is normalized with the value 1 for the ARMA(7,2) model in this figure. The PE of very poor underfitted models is limited to 1.005 times the PE of the ARMA(20,19) model, to improve the reach in the grey-scale image. This corrected value is found for the ARMA(2,1) and the ARMA(3,2) models in Figure 8.1. All models with less than seven AR parameters or less than two MA parameters are biased. The best ARMA($r,r-1$) model is the biased ARMA(5,4) model, with a prediction error PE 0.9983. This is slightly smaller than the PE value 1 of the unbiased ARMA(7,2) model, which has the same number of parameters. In other words, limiting the selection candidates exclusively to the ARMA($r,r-1$) models will not necessarily have a negative influence on the quality of estimated models. Furthermore, the lines of constant colour are more or less under 45°, for higher ARMA orders. This indicates that the normalized PE of all those unbiased models with the same number of parameters $p + q$ is about $1 + (p + q - 9)/N$ in this figure. The subtraction of nine is caused here by normalizing the PE to one for an unbiased model with nine parameters.

If more observations are available, the quality of biased models will become worse. That can be seen in Figure 8.2 where the same process has been used in simulations with 5000 observations. The PE of heavily underfitted models is

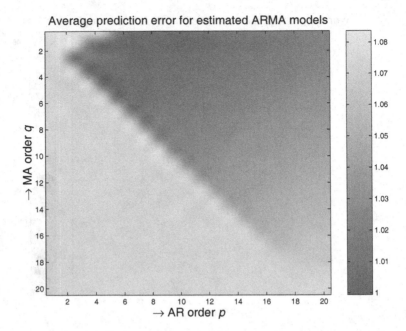

Figure 8.2. Average of 2000 simulation runs of the PE of ARMA(p,q) models estimated from 5000 observations of the ARMA(7,2) process with $\beta = 0.7$ as a function of the model orders

truncated. The darkest area with the smallest PE is the area with unbiased models with 10 to 13 parameters together. It is remarkable that the models with orders r and $r - 1$ for AR and MA, respectively, are relatively accurate in Figures 8.1 and 8.2, with a little darker colour. The ARMA($r,r-1$) model is the best of all over-fitted models with $2r - 1$ parameters.

8.4 Order Selection for MA Estimation

The theory of MA and ARMA order selection is only asymptotic. No finite-sample deviations of overfitted models from asymptotic behaviour have been reported. The reason might be that MA and ARMA models have not been studied for orders much higher than $N/10$. The discussion of possible bias due to the final parameter a_p of AR(p) models in Section 7.7 produced the preference 3 for the penalty α in GIC(p,α). Similar reasoning can also be used for the final parameter b_q of a MA(q) model. Therefore, only two possibilities are considered in simulations: GIC($p,2$) and GIC($p,3$), where the penalty 3 has been found as the optimum for AR processes in (7.75). In this way, whether the reasoning for AR selection is also applicable to MA selection can be verified experimentally.

The simulation example is chosen as a MA(11) process with the zeros at the locations of the poles of Figure 7.7. Figure 8.3 gives the autocorrelation function and the power spectral density of the process.

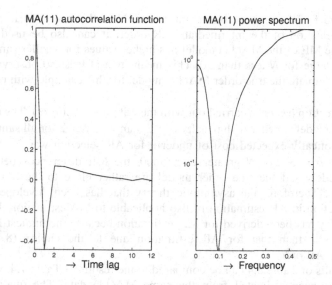

Figure 8.3. The autocorrelation function and the power spectral density of the MA(11) process with zeros at the locations of the poles in Figure 7.10

Table 8.4. Average ME of MA models estimated from data of the MA(11) process of Figure 8.3 in 1000 simulation runs. The model error (ME) is given of MA models, selected with penalty 2 and 3 the of the true order MA(11) model, as a function of sample size N.

N	$\alpha = 2$	$\alpha = 3$	MA(11)
15	17.8	15.8	94.0
20	18.0	16.3	120.4
50	17.4	16.8	22.0
100	17.4	15.6	17.8
200	16.3	14.6	15.7
500	14.6	13.7	13.3
1000	14.4	14.0	12.8
2000	14.3	14.0	11.9
5000	14.5	14.3	11.8
25000	13.5	12.4	11.1
100,000	13.2	11.8	11.0

Table 8.4 gives the ME of fixed order MA(11) models as well as the ME of selected models. MA(11) models have the value 11 as the Cramér-Rao boundary. That lower limit is obtained for the average of the MA(11) model for N equal to 100,000. This shows that the MA algorithm described in (6.24) and with the practical sliding window choice (8.1) for the intermediate AR order is asymptotically efficient. This means that the variance of the unbiased model is equal to the minimal variance that can be obtained. In this and many other examples, the asymptotic efficiency of the MA estimator (6.24) has been shown. The proposed estimator (6.24) is robust, gives invertible models, is asymptotically

efficient, and fulfills all desirable requirements. Note that the estimator includes automatic selection of the intermediate AR order. It can also be used in small samples. The ME of the MA(11) model has higher values for smaller sample sizes and is very large for N less than 50. This means that at least 50 observations are required to estimate the true order MA(11) model for this example with reasonable accuracy.

Order selection has been carried out with the criterion GIC(q, α). The quality of the selected model is better with penalty $\alpha = 3$ than with $\alpha = 2$ for all sample sizes. The asymptotically expected cost of underfit for AR selection was 2.6 for $\alpha = 2$ and 0.9 for $\alpha = 3$. For N greater than 5000, the ME differences between the selected models and the true order model are rather close to these asymptotic expectations. Therefore, the asymptotic theory that has been developed for the penalty function in AR estimation is also applicable to MA estimation. No finite-sample theory has been derived for MA estimation because the highest MA order is much lower than that for AR estimation due to the choice (8.1) of the intermediate AR order.

The results of Table 8.4 can be compared to the results of Table 7.4, where AR models have been estimated from the same MA(11) data. The quality of the selected MA model is better for $N \geq 50$. For smaller sample sizes, the quality of the selected AR model would be better. It is clear that the ME of selected MA models is less than half of the ME of selected AR models for $N \geq 500$. This demonstrates that MA models can be the best choice in practice with the algorithms that have been used.

8.5 Order Selection for ARMA Estimation

Choi (1992) has described several selection criteria developed exclusively for ARMA models. The theoretical background of special ARMA criteria relies primarily on asymptotic properties, where the quality of all biased models will certainly become poor. Then, pattern recognition methods can find the edge of unbiased models in Figures 8.1 and 8.2 in the previous section. Thus, the corner is found where both the AR order is greater than or equal to p and the MA order is at least q for a true ARMA(p,q) process. Figure 8.2 showed that 5000 observations are not always sufficient for such principles in practice. Biased models may still be good candidates. Especially in finite and small samples, pattern recognition methods are not attractive for order selection.

Simulations have been done to find the influence of the limitation to only ARMA($r, r-1$) candidates on the quality of selected models. Qualitatively, Figures 8.1 and 8.2 show that high-order AR models, though biased, are quite good. Biased ARMA($r, r-1$) models with $r < 7$ can have also acceptable accuracy. In principle, the penalty factor and the limitation of the candidate models can interfere. Therefore, the best penalty factor has been determined for selection with and without the limitation to ARMA($r, r-1$).

Table 8.5 gives the results for the average ME of the models selected with GIC($p + q, \alpha$), when α is 2 or 3. In almost all simulation examples, the average

Table 8.5. Average ME of ARMA models estimated and selected from data of the ARMA(7,2) process with the parameters of (8.3) and (8.4) in 1000 simulation runs, as a function of β. The model error ME is given of ARMA models, selected with penalty 2 and 3 from all candidates ARMA($r,<r$) and from the hierarchical ARMA($r,r-1$) models for $N = 250$.

β	ARMA($r,< r$) $\alpha = 2$	ARMA($r,< r$) $\alpha = 3$	ARMA($r,r-1$) $\alpha = 2$	ARMA($r,r-1$) $\alpha = 3$
0.1	4.2	2.7	2.5	2.0
0.4	6.6	4.8	4.5	3.6
0.7	13.2	12.4	11.8	11.5
0.9	17.2	15.5	20.4	20.7

ME of models selected with penalty α equal to 3 is smaller than the ME of penalty 2. Therefore, penalty 3 is also better for ARMA selection. This conclusion has been drawn before by Broersen and de Waele (2004) with similar simulation experiments.

Furthermore, the quality of ARMA models selected from the hierarchical, nested ARMA($r,r-1$) candidates is better than the quality of models selected from the wider class of ARMA($r,< r$) candidates. The exception to this rule is found for $\beta = 0.9$, where all biased models with less than seven AR parameters and less than two MA parameters have much greater MEs than the true ARMA(7,2) model. This means that the ARMA(7,6) model is the best possible candidate among the hierarchical ARMA($r,r-1$) candidates. This has four parameters more, which approximately explains the difference in the ME values 15.5 and 20.7 in the last row of Table 8.5. It can be concluded that, for the lowest values of β, the biased ARMA($r,r-1$) models are a good choice. The ME values are much less than nine, indicating that less than nine parameters are present in the biased model selected. Apparently, having more candidate models makes selection quality worse.

The value $\beta = 0.7$ shows some intermediate character where the number of parameters for models with a low ME is minimally nine, the size of the true process. The selection quality ME of selecting from nested candidates is still a little bit better than the ME of selected ARMA($r,< r$) models. This has been concluded for $N = 250$.

Table 8.6 gives the results of the value $\beta = 0.7$ as a function of sample size. As might be expected, the results for a large number of observations have the tendency to be most accurate if all truly nonzero parameters are included, hence for unbiased models. The differences in the ME between the results of ARMA($r,< r$) and ARMA($r,r-1$) are not so large in Table 8.6 that the accuracy gives a decisive reason to prefer the complete class of ARMA($r,< r$) models. However, the number of models in the hierarchical ARMA($r,r-1$) class is much less, and therefore the computing time is less and that is a good reason to consider only the limited class of hierarchical models as candidates for estimation and selection.

The average ME of the true order ARMA(7,2) model is 32.7 for $N = 100,000$. Apparently, the ARMA estimation algorithm is not fully efficient for this example because that would require an average ME of nine, the number of parameters in the true order model. However, a model with some more parameters is selected by

Table 8.6. Average ME of ARMA models estimated and selected from data of the ARMA(7,2) process with the parameters of (8.3) and (8.4) in 1000 simulation runs, for $\beta = 0.7$, as a function of N. The model error (ME) is given of ARMA models, selected with penalty 2 and 3 from all candidates ARMA($r,<r$) and from the hierarchical ARMA(r,r-1) models.

N	ARMA($r,< r$) $\alpha = 2$	ARMA($r,< r$) $\alpha = 3$	ARMA(r,r-1) $\alpha = 2$	ARMA(r,r-1) $\alpha = 3$
25	23.4	17.4	15.4	15.3
50	26.6	16.1	23.4	19.1
100	14.5	12.6	14.2	14.0
250	13.2	12.4	11.8	11.5
500	13.6	12.5	11.5	11.2
1000	14.0	13.5	12.3	12.1
2500	15.9	16.3	15.1	16.0
5000	16.4	17.2	16.6	19.0
100,000	15.1	13.4	16.1	16.8

order selection criteria, with better ME quality. In this example, selection gives the ME value of 16.8, which is higher than the Cramér-Rao lower limit of 9, but much lower than the outcome of the ARMA(7,2) model.

This problem has been found in several examples. The performance of the ARMA estimator with the initial stage (6.28) with reconstructed residuals did not always give the accuracy expected. The average ME of experiments with the estimated true order model often gives a less accurate result than the selection among a number of ARMA(r, r-1) candidates. This could also happen in a true ARMA(p, p-1) process. However, in all those examples, the combination with order selection was always rather satisfactory. This problem was the reason to introduce the four reduced-statistics methods for the first ARMA stage in Section 6.6. Those other methods are also not always accurate, as has been shown. However, for all examples, at least one of the initial stages gives an acceptable ARMA model with an accuracy that is not too far from the Cramér-Rao lower limit. The ARMA estimator does not yet have the same statistical efficiency that has been found for the MA estimator in Table 8.4.

The quality of selected models depends on two factors: the quality of the available models and the capacity to select one of the best models from those available. The selection capacity depends on the penalty factor. For those ARMA processes where several biased models are good candidates, penalty 3 is mostly better than penalty 2 which defines the widely used AIC criterion. For processes that fall in a critical parameter range for the given number of observations, penalty 2 can sometimes perform better in ARMA selection. Usually, penalty $\alpha = 3$ is better for an arbitrary number of observations of an unknown process and should be preferred. Therefore, penalty $\alpha = 3$ will be used in the automatic spectral estimation algorithm ARMAsel.

The limitation imposed by ARMA(r,r-1) candidate models, to only models with MA order one less than the AR order, gives an enormous reduction in computation time in comparison with the estimation of all combinations of p for AR and q for MA orders. Reducing the number of candidate models in ARMA estimation gives the possibility that the very best model is not estimated, if the MA order is not equal to the AR order minus one. Obtaining a good model may sometimes require estimating that some extra parameters with true values of zero. However, the additional variance inaccuracy of those extra parameters will generally be compensated for by reduced selection costs. Those are lower because of the limitation of the number of candidates. Generally, limiting the selection to ARMA(r, r–1) candidate models does not lead to much lower quality of the model selected. Often, the quality is even better.

8.6 Exercises

8.1 Use the ARMAsel program to estimate an AR(2) model from arbitrary input data. Arbitrary data can be found on the Internet, or they can be generated with the program "simuarma" on the computer.

8.2 Use the ARMAsel program to select the best AR model from the same data.

8.3 Compute the model error (ME) between those two AR models. Consider the selected AR model as the true process for this computation.

8.4 Use programs from the ARMASA toolbox to compute an approximating AR(100) model for the MA(2) process with $b_1 = 1.5$ and $b_2 = 0.9$.

8.5 Show theoretically that the selection criterion GIC(p,1) applied to the residuals of the true AR(∞) model of an MA(q) process selects the best AR order for prediction.

8.6 Use the ARMAsel program to select the best MA(2) model from the data of Exercise 8.1.

8.7 Use the ARMAsel program to select the best MA model from the same data.

8.8 Compute the model error (ME) between those two MA models. Consider the selected MA model as the true process for this computation.

8.9 Compute the model error (ME) between the selected MA model of Exercise 8.7 and the selected AR model of Exercise 8.2. Consider the selected AR model as the true process for this computation.

8.10 Use the ARMAsel program to select the best ARMA(4,3) model from the data of Exercise 8.1.

8.11 Use the ARMAsel program to select the best ARMA(4,1) model from the data of Exercise 8.1.

9

ARMASA Toolbox with Applications

9.1 Introduction

This chapter contains some additional topics to facilitate automatic analysis of random data. The selection of the model type is discussed, starting with the assumption that the best model order for each model type has been selected before. The easy and standard application for spectral and autocorrelation analysis will be restricted to automatically selected AR, MA, or ARMA model.

The language of random data is described to show how other interesting models are suggested by the data. By estimating the expected fit of each model to new data, the prediction error is found for all candidate models. If there appears to be an important deviation from the normally expected behaviour of prediction errors, that is a strong indication or warning that the data comprise some interesting details.

The principle of order selection with reduced statistics is described. Using as input a long AR model and the number of observations, order selection is possible. If the number of observations is not known, a good estimate of the sample size can be derived from the long AR model. The accuracy of the reduced-statistics ARMA models, computed and selected with the long AR model as first-stage input, is compared to the accuracy that can be obtained by using the data themselves in the first stage.

The application of the automatic ARMASA algorithm to harmonic data with different levels of additive noise is investigated and compared to the harmonic analysis with the periodogram.

Finally, a complete standard routine for the analysis of random data is described. Some practical applications are presented. Also the Matlab® commands are listed that are used in the ARMASA toolbox of Broersen (2002).

9.2 Selection of the Model Type

The ARMAsel algorithm computes AR(p) models with $p = 0, 1, \ldots, N/2$, using the Burg algorithm (6.12). The choice of the highest candidate AR order is limited to 1000 for $N > 2000$, but that is not necessary. It is intended to reduce the computation time required if the sample size is much greater. It has not been seen in practice that this limitation was important. However, it remains advisable to try still higher AR candidate orders if the program selected the highest available candidate. The selection of the AR order p is made with the CIC criterion of (7.77); the model order p with the minimum of CIC(p) among the candidates is selected. Mostly, all model orders are used as candidates, including the white noise model with zero parameters. For stationary random data where the global characteristics are already known from previous experiments, it might be possible to reduce the maximum candidate order without loss of accuracy. It can be taken safely as twice the highest order that ever has been selected before. The selected AR model is called the AR(K) model.

MA(q) models are estimated with $q = 1, \dots, N/5$, using algorithm (6.24). The highest candidate order is taken as 400 for $N > 2000$. The maximum MA candidate order is taken lower than that for AR. The most important reason is that a long AR model is used as an intermediate model in the MA algorithm based on (6.24). The intermediate AR order L_{MA} to estimate the MA(q) model was taken as $2K + q$ in (8.1). If the calculated L_{MA} exceeded the maximum AR order $N/2$, it is limited to that highest order $N/2$. By limiting the highest candidate MA orders, the computed intermediate AR order L_{MA} will mostly be available. Experience shows that selected MA candidates are often of a lower order than selected AR candidates and the highest MA candidate order may be limited. The best MA model order Q is selected from those candidates with the minimum GIC(Q,3) of (7.32). Use of this criterion requires computation of the residual variance of all MA(q) models as an approximation for the fit of the MA model to the data. Although this residual variance has not been used in the estimation algorithm (6.24), its computation is required as an argument of the order-selection criterion GIC(q,3) with

$$\mathrm{GIC}(q,3) = \ln\left[\mathrm{RES}(q)\right] + \frac{3q}{N} \tag{9.1}$$

The notation RES has been used for the residual variance of MA and ARMA estimation to emphasize the difference from AR estimation. To compute the residual variance, the data $x_1, \dots, x_N$ are firstly extrapolated backward with a long AR model to obtain estimates for $x_0, x_{-1}, x_{-2}, \dots, x_{-N/2}$. This elongated signal x_n is inversely filtered, and the parameters of the estimated MA(q) model are used as AR parameters, using the relation that was also exploited in (5.34):

$$\hat{\varepsilon}_n = \frac{1}{\hat{B}_q(z)} x_n, \quad n = -N/2, \cdots, 0, 1, \cdots, N \tag{9.2}$$

In Equation (9.2), the estimated parameters in $\hat{B}_q(z)$ are also found with the long AR model of the data x_n. Therefore, $\hat{\varepsilon}_n$ here has the character of a residual, not of a prediction, as in (5.35). The residual variance to be substituted in (9.1) is found as

$$\mathrm{RES}(q) = \frac{1}{N} \sum_{n=1}^{N} \hat{\varepsilon}_n^2 \tag{9.3}$$

using all N observations.

ARMA($r,r-1$) models are also estimated for $r = 2, ..., N/10$ with maximum of 200. The best ARMA order R is selected with the smallest criterion GIC($2R-1,3$). To determine residuals, the data are extrapolated backward with a long AR model, as for MA residuals. The residual variance RES($2r-1$) of the ARMA($r,r-1$) model is found from the extrapolated signal with

$$\hat{\varepsilon}_n = \frac{\hat{A}_r(z)}{\hat{B}_{r-1}(z)} x_n, \quad n = -N/2, \cdots, 0, 1, \cdots, N \tag{9.4}$$

followed by averaging the residuals in the observation interval, as in (9.3).

The AR order is selected with a finite-sample criterion CIC(p). The MA and ARMA models have been selected with an asymptotic criterion GIC($q,3$). The minimum values of those different criteria cannot be used for a mutual comparison between AR models and the other model types. Moreover, it may occur that the best models of different types have a great difference in the number of parameters. As an example, the best AR order will always increase with the number of observations for increasing sample sizes of an MA(1) process. In the derivation of the penalty factor, the possibility of selecting only one underfit order has been used as an example to derive penalty factor 3. It was clear that the same reasoning if two underfit orders are possible in (7.76) gives a smaller penalty factor. If it would have been applied to the simultaneous selection of a larger group of small parameters, penalty 2 would have been a good choice. That penalty factor will be used for the mutual comparison of different model types.

It is possible to use an estimate of the prediction error in the selection of the model type. The prediction error is estimated with the FPE of (7.30) and is asymptotically equivalent to penalty factor 2 for order selection; see (7.41). Having selected the best AR(P) model, the best MA(Q) model, and the best ARMA(R, $R-1$) model, the FPE of those three resulting models is computed as an estimate for the prediction error. For MA and ARMA models, this is given by

$$PE(m) = RES(m) \frac{1+m/N}{1-m/N} \tag{9.5}$$

where m is the number of estimated parameters in the model. The asymptotical relations (6.17) and (6.18) for AR models have been used here. Those relations have been proved only for unbiased AR models. They are applied to all ARMA and MA models with m parameters. It may be expected that RES(m) of acceptable candidates will be much smaller than the residual variance of the lower order models where the model does not include all significant parameters and the models are severely biased. The major influence on the prediction error is then caused by the high low-order values of RES. The correction term after RES(m) in (9.5) accounts for the estimation variance. It is important for higher orders.

For AR(p) models, the expression for the PE(p) is derived from expressions (6.21) and (6.22) for the finite-sample expectations of the residual variance and the prediction error, respectively:

$$\mathrm{PE}(p) = s_p^2 \prod_{m=1}^{p} \frac{1+1/(N+1-m)}{1-1/(N+1-m)} \tag{9.6}$$

The finite-sample variance coefficients (6.20) for Burg's estimation method have been substituted. The model type with the smallest estimate of the prediction error is selected. In this way, a single time series model with selected type and order can be determined for given observations. It is called the ARMAsel model. The spectral density or the autocovariance function can be computed from its estimated parameters, and that gives the best representation of the-second order characteristics of the measured data.

9.3 The Language of Random Data

In the previous section, the PE could be used to select the model type. If there are more competing models, they would have about the same PE values and that would become clear by looking at the PE of all models. In deriving (9.5), it is expected that the average of the residual variance of unbiased models decreases with $1/N$ above the true order. The average of the PE of unbiased models increases with $1/N$ for each additional parameter.

Figure 9.1 gives the PE of the models obtained from the 142 detrended global temperature data of Figure 1.5. The automatic ARMAsel program selected the AR(4) model. The figure also shows that the ARMA(3,2) model would be a possible choice; MA models are certainly less accurate. The MA models with orders between 5 and 13 all have about the same estimated PE, but the accuracy is

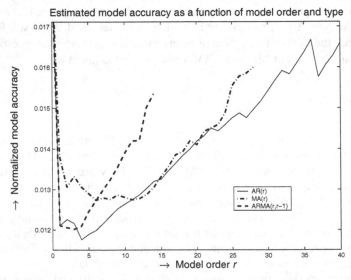

Figure 9.1. The estimated accuracy of AR, MA, and ARMA candidates for selection, estimated with ARMAsel from the 142 detrended global temperature data shown in Figure 1.5

much less than that of the selected AR(4) model. The similar behaviour of the higher order AR and MA models is very remarkable. All unbiased high-order models with the same number of parameters have almost the same PE. AR and MA models are very close. The slope of the ARMA models is two times steeper because each higher order has two more parameters.

If the slopes of the higher model orders are regular, it is almost certain that the parameters are all small and not statistically significant. In this example, only one candidate seems attractive: the AR(4) model that has been selected by the ARMAsel program. The AR model of order 37 gives a sudden decrease in the PE. The AR(37) reflection coefficient was –0.27; the standard deviation for all reflection coefficients above the true order would be $\sqrt{1/N}$, which is 0.084. This means that the largest of the estimated AR high-order reflection coefficients had a value about three times the standard deviation for a true parameter value of zero, which is not impossible. Therefore, the detrended temperature data do not deviate significantly from an AR(4) process. However, if the decrease in the PE at order 37 were much stronger, that would indicate that something interesting would happen at that order and it would be advisable to compare the AR(4) and the AR(37) spectral densities and autocorrelation functions. ARMAsel selects one single model for the data, but inspection of the PE of all candidates reveals possibly interesting details, if present. In this respect, ARMAsel gives a language to random data.

9.4 Reduced-statistics Order Selection

Order selection is usually based on the reduction of the residual variance as a function of the model order with an additional penalty for every estimated parameter. Reduced-statistics MA and ARMA estimators do not use the data; hence no direct estimate for the residual variance is available. Broersen and de Waele (2002, 2005) compared the different AR, MA, and ARMA models with the very long AR model $C_M(z)$. The order of this very long AR model is determined by the available size M of the reduced statistic. Residuals $\eta_{n,r}$ for an ARMA($r,r-1$) model can be related symbolically to the unknown observations by the equation

$$\eta_{n,r} = \frac{\hat{A}_r(z)}{\hat{B}_{r-1}(z)} x_n \approx \frac{\hat{A}_r(z)}{\hat{B}_{r-1}(z)} \frac{1}{C_M(z)} \varepsilon_n \qquad (9.7)$$

where the reduced-statistics AR model $\varepsilon_n/C_M(z)$ is substituted for the unknown process x_n. Hence, $\eta_{n,r}$ can be expressed approximately as a filtered version of the white noise process ε_n. The ratio of the output and input variances is given as the power gain $P_g\left[\hat{A}_r(z), C_M(z)\hat{B}_{r-1}(z)\right]$ of (5.47). A simple computation of the ARMA power gain is found by separating the AR and MA filter operations. Thus, a value for the power gain or the *relative* residual variance of the ARMA($r,r-1$) model can be calculated from the estimated parameters and $C_M(z)$, without knowledge of η_n or ε_n, using a simple expression in the time domain:

$$\text{RES}(2r-1)/\sigma_\varepsilon^2 = \sigma_\eta^2/\sigma_\varepsilon^2 = P_g\left[\hat{A}_r(z),C_M(z)\hat{B}_{r-1}(z)\right] \tag{9.8}$$

This relative residual variance can be used in an order-selection criterion based on the logarithm of the residual variance. Penalty factor 3 is preferred for order selection.

The reduced-statistics order-selection criterion for MA(q) models becomes

$$\text{GIC}_{RS}(q,3) = \ln\left\{P_g\left[1,C_M(z)\hat{B}_q(z)\right]\right\}+\frac{3q}{N} \tag{9.9}$$

The unknown constant $\ln(\sigma_\varepsilon^2)$ has been subtracted in this reduced-statistics criterion in comparison with the usual definition of GIC in (7.32):

$$\begin{aligned}
\text{GIC}(q,3) &= \ln\left[\text{RES}(q)\right]+\frac{3q}{N}\\
&= \ln\left\{P_g\left[1,C_M(z)\hat{B}_q(z)\right]\right\}+\ln(\sigma_\varepsilon^2)+\frac{3q}{N}\\
&= \text{GIC}_{RS}(q,3)+\ln(\sigma_\varepsilon^2)
\end{aligned} \tag{9.10}$$

This unknown constant is the same for all model orders. Hence, the order at which the minimum of the selection criterion is found is not altered by this subtraction. Even the numerical values of the two selection criteria are equal for $\sigma_\varepsilon^2 = 1$.

The selection criterion for ARMA($r,r-1$) models becomes

$$\text{GIC}_{RS}(2r-1,3) = \ln\left\{P_g\left[\hat{A}_r(z),C_M(z)\hat{B}_{r-1}(z)\right]\right\}+\frac{6r-3}{N} \tag{9.11}$$

This order-selection criterion can be used to select the best order for each of the four reduced-statistics first-stage ARMA estimators in chapter 6.6, but it can also be used to select *among* those four estimators. In this way, the best, first-stage ARMA algorithm can be selected automatically.

The same type of criterion can also be used to select the order of the best AR model order by substituting $b_0 = 1$ for $\hat{B}_{r-1}(z)$ and by using the number of estimated parameters in the penalty function. In this way, the criterion for the AR order p becomes

$$\text{GIC}_{RS}(p,3) = \ln\left\{P_g\left[\hat{A}_p(z),C_M(z)\right]\right\}+\frac{3p}{N} \tag{9.12}$$

For AR order selection, this criterion can be used, but a finite-sample criterion is preferred if the highest candidate order for selection is greater than $N/10$. The reduced-statistics variant of the CIC criterion of (7.77) becomes

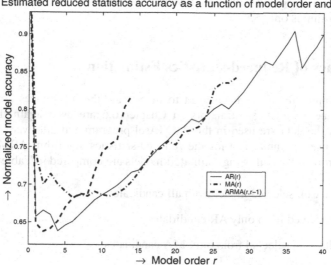

Figure 9.2. The estimated accuracy of AR, MA, and ARMA candidates for selection, estimated with ARMAsel-rs from the 142 detrended global temperature data shown in Figure 1.5

$$\mathrm{CIC}_{RS}(p) = \ln\left\{P_g\left[\hat{A}_p(z), C_M(z)\right]\right\} + \max\left[\frac{\prod\limits_{i=0}^{p}\left(1+\dfrac{1}{N+1-i}\right)}{\prod\limits_{i=0}^{p}\left(1-\dfrac{1}{N+1-i}\right)} - 1, 3\sum_{i=0}^{p}\frac{1}{N+1-i}\right] \quad (9.13)$$

where the approximations in (7.45) for the Burg method have been substituted for the finite-sample variances.

The P_g values can also be used for a survey of the accuracy of all candidate models in a reduced-statistics ARMAsel program. Figure 9.2 gives the results obtained by using the long AR(71) model of the data of Figure 9.1 as input for reduced-statistics estimation and selection. The numerical values for the estimated accuracies are different, but the shapes of the curves in Figures 9.1 and 9.2 are similar.

The results for AR models have to be identical, apart from scaling. They are derived from the same computed values of the residual variance, given by (6.14). The results for MA models are only slightly different because the residual variance is calculated by filtering the data in (9.3) in ARMAsel and with P_g for the reduced-statistics estimator. The results for ARMA models can be more different when using the reduced-statistics method because all four first-stage estimates are used in the second-stage ARMA estimation and the best of those four is selected. As an example, the ARMA(2,1) that was worse than the AR(4) model in Figure 9.1 turns out to be a little better in Figure 9.2, and the ARMA(2,1) model is selected by the ARMAsel-rs algorithm. The difference between those models is small: the ME of the AR(4) model compared with the ARMA(2,1) model estimated by the reduced-statistics algorithm is only 5.8.

9.5 Accuracy of Reduced-statistics Estimation

The four examples that have been used to investigate the differences among the first-stage reduced-statistics estimators in Chapter 6.6 are used again. In simulations, firstly the data are used in the ARMAsel program and afterward the long AR model is used as input data for the reduced-statistics algorithm ARMAsel-rs. The model errors of the following estimated models are compared in Table 9.1

- ARMAsel, selected model from all candidates

- AR, selected from only AR candidates

- results MA, selected from only MA candidates

- ARMA, selected from only ARMA$(r,r-1)$ candidates

- True order, fixed true order ARMA model, no selection.

In all four examples, the AR results for the ME are identical for the data and for the reduced-statistics algorithm. The average quality of selected MA models was only slightly different because in some runs the model selected was not the same.

The quality of ARMA models selected can be better and worse for ARMAsel or ARMAsel-rs. The data algorithm ARMAsel had numerical problems for less than 250 observations in the ARMA(5,4) example. That turned out to be the most difficult example in the simulations for ARMA algorithms. The ARMAsel-rs algorithm sometimes had numerical problems and created warnings for 1000 or more observations of that same example ARMA(5,4) process but only with the long COV initial-stage method.

Unfortunately, numerical problems sometimes have as a consequence that automatic, order-selection criteria will select precisely the model for which a warning was given. That problem is also easily identified by looking at the expected model accuracies, as plotted in Figure 9.2. Dangerous numerical problems cause an isolated low prediction error for one single ARMA model. That is impossible because, without numerical problems, the residual variance cannot become greater if more parameters are included. It can only stay the same. Hence, the asymptotic increase of the prediction error can never be more than $1 + 2/N$ with (9.5), and isolated deep holes are impossible in figures such as 9.2. Those numerical problems are avoided by limiting the highest candidate order or by excluding candidates with a warning from the selection. For Table 9.1, the highest ARMA candidate was taken as ARMA(10,9). If the highest candidate order was ARMA(30,29), the average ME of the ARMAsel-rs and also of the ARMA selected would become 177.8 for the ARMA(5,4) example. That means that the candidate with numerical problems would be selected if the warnings were neglected.

Table 9.1. Average of 2500 simulation runs of the ME of models estimated from 1000 observations of the four different ARMA processes of Section 6.6.7. The selection is ARMAsel with type selection included, only for AR order, only for MA order, only for ARMA order, and for the fixed, true order ARMA model, for estimates obtained with the ARMAsel algorithm that uses the data and the reduced-statistics ARMAsel-rs algorithm that uses only a long estimated AR model of the data for MA and ARMA estimation and for order and type selection.

	ARMA(2,1)	ARMA(3,2)	ARMA(4,3)	ARMA(5,4)
		Directly from data		
ARMAsel	5.6	7.1	21.0	28.0
AR	27.5	49.2	30.2	81.4
MA	7.7	22.3	80.9	711,587
ARMA	3.6	5.3	17.3	29.7
True order	3.5	5.2	179.2	6973.4
		Reduced statistics		
ARMAsel-rs	4.7	7.4	9.1	19.7
AR	27.5	49.2	30.2	81.4
MA	7.7	22.3	81.1	711,704
ARMA	3.4	5.8	8.4	19.7
True order	3.1	5.4	8.2	6324.3

For some examples given here, fixed order ARMA estimation is no problem. Then the fixed order ARMA results are better than the results of selection. This is representative of many practical data. However, ARMAsel for the ARMA(4,3) example and both algorithms for the ARMA(5,4) algorithm give poor averages for fixed order models. Sometimes, the algorithm does not estimate a good model for some model order, and other orders are estimated better. The average fixed order result becomes worse than the average selection result. However, in all four examples, the automatically selected ARMAsel and ARMAsel-rs models were much better than they would have been without ARMA candidates. For the ARMA(5,4) example, it was necessary to limit ARMA candidate orders.

The Cramér-Rao lower bound for the ME of unbiased models is asymptotically equal to the number of parameters estimated. The AR and the MA estimators are asymptotically close to that bound; Tables 7.1 and 8.4 show this for AR and MA, respectively. For some ARMA examples in Table 9.1, the fixed order ME was close to that bound. Selection with only ARMA candidates could also give an average ME close to the lower bound. This shows that the loss due to overfit is less for ARMA than that for AR or MA because the first overfitted model has two parameters more, instead of one. Unfortunately, the good performance of the ARMA algorithm is not found in all examples. Sometimes, the fixed order estimates are poor in the average, and the model errors of the selected models are as much as three times higher than the Cramér-Rao lower bound. A long search did not provide a better ARMA estimator. Maximum likelihood estimators have many more numerical problems than the algorithms of ARMAsel and ARMAsel-rs.

Another candidate for ARMA algorithms of Bauer and de Waele (2003), based on subspaces, has much greater errors for examples where ARMAsel is close to the Cramér-Rao lower bound. The model error of the subspace method for the

ARMA(2,1) example is two times higher, and for the ARMA(5,4) example, it is more than 1000 times higher than in Table 9.1. Those errors are caused by the triangular bias of the subspace procedure, so that method cannot be recommended. ARMAsel and its reduced-statistics variant are the most accurate so far.

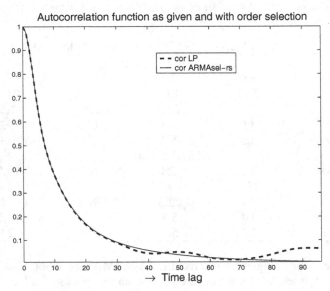

Figure 9.3. The measured autocorrelation function of a numerical statistical turbulence experiment and the autocorrelation function selected with reduced statistics

A practical application of the reduced-statistics estimator is given in Figure 9.3. From computational flow experiments solving the Navier-Stokes equation, a lagged product autocorrelation function is measured for turbulent channel flow. This autocorrelation gives information about the characteristic eddies formed in the channel. To identify turbulent structures, the maximum contribution to turbulent kinetic energy should be resolved. A technique called proper orthogonal decomposition is capable of determining these turbulent structures, provided an accurate estimate of the autocorrelation function is available. This autocorrelation function of the flow speed is the average computed from a number of parallel tubes in an artificial quasi-periodic turbulent flow. The dependence of the computations between neighbouring parallel tubes is high. The total number of observations that contributed to the given autocorrelation function was about 1,500,000. However, those observations from parallel tubes and different periods are not statistically independent.

Therefore, the *effective* number of observations is unknown. The 96-point, measured, lagged product autocorrelation function is transformed into the reflection coefficients of an AR(96) model by the Levinson-Durbin recursion (5.24). Supposing that the best AR order is lower than half the length of the given autocorrelation, the effective number of observations is approximated with the average of the squared reflection coefficients:

$$N_{\text{eff}} = 1/\sqrt{\frac{1}{48}\sum_{m=49}^{96}\hat{k}_m^2}$$
(9.14)

Equation (6.16) is applied to the higher order reflection coefficients with the assumption that those will be comparable statistically to reflection coefficients estimated from white noise. The effective number of observations was about 58,500 in this example, less than 4% of the number of contributions to the given autocorrelation. It is also possible to approximate the effective number with the median of the squared higher order reflection coefficients or with their standard deviation. If the three approximations are close, the resulting value for N_{eff} can be considered satisfactory. If they are quite different, some further analysis is required to determine why.

The selected model was the ARMA(8,7) model. The best AR model was AR(31) and the best MA candidate was MA(44). A plot of all model accuracies as in Figure 9.2 can be used to investigate the expected quality of the other candidate models. It turns out that the estimated model accuracy for AR orders greater than 35 increases linearly in a way that agrees with a true AR(31) process. In other words, the use of (9.14) to compute an effective number of contributions was allowed in this case. The autocorrelation function obtained with order selection is much more in agreement with the theoretical expectations than the given autocorrelation. The convergence to zero at the end had been especially expected theoretically. That convergence was even necessary for the validity of the experiment.

9.6 ARMASA Applied to Harmonic Processes

Figures 9.4 and 9.5 give the spectra of two sine waves with different levels of additive white noise. An integral number of periods fits precisely in the observation interval in Figure 9.4. The signals are periodic with the observation interval. This is favourable for periodogram estimates that always suppose this periodicity. The periodogram is more accurate for the magnitudes of the amplitudes than the time series spectrum, if the signal-to-noise ratio (SNR) is better than 0 dB. However, the ARMAsel spectra also show two peaks at the correct frequencies. For the 0-dB result, time series still give two peaks whereas the periodogram seems to have a valley at $f = 0.2$ or $f = 0.47$. For -10 dB, both the periodogram and the time series model do not detect the peaks.

In the lower part of Figure 9.5, the observation length was 10.5 periods of a very weak periodic signal and 30.5 periods for the second frequency involved. The ARMAsel solution detects the frequencies of the two peaks well, but the amplitudes are not accurate. The raw periodogram of the data also presented misses the weaker peak completely. It is clear that the background level for the SNR 120-dB spectra in Figures 9.4 and 9.5 is the same for the time series models selected by ARMAsel and very different for the periodograms. That is the reason that this small sine wave is not seen in the periodogram.

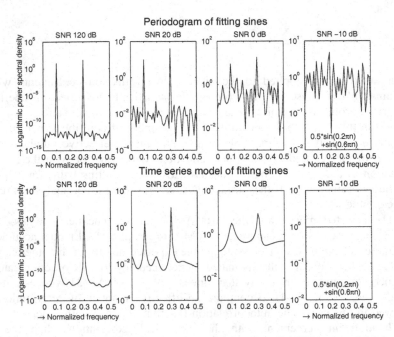

Figure 9.4. Raw periodogram above and spectrum of the selected ARMAsel model below from 100 observations of a periodic signal that fits in the observation interval with an integral number of periods

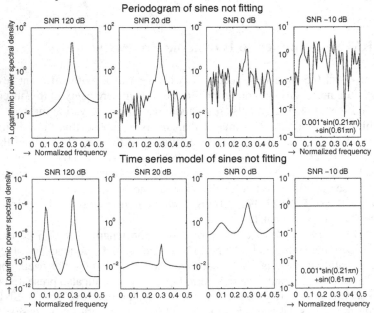

Figure 9.5. Raw periodogram above and spectrum of the selected ARMAsel model below from 100 observations of a periodic signal that does not fit in the measurement interval; the length is an integral number of periods plus a half period for both frequencies

This example shows that the performance of the automatic ARMAsel selection is still valuable if the true process is not a stationary stochastic process but a sine wave with a small amount of noise.

Of course, the behaviour of the periodogram can be improved by using tapers and windows. However, no automatic best solution is available, but many different approaches have been used. In practice, periodic data are multiplied by a taper or data window before they are transformed. Harris (1978) gives a survey of the improvements for periodic signals that do not have to be synchronized with the observation interval. With a suitable taper, the amplitude of the smaller sine wave would also have been detected in the low noise example of Figure 9.5.

9.7 ARMASA Applied to Simulated Random Data

The MA(13) example process, whose autocorrelation and spectral density are shown in Figures 3.2 and 3.3, respectively, has been used to generate 10,000 observations. The ARMAsel algorithm is used for the first N of those observations. Table 9.2 gives the selected model order and model type as well as the ME of the model selected, as a function of sample size N.

Table 9.2. The type, order, and ME of models estimated from a single realisation of the MA(13) process of Figure 3.2, as a function of the number of observations N.

N	Type	ME
10	AR(4)	549.0
20	AR(7)	410.8
50	AR(15)	195.2
100	AR(33)	63.8
200	AR(41)	57.0
500	MA(13)	21.3
1000	MA(13)	12.5
2000	MA(13)	19.3
5000	MA(13)	22.1
10,000	MA(13)	14.6

Repeating the experiment with newly generated data will often give the same selection behaviour. For 500 and more observations, the MA(13) model is also selected in a number of repeated runs. If enough observations are available, the true type of the process will be selected, mostly with the true order. Using penalty 3 for order selection still gives a small probability of one order overfit and a very small probability of two or more orders overfit. Underfit becomes unlikely if the last parameter is sufficiently greater than $1/\sqrt{N}$.

The values of the ME for the single simulation run vary considerably in Table 9.2. It should be realised that the standard deviation of the model error (ME) is rather large. For $N = 1000$, the experimental standard deviation of the ME was 7.5 in this MA(13) example. The results for larger sample sizes where the unbiased

MA(13) model is chosen are close to the Cramér-Rao lower bound, which is 13 for this example.

9.8 ARMASA Applied to Real-life Data

9.8.1 Turbulence Data

In the numerous simulations where the algorithms have been tested on simulated data, no difficulties or abnormal behaviour have been found. All overcomplete unbiased models have an average growth of the prediction error of $1/N$ in plots where the accuracy of the ARMAsel candidates is given as a function of the number of estimated parameters. Data available on the Internet have also been used as examples. In the majority of examples, the automatically selected single ARMAsel model gives all necessary or relevant information. On some occasions, questions arise about the real true spectrum, but they cannot be answered for those unique data because the truth is unknown. Irregular behaviour in those data could be caused by careless experiments or errors in the registration or the transmission of the data.

Turbulent flow is a challenging application area for signal processing. Legius (1997) used AR models as references for different flow regimes. He was successful in detecting a bubbly or a slug flow regime and also the transition between both. Pavageau *et al.* (2004) studied the application of AR models to isotropic turbulence. They found that AR models are a nonambiguous way to identify spectral ranges of constant slope. Here, an example from computational physics is treated, where data for turbulent flow behind a cylinder have been produced; see Sobera

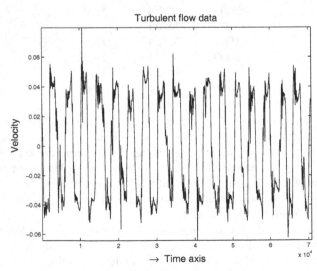

Figure 9.6. A computer simulation of turbulent flow, consisting of 70,814 observations, with the mean subtracted

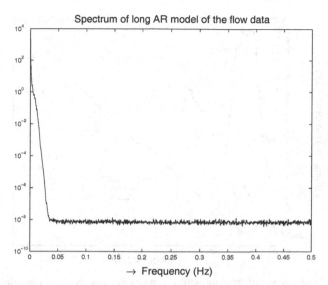

Figure 9.7. Spectrum of the 70,814 observations of turbulent flow, estimated with an AR(1000) model

et al. (2003, 2004). This study focused on heat and mass transfer to a clothed human body in outdoor conditions. It is expected that a turbulent von Karman street is found, and slopes of $f^{-5/3}$ and f^{-7} are in the double logarithmic spectrum. Those specific data are used because they are generated outside the framework of time series analysis and also because new questions evoked by the analysis with ARMAsel can be answered, if necessary, by adding new experimental data. Moreover, those data may illuminate some practical aspects of the selection and the analysis of additional models that are worth examining.

The first stage in analysing new data is always to look at a plot of the data. Figure 9.6 gives the flow data. At first sight, there are no strong objections to the hypothesis that the signal can be treated as stationary stochastic. At this stage, trends, nonstationary variance, seasonality, and other obvious deviations have to be treated.

The next stage is never necessary for simulated data, but it might be important for real-life data, as in the example. If the sampling frequency is extremely high, the correlation between the successive observations can become very close to one, and that may cause numerical problems in the analysis. This can be detected by estimating a high-order AR model and plotting the spectral density. Figure 9.7 gives the spectrum that contains hardly any power or information for frequencies above 0.033 Hz. Nevertheless, AR parameters are estimated to approximate this spectrum that is flat over more than 90% of the whole frequency range. Many parameters are used to describe behaviour at high frequencies. Spectral approximation can be more accurate with the same number of estimated parameters if it is restricted to the first small part of the frequency domain. Downsampling with a factor of 15 is a good way to continue. If possible, no antialiasing filters should be used in downsampling. A big advantage is that the time domain is not distorted if

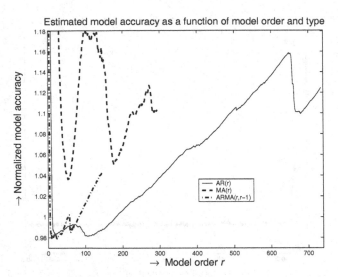

Figure 9.8. The language of the data with estimated accuracies of all candidate models in ARMAsel, applied to the 4721 downsampled turbulence data

if those filters are not used. The distortion in the frequency domain is a redistribution of the original power at frequencies above half the new resampling frequency over the smaller, new, frequency domain. Broersen and de Waele (2000b) have shown that antialiasing filters often give much more distortion than simply downsampling without filters in many cases. If filters are absolutely necessary, the new downsampled frequency range should at least be completely within the passband of the antialiasing filter. In other words, the cutoff frequency of the antialiasing filter should be higher than half the resampling rate. Otherwise, time series analysis will describe mainly the characteristics of the transition band of the filter, not the spectrum of the data.

Figure 9.8 gives the estimated accuracies of all models that have been estimated from the signal after downsampling with a factor of 15. The selected ARMAsel model was ARMA(9,8). MA models are not attractive for those data, but some AR models are close competitors. The best AR order was AR(18) with a spectral estimate that very much resembles the ARMA(9,8) spectrum.

Two very remarkable features are seen in Figure 9.8: the higher order minimum at AR(107) and a very strong dip at AR(679). The local minimum of ARMA(54,53) has the same number of parameters as the AR(107) minimum. The spectra of the ARMA($r,r-1$) models are very close to the spectra of AR($2r-1$) models with the same number of parameters. This could be expected from Figure 8.1. Therefore, only AR models are considered further for this example. This figure represents the language of the data, and the reason for the local minima of the AR models will be investigated.

After downsampling or decimating by a factor of 15, 15 different downsampled segments can be obtained, starting with $t = 1, 16, \ldots$ and $t = 2, 17, \ldots$ and so on. De Waele and Broersen (2000) described a Burg algorithm for using segments of data simultaneously in estimating an AR model. The effective number of observations

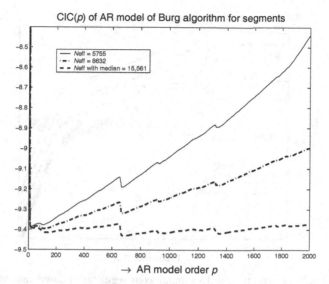

Figure 9.9. Order-selection criteria for AR models obtained from 15 simultaneous segments of downsampled data with minima at orders 18, 107 and 679

can be approximated by the method used in (9.12). The average of the estimated squared reflection coefficients at higher orders is taken as the inverse of the number of independent data. The effective number found for the Burg algorithm applied to 15 segments was 5755 for the reflection coefficients at AR orders between 1000 and 2000. Each segment has a length of 4721. There is only an advantage of about 20% in accuracy by using the 15 segments simultaneously in this example. Downsampling by a factor of 15 and forgetting all other segments gives almost the same accuracy. The correlation between the segments is so high that they all contain the same information. This agrees with the very flat spectrum in Figure 9.7.

Instead of using the average squared estimated reflection coefficients, it is also possible to use the median of the squared reflection coefficients to estimate the effective number of observations. Unfortunately, 15,561 was found by using the median. This difference between average and median indicates that the magnitude of high-order reflection coefficients is greater than the average expected. That behaviour is explained partially by the finite-sample behaviour of reflection coefficients. Estimating 2000 reflection coefficients from an effective number of 5755 observations gives values greater than $1/N$ for the variance of high-order reflection coefficients with the finite-sample variance coefficients of (6.20). Having obtained these guesses for the effective number of observations, the selection criterion CIC can be used for order selection. Due to the uncertainty in the effective number of observations, the values of CIC(p) have been calculated with $N_{\text{eff}} = 5755$, 8632, and 15,561. The middle value is 1.5 times the first. The higher values of N_{eff} have the same influence as a smaller penalty for each additional parameter in CIC(p). The three results are presented in Figure 9.9 to demonstrate the influence of N_{eff} on estimated model accuracy.

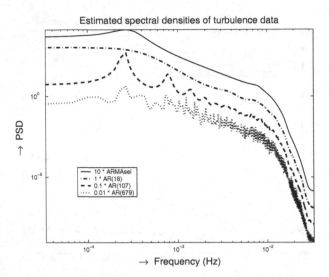

Figure 9.10. Spectra estimated from 15 simultaneous segments of downsampled turbulence data, calculated for the AR orders 18, 107, and 679 that were the minima in Figure 9.9. The AR(679) is shifted by a factor of 10 to enhance visibility. As a comparison, the ARMAsel spectrum of the ARMA(9,8) model, that had been selected in Figure 9.8 is given.

The flat shape of the lower curve in Figure 9.9 is what can be expected if the effective number of observations is replaced by a twice too high value. The estimated prediction error becomes a constant then. Therefore, the first two values of N_{eff} are probably good guesses. The minimum of the lower curve is at order 679. It is remarkable that both other curves also show a irregular shape at that order. The reason for this will be investigated. At an order around 1350, all criteria also have a local valley. This order is about twice the order 679 of the minimum. The global minimum at order 107 is found for the middle curve. This is also a local minimum in the other curves. The upper curve has its minimum at order 18. In the derivation of the penalty and of the selection of the model type, it has been discussed that the value for penalty 3 is a compromise to prevent a couple of orders overfit and that the penalty 2 would become interesting for comparison of models with greater differences in model order. For that reason, the AR(107) model is considered the most interesting model order.

The whole discussion of the effective number of observations after decimating could have been omitted here. It just showed that three different AR orders could be selected, depending on the assumptions. However, inspection of Figure 9.8 immediately shows that there are three interesting AR models at orders 18, 107, and 679. The spectra of those three AR orders are presented in Figure 9.10 on a doubly logarithmic scale, which is usual for turbulence spectra. The AR(18) spectrum gives global appearance with three different slopes: flat for very low frequencies, a moderate slope in the middle, and very steep for the highest frequencies in the signal. The AR(107) spectrum shows some wiggles around this curve, and the AR(679) spectrum shows a very irregular pattern around the same global spectral shape. The ARMA(9,8) spectrum, selected by ARMAsel, is also

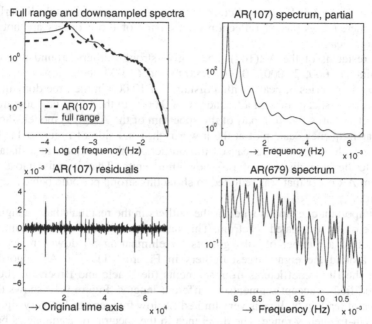

Figure 9.11. Spectral details. The upper left figure compares the downsampled AR(107) spectrum with the first part of the full range spectrum of Figure 9.7. Upper right gives a part of the AR(107) spectrum on a linear frequency scale. The lower part gives the residuals of the AR(107) spectrum and the equidistant peaks in the AR(679) spectrum caused by a few of the largest residuals.

presented. This looks very much like the AR(18) spectrum, except for the very wide peak around $f = 0.00025$.

The upper left Figure 9.11 shows that the low-frequency part of the full range spectrum of Figure 9.7 is almost completely conserved by downsampling. Only in the very highest part of the new frequency range, can it be seen that a visual redistribution of power makes a noticeable difference. The white noise of higher frequencies has very little influence. The differences around 0.001 Hz can be attributed to using different AR orders for the spectra. Note that the area below 0.001 Hz is very small in Figure 9.7 with a linear frequency scale.

The linear frequency plot on the upper right hand side shows that the selected AR(107) model gives a number of almost equidistant spectral peaks. This phenomenon is characteristic of turbulence data that has strong periodicity. The periodic behaviour is fully developed in the AR(107) model, and it was completely absent in the AR(18) spectrum. The AR(107) model is on the boundary of statistical significance for the given amount of data. It would probably be selected if many more data had been collected. The language of the data in Figure 7.8 gives a strong indication that this model has interesting properties. The periodicity is less developed in lower order AR models. Higher order models, until order 679, do not require special attention. The plot of Figure 7.8 is what would be expected for a white noise signal in the range between 120 and 650. In asymptotic theory, that is a linear increase in the prediction error with model order. Finite-sample effects are shown in Figure 7.4. If the prediction error increases approximately linearly with model order, the estimated reflection coefficients of those orders are not statistically significant.

The residuals of the AR(107) model give strong outliers around the original time points 10,000, 20,000, 30,000, 55,000 and 65,000 at the lower left-side of Figure 9.11. A series of peaks with a distance of 10,000 in the time domain would give a periodic spectrum with a frequency of 0.0001 in the frequency domain. That is precisely what is found in part of the spectrum of the AR(679) model, shown on the linear frequency scale plot at the lower right-hand side of Figure 9.11. Downsampling with a factor of 15 caused the outliers in the residuals at a distance of 10,000 to lie at a distance of 667 new time units. That is quite close to the minimum AR order that was required to show this strong periodic behaviour in the spectrum.

An important question is whether the outliers in the residuals have a significant influence on the estimated spectrum. This has been investigated by dividing every segment into a number of subsegments by eliminating 50 downsampled observations around the eight largest outliers in Figure 9.11. The AR algorithm to estimate the Burg coefficients from segments (de Waele and Broersen, 2000) has been modified to accept segments of different lengths. Figure 9.12 shows that the influence of leaving outliers is very limited for low frequencies but very significant in the high-frequency range. The difference in the spectra for frequencies between 0.01 and 0.03 Hz is more than a factor of 10. About the same spectrum without outliers is found by using only the first 10,000 observations with downsampling or by using any contiguous part where no large outliers are found. The few outliers have a great influence on the magnitude and shape at high frequencies. Therefore, the signal properties are not stationary after all. The spectrum estimated without

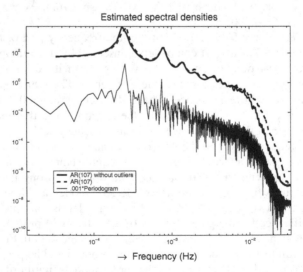

Figure 9.12. Spectra of the AR(107) model estimated from 15 complete segments and from fragmented segments where the parts with large residuals have been eliminated. The periodogram of one downsampled segment is also given, shifted vertically to increase visibility.

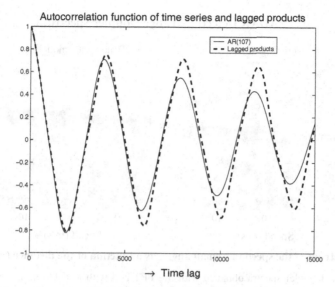

Figure 9.13. Autocorrelation function of the AR(107) model and the lagged product estimates estimated from the decimated data. The frequency 0.00025 of the spectral peaks in Figure 9.12 is equivalent to a period of about 4000 in the autocorrelation function.

outliers is considered the best estimate. Finally, knowing that the outliers at a distance of 10,000 have such a strong influence on the spectrum, it is also possible to detect those outliers visually in the original data, as presented in Figure 9.6. As a comparison, the periodogram is shown in Figure 9.12.

Two estimated autocorrelation functions are presented in Figure 9.13. The Fourier transform of the lagged product estimate is given by the periodogram in the previous figure. The details and the differences visible in the two estimated auto-correlation functions are almost exclusively due to the differences around the peak at 0.00025 Hz in Figure 9.12. Spectral details for the frequency range above 0.001 Hz are invisible in a plot of the autocorrelation function. The absolute sum of squares of those frequencies in the spectrum is very small, and hence they do not influence the sum of squares of the autocorrelation function. This shows why the cepstrum is necessary as a measure for the accuracy of autocorrelation functions.

9.8.2 Radar Data

The detection of moving objects with radar is a practical application of ARMAsel. De Waele (2003) has compared the performance of the usual detection criteria based on periodograms and lagged product autocorrelations, on the one hand, and fixed order or selected time series models on the other hand. He used frequency-modulated, continuous-wave (FMCW) radar data.

Target detection is an important step in radar processing. The aim of a detector is to distinguish between reflections from a target and background reflections. An example of a target is a small boat, which is to be detected amid reflections of waves and measurement noise. The statistical theory is hypothesis testing. The idea

RADAR Detection, range Doppler spectrum

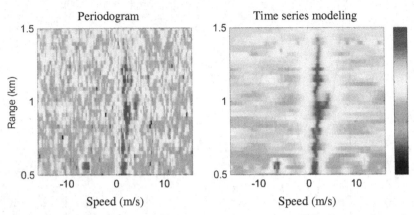

Spectrum of the speed in a small angle, as a function of the distance (range)

Figure 9.14. Doppler spectra obtained with the FFT and with ARMAsel, as a function of the range. A horizontal line in the range-Doppler spectrum shows the distribution of velocities in a particular range cell. The reflections around a speed of 2 m/s represent sea clutter, reflections of waves. Possible targets are found in range cell 0.55 at −7 m/s and a small one at range cell 1.4 at −6 m/s. The ARMAsel estimate shows these targets more clearly than the FFT estimate.

is to estimate a model of the surrounding environment without the boat. If that is the sea, waves with low velocities will be present, and the prototype model is the sea clutter. Figure 9.14 illustrates his conclusions. The spectrum of the speed as a function of the distance shows slowly moving waves and some moving targets. It is easier to detect them in the time series spectra than in periodograms. In general, the FFT plot shows many more spurious details than the time series models.

9.8.3 Satellite Data

The development of space-borne and aircraft-based sensors and powerful computer hardware in combination with improvements in physical and mathematical modeling give the possibility of determining the earth's gravity field with increasing resolution and accuracy. Satellite sensors often provide data corrupted by noise that is strongly correlated along the orbit and trajectory. Prominent examples are the Gravity Recovery and Climate Experiment (GRACE) and the Gravity Field and Ocean Circulation Explorer (GOCE) satellites. Accelerometers onboard the GOCE satellite will be subject to a strong, coloured, noise signal. The intention is to describe the geoid, the height of the equigravity plane around the world, as accurately as possible. Using a simple least-squares evaluation of the observations gives large geoid errors. Due to the coloured noise, the least squares solution is not optimal, and the orbit of the satellite can be recognized erroneously in the gravity field. The data have to be multiplied by the inverse of the covariance matrix of the noise in (2.43) to give a better estimate. However, too many data are available to invert that covariance matrix. Klees *et al.* (2003) describe how the data can be filtered so that the errors become white noise. Then, the least-squares solution of (2.41) gives the best solution. The errors in the geoid map are very much reduced.

9.8.4 Lung Noise Data

This example demonstrates the use of ARMAsel for feature extraction in a medical context. Many applications use the recognition of electrical signals that can be measured on living people or animals. Examples are the electro-encephalogram (EEG) signal for brain activities, the electrocardiogram (ECG) signal for heart activities and the electromyogram (EMG) for muscle activities. Liley *et al.* (2003) describe how effects of tranquilliser drugs can be detected in an EEG with ARMA models. Yang *et al.* (2001) use AR modeling with order selection to investigate the properties of transplanted kidneys of rats. Friedman *et al.* (2004) describe how time series analysis can be used for continuous monitoring of blood pressure. Kim *et al.* (2004) recognize emotions from physiological measurable signals. Pfurtscheller (2004) gives a survey of brain-computer interface prospects. Feature extraction and classification use new time series techniques to improve the interaction. The EMG signal can be used for prosthesis control. Muscle activity can be detected in electrodes on the skin, and the signal specifications can be used for various commands. ARMAsel could very well improve the detection of the signal type and hence make revalidation with artificial limbs somewhat less difficult.

A first step toward the detection of asthma in lung sounds is the detection of the presence of methacholine, a drug that has an influence on the properties of the lungs similar to that of asthma. This can be done by measuring the lung noises of the same person with and without methacholine; see Broersen and de Waele (2000). Figure 1.2 gives two cycles of the inspiration and the expiration signal. The signal is not stationary, but using only a finite part of the cycle that is synchronized with the beginning of the cycle will give data that can be considered more or less stationary. The experiment starts by defining two prototype spectra with known conditions: one with methacholine and one without methacholine. The parameters of the time series models of both prototypes are estimated. Later, a new test signal is recorded, and the experimenter does not know whether methacholine is present. Now, ARMAsel is used to estimate the parameters of the test signal. Figure 9.15 gives an example of spectra of the two prototypes and a test signal. The model error is used as a measure of the difference between the test signal and the prototypes. In this example, the ME between the test signal and the prototype without methacholine was 94 and with methacholine 437. Hence, it may be concluded that the test was measured without methacholine.

This example shows how ARMAsel can be used in the detection of diseases, in the classification of EEG signals, and in the control of prostheses. The principle is to measure prototype time series models under known circumstances. The prototypes can be established as the average of many experiments. For new test data, ARMAsel is used to estimate the time series model, and the model error is used to measure the difference between the test data and all prototypes. More than two prototypes can be used. The prototype with the smallest difference is selected. If desired, a maximum value can be given for the ME. If the test is not closer to

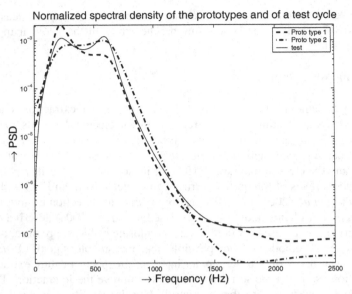

Figure 9.15. Normalized spectral density of two prototype spectra and a test spectrum that must be classified as belonging to one of the prototypes. Each spectrum is computed with 500 observations, starting at the beginning of the expiration phase. The ME between test and prototype 1 without methacholine was 94; with prototype 2 with methacholine, it was 437, and the test signal belongs to the class without methacholine.

one prototype than the prescribed maximum allowable difference, the experiment is undecided. For instance, a prosthesis can be commanded to move forward, to move backward or not to move at all. A maximum ME difference will reduce unintentional motions.

9.8.5 River Data

The importance of an accurate early warning system for river flooding is obvious. Improving the forecast of river discharge enhances the safety of riverside residents and can prevent damage. Moreover, an improved forecast of river discharge or water level has economic advantages because the maximum loading of commercial vessels in inland navigation depends on the actual water level in rivers and ports. Therefore, much attention has been given to obtain better forecasts of river discharges.

A practical application is improved prediction of the water level of rivers. The height can be predicted days in advance with a deterministic model that contains knowledge of the amount of rainfall in the river area in the past days. However, measuring the level will give a difference from the deterministic prediction. If the actual level exceeds the predicted level at a certain moment, it will probably also exceed the level predicted one hour later. The difference between the deterministic predictions and the actual observations of the water level is a stochastic signal. That signal is used to correct the deterministic predictions. Figure 9.16 gives the observed and the predicted hourly water levels of the Mosel discharge.

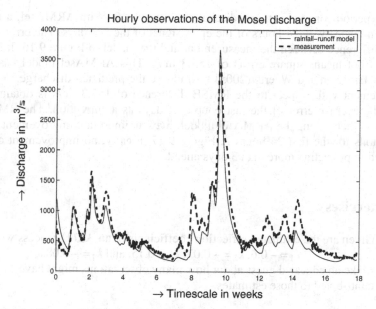

Figure 9.16. Measured discharges of the Mosel and simulated discharges with a rainfall runoff model around the end of 1998. The RMSE of the difference is 155.3 m^3/s.

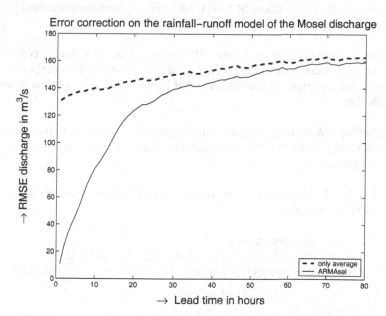

Figure 9.17. Average forecast accuracy for a period of 80 hours for the Mosel catchment. Three hundred updates have been made at dayly intervals with a model estimated from the error signal in the last eight days. The error correction is made with the selected ARMAsel model and with the mean value of the error in the last 8 days. The ARMAsel correction is a considerable improvement for the first couple of hours. After 1 day, the difference becomes small. The average discharge was 647 m^3/s.

The period contains extreme discharges in week 10. With ARMAsel, a model has been made every 24 hours of the error signal of the last days. The error signal is the difference between the measurement and the model in Figure 9.16. It has an RMSE (root means square error) of 155.3 m^3/s. This ARMAsel model has been used by Broersen and Weerts (2005) to improve the predicted discharge. A slight improvement with respect to the RMSE difference of 155.3 m^3/s is obtained by using the average error of the last couple of days as a correction. The RMSE is about 140 then. Using the ARMAsel model gives an important improvement in the predictions for the first 24 hours in Figure 9.17. Finally, no improvement can be obtained in predicting more than 3 days ahead.

9.9 Exercises

9.1 Given are the estimated reflection coefficients of an AR(1) process with
$$k_1 = -0.6, k_2 = -0.03, k_3 = 0.15, \text{ and } k_4 = -0.08.$$
Make an educated guess about how many observations might have contributed to those estimates.

9.2 Given are the two observations -1 and $+1$. Use ARMAsel to compute the parameters of the AR(1) model. The download address of ARMAsel is given in the next section. MATLAB® is the required computational environment.

9.3 Given are the two observations -101 and $+1$. Use ARMAsel to compute the parameters of the AR(1) model. Compare the result with the previous exercise and explain. Think about what happens as a rule with the mean of the data.

9.4 Run the MATLAB® program "simple_demo" to become familiar with the ARMAsel instructions to get a spectrum or an autocorrelation function from simulated data.

9.5 Run the MATLAB® program "demo_armasa" to learn more about the ARMAsel program.

9.6 Given are eight observations:
$$x = [0.6, -0.7, 0.9, -0.3, 0.8, -1.2, 1.1, \text{ and } -0.9].$$
Use ARMAsel to compute the parameters of the various models with

$$[a,b,\text{sellog}]=\text{armasel}(x)$$

AR(1) will be selected. Verify with the results in sellog that this model was the best candidate.

9.7 Verify that the model error (ME) between the best AR model and the best MA model was only 0.09 in the previous example.

9.8   Given are eight observations
        $x = [-0.9, \ 1.1, \ -1.2, \ 0.8, \ -0.3, \ 0.9, \ -0.7, \text{ and } 0.6\ ]$.
This is the reversed sequence of Exercise 9.6. What will be the estimation result of the reversed data? Realise that the autocorrelation function of arbitrary data will not change if their sequence is reversed.

9.9   Use the ARMAsel program to analyse some data sets, from the Internet. Many historical data sets can be found with the search term "climate data". Also economic data and industrial production data can easily be found.

ARMASA TOOLBOX

MATLAB® programs for automatic time series analysis

Download address:

http://www.mathworks.com/matlabcentral/
then select: File Exchange, Signal Processing, Spectral Analysis,

ARMASA   contributed by Piet M.T. Broersen
Automatic Spectral Analysis   contributed by Stijn de Waele

name	- function
armasel	- ARMAsel model identification from data
sig2ar	- AR model identification from data
sig2ma	- MA model identification from data
sig2arma	- ARMA model identification from data
armasel_rs	- ARMAsel identification from long AR model
arh2ar	- AR model identification from long AR model
arh2ma	- MA model identification from long AR model
arh2arma	- ARMA model identification from long AR model
arma2cor	- parameters to autocovariance function
arma2psd	- parameters to power spectral density
arma2pred	- parameters to prediction
moderr	- Model Error, difference between models
ar2arset	- AR parameters to lower-order AR models
rc2arset	- AR reflection coefficients to AR models
cov2arset	- autocovariances to AR models
cov2arset + arh2ma	- autocovariances to MA models
cov2arset+arh2arma	- autocovariances to ARMA models
armafilter	- calculation of residuals or predictions of data
Burg	- Burg type AR estimator
Burg_s	- Burg for segments
CIC	- CIC finite-sample order-selection criterion
cor2ma	- MA(q) covariance function to MA(q) parameters
simuarma	- generates time series data
psd2ar	- power spectral density to AR model
sumarma	- sum of ARMA processes with different variances
demo_armasa	- demonstration of some ARMASA toolbox features
simple_demo	- demonstration of some ARMASA toolbox features

10

Advanced Topics in Time Series Estimation

10.1 Accuracy of Lagged Product Autocovariance Estimates

Raw periodograms have never been considered useful spectral estimates for random processes. The standard deviation is equal to the expectation of the spectrum and does not become smaller if more observations are available. Only the number of frequencies with an independent spectral estimate increases. Nevertheless, the inverse Fourier transform of the periodogram, the mean lagged product estimator, is still often considered the natural estimator for the autocovariance function.

Maximum likelihood is a reliable principle for deriving efficient estimators in ordinary linear regression problems. However, no relation can be established between the lagged product estimator for the autocovariance function and the maximum likelihood principle. Simply taking the average of a number of estimates with the same expectation is only an optimal estimator if the estimates are uncorrelated and all have the same variance. In other cases, weighted least squares gives a better estimate.

Porat (1994) proved that only sample autocovariances until lag $p - q$ are asymptotically efficient in an ARMA(p,q) process. All further lagged product estimators are not efficient, and their variances are greater than minimally obtainable. However, the autocovariance function can be calculated efficiently from efficiently estimated parameters of the proper time series model. The theoretical background of this statement is the invariance property, found in Zacks (1971): the maximum likelihood estimate of a function of parameters is found by substitution of the maximum likelihood estimates of the parameters in that function.

Some expressions for the variance of estimated autocovariance functions have been given in Section 3.5. However, they could not be evaluated then because no standard expressions for infinitely long autocovariances had been developed at that stage. That requires the time series results of Chapter 4. Some explicit formulas are derived here for simple processes. The variance of the autocorrelation function has been given in (3.40). The asymptotic expression for the covariance of the autocorrelation with the lagged product estimator at lags k and $k + v$ is found with the Taylor approximation (2.32) for the quotient of two stochastic variables (3.31) with lag k and lag 0 as

$$N \operatorname{cov}\{\hat{\rho}_{LP}(k), \hat{\rho}_{LP}(k+v)\}$$

$$\cong \sum_{m=-\infty}^{\infty} \begin{bmatrix} \rho(m)\,\rho(m+v) + \rho(m+k+v)\,\rho(m-k) \\ +2\,\rho(k)\,\rho(k+v)\,\rho^2(m) - 2\,\rho(k)\,\rho(m)\,\rho(m-k-v) \\ -2\,\rho(k+v)\,\rho(m)\,\rho(m-k) \end{bmatrix} \qquad (10.1)$$

The variance of $\hat{\rho}_{LP}(k)$ follows from this expression for $v = 0$ and is given already in (3.40). The formulas are inversely proportional to sample size N, so the statistical variations will finally approach zero for larger sample sizes. For larger lags k where the true correlation $\rho(k)$ becomes approximately zero, all products in (10.1) vanish if the indexes differ by more than k. It follows that

$$N \operatorname{var}\left[\hat{\rho}_{LP}(k)\right] \cong \sum_{m=-\infty}^{\infty} \rho^2(m), \qquad k \to \infty \qquad (10.2)$$

The practical importance of this latter formula is that the statistical inaccuracy of lagged product estimates of decaying autocorrelation functions is a constant for all lags where the true autocorrelation $\rho(k)$ has died out. Substitution of lagged product estimates for the autocorrelation in (10.2) does not generally give a good idea of the actual accuracy. Take white noise as an example; only $\rho(0)$ is 1 and all other lags have zero expectation. The infinite summation in (10.2) has the true value 1, but each lag would give a contribution of about $1/N$ if estimated lagged product autocorrelations were used.

A further example is the MA(1) process with parameter b. The variance expression (3.40) for lagged products gives

$$N \operatorname{var}\left[\hat{\rho}_{LP}(1)\right] = 1 - 3\rho^2(1) + 4\,\rho^4(1)$$
$$N \operatorname{var}\left[\hat{\rho}_{LP}(k)\right] = 1 + 2\rho^2(1), \qquad k \ge 2 \qquad (10.3)$$

with the true autocorrelation values $\rho(0) = 1$, $\rho(1) = b / (1+b^2)$ and $\rho(k) = 0$ for $k > 1$. An old expression of Whittle (1953) for the variance of the time series model based autocovariance estimate follows from a Taylor approximation of $\rho(1)$:

$$\operatorname{var}\left[\hat{\rho}_{MA}(1)\right] = \operatorname{var}\left[\frac{\hat{b}}{1+\hat{b}^2}\right] = \frac{(1-b^2)^2}{(1+b^2)^4}\operatorname{var}\left(\hat{b}\right)$$
$$\operatorname{var}\left[\hat{\rho}_{MA}(k)\right] = 0, \qquad k \ge 2 \qquad (10.4)$$

The variance of a MA(1) parameter follows from (6.45) as $(1-b^2)/N$ and

$$\frac{\operatorname{var}\left[\hat{\rho}_{LP}(1)\right]}{\operatorname{var}\left[\hat{\rho}_{MA}(1)\right]} = \frac{1+b^2+4b^4+b^6+b^8}{(1-b^2)^3} \qquad (10.5)$$

This ratio is greater than one for all values of b other than zero because the numerator is greater than one and the denominator is smaller than one. The quotient of the variances becomes ∞ for $k \geq 2$, where the MA estimate equals zero. Even though the variance of the mean lagged product estimator converges to zero for increasing N for all lags, the *ratio* of both estimates remains ∞.

Monte Carlo simulation runs have been made with a MA(2) process with poles at an equal radius β. In each run, three different autocorrelation estimates were determined for lags 2 and 4. The first is measured with the estimated true order MA(2) model, the second with the selected ARMAsel time series model (TS), and the third with mean lagged products (LP). The average of N times the mean square error (MSE) is given in Table 10.1.

Table 10.1. Average of N*MSE of 1000 simulation runs of 100 observations of the quantities $\hat{\rho}_{MA}(2), \hat{\rho}_{TS}(2), \hat{\rho}_{LP}(2), \hat{\rho}_{MA}(4), \hat{\rho}_{TS}(4)$, and $\hat{\rho}_{LP}(4)$ of a MA(2) process, as a function of the radius β

	Lag $k = 2$			Lag $k = 4$		
β	MA(2)	TS	LP	MA(2)	TS	LP
−0.9	0.005	0.52	1.53	0	0.33	1.58
−0.6	0.21	1.18	1.33	0	0.94	1.65
−0.3	0.77	0.89	1.11	0	0.23	1.20
0	0.97	0.26	0.97	0	0.07	0.95
0.3	0.81	0.91	1.13	0	0.22	1.11
0.6	0.20	1.36	1.49	0	1.13	1.73
0.9	0.005	0.56	1.47	0	0.38	1.47

According to asymptotic theory, no autocorrelations can be estimated efficiently with lagged products for any lag. All columns for TS give errors smaller than LP. The much better accuracy of time series estimates in practice is clear in the results for lag 4.

With some manipulation, the asymptotic expressions (3.40) and (10.2) for lagged products yield, for an AR(1) process with true parameter $-a$, hence with $\rho(1) = a$

$$N \operatorname{var}\left[\hat{\rho}_{LP}(k)\right] = \frac{1 + a^2 - (2k+1)a^{2k} + (2k-1)a^{2k+2}}{1 - a^2}$$

$$N \operatorname{var}\left[\hat{\rho}_{LP}(\infty)\right] = \left(1 + a^2\right)/\left(1 - a^2\right)$$

$$\frac{\operatorname{cov}\left[\hat{\rho}_{LP}(1), \hat{\rho}_{LP}(2)\right]}{\sqrt{\operatorname{var}\left[\hat{\rho}_{LP}(1)\right]\operatorname{var}\left[\hat{\rho}_{LP}(2)\right]}} = \frac{2a(1-a^2)}{\sqrt{1 + a^2 - 5a^4 + 3a^6}} \qquad (10.6)$$

The last line demonstrates that the correlation coefficient between the estimated autocorrelations for two successive lags (1 and 2 in this example) can be greater than the correlation between two successive observations. In other words, the autocorrelation function estimated with lagged products can look more correlated or regular than the observations.

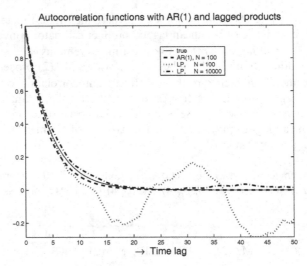

Figure 10.1. True and estimated autocorrelation functions for an AR(1) process with $a_1 = -0.8$

Another approach for determining the autocorrelation of an AR(1) process is using the estimated AR(1) parameter $\hat{a}$ to compute the autocorrelation function as

$$\hat{\rho}_{AR}(k) = (-\hat{a})^k \qquad (10.7)$$

An expression for the variance of this parametric approach can be found with a Taylor approximation (2.29) and the asymptotic variance expression (6.42) for $\hat{a}$. This gives

$$N \operatorname{var}\left[\hat{\rho}_{AR}(k)\right] = k^2 a^{2k-2}(1-a^2)$$
$$N \operatorname{var}\left[\hat{\rho}_{AR}(\infty)\right] = 0 \qquad (10.8)$$

A comparison of (10.6) and (10.8) reveals that the two variance expressions are identical for $k = 0$ and $k = 1$, where both are equal to 0 and $1 - a^2$, respectively. For other k, the variance expressions give different outcomes.

Figure 10.1 shows that the time series ARMAsel estimate for the autocorrelation function, obtained from 100 observations, may be more accurate than the lagged product estimate that is obtained from 10,000 observations. The difference is often large at higher lags where the true autocorrelation approaches zero because the AR poles damp out and the variance of lagged products is still given by (10.2). The LP results for $k = \infty$ in (10.6) are good approximations for all values of k where the true autocorrelation becomes negligible.

The ratio between the variance expressions (10.6) and (10.8) is shown in Table 10.2. For $k > 1$, the time series estimator (10.8) has better accuracy than the mean lagged product estimator (10.6). The confidence that has been given to lagged product estimates in the past is based on the fact that the variance of mean lagged products finally converges to zero if N is ever increasing. However, the time series estimates already converge for much smaller sample sizes. Table 10.2 shows that the qualification "not asymptotically efficient" in practice means very inefficient.

Table 10.2. Theoretical ratio $\mathrm{var}[\hat{\rho}_{LP}(k)]/\mathrm{var}[\hat{\rho}_{AR}(k)]$ for an AR(1) process, as a function of the AR parameter $\rho(1) = a$ and of k

k	$a = 0.25$	$a = -0.5$	$a = 0.75$	$a = 0.9$
1	1	1	1	1
2	4.75	1.75	1.19	1.06
5	3169	22.6	2.56	1.33
10	83 10^7	5825	14.2	2.08
∞	∞	∞	∞	∞

The AR(1) spectrum can be computed directly with a true or an estimated value of the AR(1) parameter. It is equal to the exact Fourier transform of an *infinite* length of autoregressive autocovariances, as estimated from the single true or efficiently estimated AR(1) parameter.

Whatever windows are used, it will not be possible to obtain an efficient spectral estimate with periodograms, with an accuracy that is comparable to that of the AR(1) spectrum, because the periodogram is based on inefficient auto-correlation estimates for all lags greater than one. This result can be generalized to AR(p) processes where only p autocorrelations can be estimated efficiently with lagged products. The Fourier transform of more than p lagged product correlations will give an inefficient spectral estimate because of inefficient higher lag estimates.

Table 10.3. Average of N*MSE of 10,000 simulation runs of $\hat{\rho}_{ARMA(2,1)}(k), \hat{\rho}_{TS}(k), \hat{\rho}_{LP}(k)$ of an ARMA(2,1) process with $a_1 = -0.625, a_2 = 0.25, b_1 = -0.9, N = 10,000$

k	ARMA(2,1)	TS	LP
1	0.73	0.74	0.73
2	0.66	0.79	0.86
3	0.44	0.82	1.18
6	0.07	0.58	1.20

The results of a simulation experiment with an ARMA(2,1) process are given in Table 10.3. The empirical accuracy of the three autocorrelation estimators is equal for $k = 1$. For higher lags, the autocorrelation can better be estimated with a time series model. The estimated ARMA(2,1) model of the true order gives better accuracy in this example than the selected time series model TS. Only the first lag is estimated with the same inaccuracy using the lagged product estimator (LP). This confirms the validity in finite samples of the asymptotic theoretical result of Porat (1994) that only $p - q$ lagged product estimates are efficient for the autocorrelation of an ARMA(p,q) process.

A general but inaccurate method for a rough comparison of estimated auto-correlation functions has been shown in Figure 3.9, for lagged product estimates

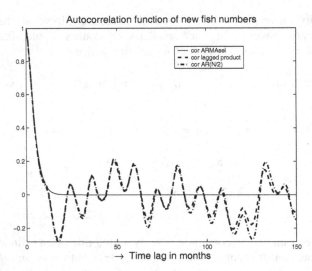

Figure 10.2. Three estimated autocorrelation functions from the fish data of Figure 3.8.

only. Divide the data into two halves and estimate the autocorrelation functions separately for the first and the second half of the data. This comparison is rough because it is based on looking at absolute differences and not at the cepstrum (5.57).

Figure 10.2 gives three estimates of the autocorrelation function of all 445 fish number observations. All three estimates almost coincide for the first 10 lags. Afterward, the ARMAsel estimate dies out. The lagged product estimate and the AR($N/2$) estimate remain close to each other. The influence of the triangular bias of (3.32) on the lagged product estimator can be seen; the values of the AR($N/2$) estimate are slightly further away from zero. It is a general property that high-order AR models and lagged products produce similar estimates of the autocorrelation function. The great difference is that time series models allow order selection to find out which part of the high-order AR autocorrelation function is statistically significant and which part can better be extrapolated. Figure 10.2 shows that the selected ARMAsel model gives a small autocorrelation for lags greater than 10. The lagged product autocorrelation estimate also illustrates the middle formula of (10.6) for lags where the true correlation is zero. If that is true, the variance of lagged product estimates for all higher lags is the same. Applied to Figure 10.2, all lagged product estimates between + 0.2 and – 0.2 are probably due to statistical variability. The high negative peak at lag 18 might also be due to the same uncertainty.

Figure 10.3 shows the estimated model accuracy of all ARMAsel candidates of the fish data. The high-order AR models have the worst estimated accuracy. Lagged product estimates give an autocorrelation function that is very similar to high-order AR models, but those high-order AR models are poor time series candidates. AR(2) has been selected with the usual selection criterion with penalty 3. Selection with penalty 2 would have selected the AR(37) model in this case. MA models are less attractive for those data, and ARMA models are also less accurate

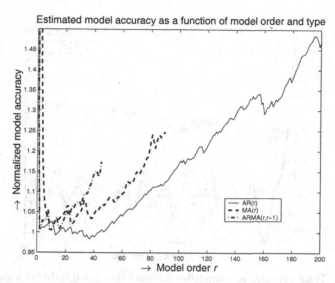

Figure 10.3. The language of the data with estimated accuracies of all candidate models of the fish data in ARMAsel

than AR candidates. For the given number of observations, many AR models with orders between 2 and 40 are almost equally attractive as candidates. Their estimated accuracies are almost the same. The parameters of those orders are on the edge of statistical significance. It is quite certain that the best model would become AR with order about 37 if many more observations were available.

It is almost equally certain that AR(200) will never become an attractive candidate, no matter how many observations are available. The accuracy of high-order AR models in Figure 10.3 is representative of many examples. High-order AR models have an autocorrelation that is similar to the lagged product result, but the estimated model accuracy is very poor. If the slope of the estimated accuracy of the AR models is a constant that resembles r/N, the models contain only insignificant parameters. That applies to AR models of order higher than 40 in Figure 10.3.

Figure 10.4 gives the autocorrelation function of the AR(37) model, as well as the selected AR(2) and the lagged product estimates. As AR(37) models use the Burg estimates for the first 37 correlations, it is clear that the lagged product and the AR(37) estimate stay close for the first 37 lags. For higher lags, the AR(37) estimate is obtained by extrapolation with (5.3). It damps out, but the lagged product estimate keeps heavy wiggles with constant periodicity. The periodicity of 12 months or 1 year in the correlations is not surprising for ocean data.

However, it is surprising that a real periodicity would have given a first minimum at lag 6 and a first maximum at lag 12 in the autocorrelation function. Those peaks are not present in any estimated function in Figure 10.4. The first sign of this yearly periodicity is the negative peak at lag 18. The periodicity is clearly visible at higher lags. The absence of the peak at lag 12 is probably the reason that the order-selection procedure is not very decisive for low-order AR models. The periodicity is on the boundary of statistical significance for this sample size.

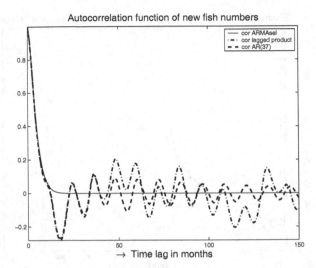

Figure 10.4. Three estimated autocorrelation functions from the fish data of Figure 3.8. The estimates are from all 445 data together, with ARMAsel, the AR(37) model, and obtained with lagged products.

Dividing the data in two parts brings one uncertainty to an end. The reduced number of observations is not enough to support high-order models in an order-selection procedure for one-half of the data. ARMAsel has been applied to the first and second halves of the data separately. The selected ARMAsel model was MA(8) for the first half and AR(2) for the second half of the observations. However, the autocorrelation functions belonging to those selected models are very close. For all observations together, the AR(2) model had also been selected by ARMAsel. It should be realised that certain details that can be statistically significant for all data are not always significant if half of the data is used. It is quite remarkable that the first 25 lagged product estimates are similar in the upper half of Figure 10.5 that gives the 50 first lags of Figure 3.9. However, there is no reason to conclude in Figure 10.3 that there is something statistically special in the AR model of order 25. The conclusion from Figure 10.3 is that the two models AR(2) and AR(37) deserve special interest because those two are selected by the data with penalties 3 and 2, respectively. The conclusion from the upper part of Figure 10.5 can be only that all details between plus and minus 0.3 are not reliable for the halved sample size. Therefore, applying a rule of thumb, only the first five lags in the lagged product autocorrelation function can be trusted as reliable because they are outside the uncertainty boundaries. This shows once more how the data can speak for themselves with the accuracy plot of Figure 10.3 for estimated time series models.

A special property of those practical fish data is that there is not enough information to decide on the best model order. Automatic selection with ARMAsel, with penalty factor 3 for each parameter, selects the AR(2) model. Figure 10.3 shows that almost all AR models with orders between 2 and 37 would be good candidates. For AR orders greater than 37, the performance in Figure 10.3 is more or less as expected when all higher order parameters are not significant.

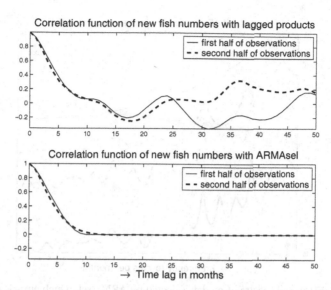

Figure 10.5. Estimation of the autocorrelation function of practical data on the amount of fish in the Pacific Ocean. The number of observations was 445. The estimates are from the first and second halves of the data.

In comparing models, penalty 3 is favourable for a comparison of neighbouring models whereas penalty 2 is better if there is a large difference in the number of parameters. Therefore, the type selection of Section 9.2 uses penalty factor 2 for a mutual comparison of AR, MA, and ARMA models.

To study what happens if not enough data are available to make all details statistically significant, Broersen (2005c) used the estimated AR(37) model of the 445 fish data in Figure 3.8 to generate 10,000 new observations with the AR(37) parameters as the true process. According to Figure 10.3, all AR models between orders 2 and 37 are acceptable candidates for selection from about 450 observations. With much less, AR(2) would be the best and with much more, AR(37) has the smallest ME. From those artificial data, sequences of 100, 500, 1000, and 10,000 observations have been analysed. Estimated spectra are given in Figure 10.6, together with the true AR(37) spectrum. Selected by ARMAsel are the AR orders 2, 15, 37, and 37 for the increasing N of Figure 10.6. AR(37) is almost always selected if enough data are available, sometimes AR(38), and almost never any other model, no matter how many candidates are taken. For $N = 1000$ with AR(37) selected, it is visible that the amplitude at a yearly interval of 0.083 cycles per month is a bit smaller than in the true spectrum, but all true details are present. In a linear plot, almost no details at all would be visible for frequencies above 0.2, and the three peaks between 0.1 and 0.2 would just be visible with amplitudes around 4% of the full scale.

The autocorrelation functions of the models selected are presented in Figure 10.7. Being the inverse Fourier transform of the spectrum, the sum of squared differences of spectrum and autocorrelations are the same. Therefore, all details that were not visible in the linear spectrum are equally invisible in the representation of the autocorrelation functions. In other words, if the spectra of

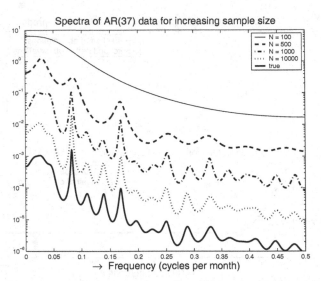

Figure 10.6. Estimated spectra of models selected by ARMAsel, which are AR of orders 2, 6, 37, and 37 for the four sample sizes

Figure 10.6 had been made equal to zero for the frequency range from 0.1 to 0.5 and had been transformed to autocorrelation functions afterward, the new autocorrelations would have been almost identical to those in Fig. 10.7. The removal of all weak details from the spectrum has no significant visual influence on the linear spectrum or on the autocorrelation representation. Differences can be found only in the decimals farther behind the decimal point.

Evaluating the first and the second halves of the data is simple. This rough comparison of the quality of autocorrelations and spectra of time series models with the lagged product estimates and periodograms has been applied to many artificial and real-life data; see Figure 3.9. One would like to have a better approximation of the truth if more data become available. Without using any time series measure or argument, intuitively the method with the closest results for both halves is preferred. In almost all situations, time series with the selected ARMAsel model outperform lagged product estimates.

The poor properties of the periodogram for spectral estimates and of lagged products for autocorrelation estimates are also encountered in simple transfer function estimators. Broersen (1995, 2004) proved that the variance of the popular empirical transfer function estimator (ETFE) is infinite. Therefore, the estimation of the combined properties of two or more signals also has some peculiarities if the signals are random. The input signal before the measurement interval has influence on the output signal during the interval. That influence of transient responses does not become negligible for increasing sample sizes. It is always present for stationary stochastic processes, which cannot be defined as zero before the observation interval. That would be in conflict with the concept of stationarity. Another well-known peculiarity for the joint properties of two random variables is the outcome of the coherency estimate that is identical to one if N coherency points are estimates from N joint observations.

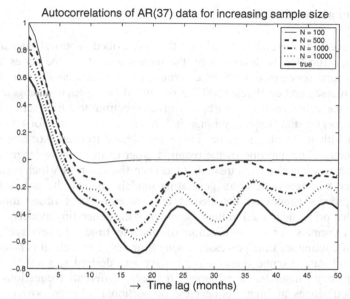

Figure 10.7. Estimated autocorrelations of models selected by ARMAsel

ARMAsel analyses only a single stochastic signal and automatically selects one model from all candidates. It estimates autocorrelations of one signal, not the cross correlation between two signals. That would be a welcome addition to the ARMAsel toolbox, but cross correlations, cross spectra, and transfer functions have no simple and general classes of nested models. The equivalents of AR, MA, and ARMA are not sufficiently rich to describe all possible model types for more signals simultaneously.

Ljung (1987) gives ARMAX structures with an input signal, an output signal, and noise. In principle, the model orders for input, output, and noise can be chosen freely. He also described several other structures, which are not nested either. Therefore, these model types are difficult as candidates for order selection. Automatic selection of the best model type for multivariable signal processing is not yet feasible. Neither is a description of a complete and compact class of linear models available for all possible interactions of two stochastic processes.

Vector autoregressive models are a limited class of models for treating more signals simultaneously. They are nested artificially by taking the same order for all components. That has been done before in ARMA estimation where the unstructured class of ARMA(p,q) models has been replaced by the nested class of ARMA($r,r-1$) models. The influence of this nesting on the achievable model accuracy was small in Table 8.5. De Waele and Broersen (2002, 2003) give details of finite-sample effects on vector autoregressive order selection. The accuracy of vector autoregressive models for unknown cross correlations and cross spectra has not been compared with the accuracy of other possible types of candidate models.

No computer program with the automatic selection of the best model type and model order is available for the joint analysis of more stochastic signals.

10.2 Generation of Data

Data generation is straightforward if the prescribed spectral density or autocovariance function is described by the parameters of a time series model. Otherwise, a time series model can be determined to generate equidistant discrete time data. Broersen and de Waele (2003) give several examples of data generation and describe the transition from an arbitrary given spectrum to a time series model. For discrete spectra, the frequency range is limited naturally, and for continuous time spectra with an infinite frequency range, the highest frequency of interest has to be determined. The question of the required order of the time series model for sufficient accuracy has also been treated. It has been shown under which conditions the difference between finite-order generating models and infinite orders for the generating process cannot be detected if only N equidistant observations are available. That principle is used to limit the order of the generating models.

Special examples for the generation of discrete time data emerge in the simulation of continuous time processes. Using time series models, it is possible to generate equidistant discrete time data that have any desired spectral slope in a limited frequency range. For continuous processes, differential equations allow only specified slopes at high frequencies, proportional to even powers of the frequency. For the generation of discrete time data, spectral slopes with odd or broken powers of the frequency can be approximated with every desired degree of accuracy. Examples are turbulence, where the spectrum is proportional to $f^{-5/3}$ or f^{-7}, or chaotic physical problems with $1/f$ noise, and many other problems with broken or uneven powers as asymptotes for the high frequency range.

As an example, the desired spectrum of $1/f$ noise would increase without limits for f going to zero due to the singularity at $f = 0$. The lowest frequency that can be detected in N equidistant observations is around $1/N$. Hence, the generated spectrum has to be accurate only for frequencies between $1/N$ and half the sampling frequency. The target spectrum to be generated at frequencies lower than $1/N$ is simply extrapolated from the frequencies just above $1/N$. Broersen and de Waele (2003) used a parabola to find a value for $f = 0$ which gives an easily realisable time series model for the whole frequency range of equidistant data.

The model error (ME) of (5.40) has been used by Broersen and de Waele (2003) to limit the order of generating AR models. First of all, the desired spectrum is approximated by a very high-order AR model. This can be approximated with any desired accuracy by taking the AR order high enough. Afterward, the ME between the very high-order AR model and an AR(L) model can be calculated for every value of N and L. For a given value of N, the order L must be such that the ME value is less than one.

Generating stationary data for an ARMA process requires some care with the initialisation. If zeros or any arbitrary values are used as initial values, the generated signals become strictly stationary only after the duration of the impulse response. Unfortunately, the impulse response is only finite for a MA process. It is infinitely long for AR and ARMA processes. Therefore, this primitive method of data generation without care for initial conditions is exact only for MA processes. It is at best an approximation for AR or ARMA processes. A much better method is found by separating the generation into AR and MA parts. Consider the joint probability density function of a finite number of AR observations with a prescribed correlation function. Data can be generated that obey that prescribed AR correlation. A realisation of the ARMA process is obtained by filtering AR data with the MA polynomial.

The joint probability density function of N normally distributed observations X is given by (2.22). That is a general expression that can be applied to AR, MA, or ARMA processes by expressing the Toeplitz matrix R_{xx} in the parameters. The probability density function of N observations can also be written as a conditional product of the last $N - k$ observations, given the first k with (2.25). By using conditional densities, it follows that a joint distribution of N observations can be written as a product of N conditional distributions in (2.26). The probability density function of N observations with (2.25) becomes

$$f(x_1, x_2, \cdots, x_N) = f(x_{p+1}, \cdots, x_N \mid x_1, \cdots, x_p) f(x_1, \cdots, x_p) \qquad (10.9)$$

The first part of the right-hand side for the last $N{-}p$ observations for an AR(p) process with polynomial $A_p(z)$, with normally distributed zero mean innovations ε_n, is given by

$$f(x_{p+1}, \cdots, x_N \mid x_1, \cdots, x_p) = \left(\frac{1}{2\pi\sigma_\varepsilon^2}\right)^{\frac{N-p}{2}} \exp -\left\{\frac{1}{2\sigma_\varepsilon^2} \sum_{n=p+1}^{N} \left[A_p(z)x_n\right]^2\right\} \qquad (10.10)$$

The second part describing the first p observations can also be derived. By using conditional densities in (2.26), it follows that

$$f(x_1, x_2, \cdots, x_p) = \left[\prod_{q=2}^{p} f(x_q \mid x_1, \cdots, x_{q-1})\right] f(x_1) \qquad (10.11)$$

The Levinson-Durbin algorithm (5.24) can be used to evaluate this expression. The polynomial $A^{[K]}(z)$ is made with the parameter vector $\alpha^{[K]}$ of (5.12) without the parameter of order zero. The polynomial $1 - A^{[K]}(z)$ is the best linear predictor of order K or the best linear combination of K previous observations to predict the next observation x_K. Hence, the conditional probability density of x_q, conditional on $q - 1$ previous observations has $[\, 1 - A^{[q-1]}(z)\,]\, x_q$ as expectation with variance s_{q-1}^2. Using this in the conditional expectations (10.11) gives

$$f(x_q \mid x_1, \cdots, x_{q-1}) = \left(\frac{1}{2\pi s_{q-1}^2}\right)^{\frac{1}{2}} \exp -\left\{\frac{1}{2s_{q-1}^2}\left[A^{[q-1]}(z)x_q\right]^2\right\} \qquad (10.12)$$

The recursive variance relation for intermediate AR orders can be expressed with an increasing or decreasing index, which gives

$$\sigma_q^2 = \sigma_\varepsilon^2 \Big/ \prod_{i=q+1}^{p}\left(1-k_i^2\right) = \sigma_x^2 \prod_{i=1}^{q}\left(1-k_i^2\right) \tag{10.13}$$

Now, the conditional density (10.12) becomes

$$f(x_q \mid x_1,\cdots,x_{q-1}) = \frac{\left[\prod_{i=q}^{p}\left(1-k_i^2\right)\right]^{1/2}}{\sigma_\varepsilon\sqrt{2\pi}}\exp-\left\{\frac{1}{2\sigma_\varepsilon^2}\prod_{i=q}^{p}\left(1-k_i^2\right)\left[A^{[q-1]}(z)x_q\right]^2\right\} \tag{10.14}$$

All ingredients for the probability density function $f(x_1, x_2,\ldots, x_N)$ have been given and this derivation is sufficient to be used in generating data for an AR(p) process.

Data generation is strictly separated in MA and AR generation. For ARMA processes, it is essential that the AR part is used first. The AR observations are denoted v_n. The first observation v_1 has only the requirement that it has expectation zero and variance $r(0)$ or σ_v^2. That is found with a random number generator with normal or Gaussian density and the prescribed variance. The second observation v_2 follows as a normally distributed random variable with mean $-a_1^1 v_1$ and variance s_1^2. The third observation v_3 has mean $-a_1^2 v_2 - a_2^2 v_1$ and variance s_2^2. The first p observations are generated in this way. All further observations can simply be generated with

$$v_n + a_1 v_{n-1} + \cdots + a_p v_{n-p} = \varepsilon_n, \ n = p+1,\cdots,N \tag{10.15}$$

This is a filter procedure with a Gaussian, random, white noise signal as input signal and the first p observations as initial conditions. For AR processes, the p initial observations and the $N - p$ filter results with (10.15) together are the N observations.

For MA(q) data, the input signal v_n is a Gaussian white noise ε_n and length $N + q$. For ARMA(p,q) processes, the input to the MA filter will be the output v_n of the AR filter (10.15) of length $N + q$. The data x_n are computed with

$$x_n = v_n + b_1 v_{n-1} + \cdots + b_q v_{n-q} \tag{10.16}$$

The first q data require a negative input index. Therefore, to generate N MA or ARMA observations with this method, the input sequence v_n has to be $N + q$ long. The first q points of the filter output are disregarded.

10.3 Subband Spectral Analysis

Standard time series analysis estimates the power spectral density over the full frequency range until half the sampling frequency. Processing of a time series spectrum in a subband may be useful if observations of a stochastic process are analysed for the presence or multiplicity of spectral peaks. Separating the heartbeat of mother and child is a medical example. Magnetic resonance spectroscopy often analyses absorption spectra where different chemical components produce narrow peaks in a small frequency subband (see Tomczak and Djermoune, 2002). If two close spectral peaks are present, a minimum number of observations is required to observe two separate narrow peaks with sufficient statistical reliability. With fewer data, a model with one single broad peak might be selected.

A high-order autoregressive model or a periodogram will often indicate the separate peaks in power spectral density, together with many other similar details that are not significant. However, order selection among full range models may select a model with a single peak. By using subband order selection, it is sometimes possible to detect the presence of two peaks from the same data by using subband analysis.

Makhoul (1976) described a simple time series subband model estimation method where an AR(p) model is fitted to a subband of a periodogram. This method uses the inverse Fourier transform of a subband of the periodogram as if it were a new autocorrelation function. The AR parameters are found from that function with the Yule-Walker relations. Another approach of de Waele and Broersen (2001) is applicable only to the lowest frequency range from zero to a small fraction of the sampling frequency. The signal is downsampled at a lower sampling rate that automatically reduces the highest frequency in the spectrum, as in Section 9.8.1. If no filtering is applied, the downsampled spectrum becomes aliased, and the autocovariance function remains undistorted. It turned out that a narrow low-frequency peak in the spectrum of experimental wind tunnel data was found only after resampling at a reduced rate. Order selection for the whole frequency range did not select a model with the peak in the wind tunnel spectra, but order selection for only the first 1% of the frequency range showed the statistical significance of the peak.

To evaluate the quality of subband estimation methods, it is necessary to have an objective quantitative measure that applies to part of the frequency range. Error measures have been developed by de Waele and Broersen (2000b) for the lowest part of the frequency range, with and without taking into account the possibility of spectral aliasing. Those measures can be transformed to arbitrary parts of the frequency range. It turned out that the dedicated order selection is especially important for the accuracy of frequency-selective models. Frequency-selective analysis for the lowest frequency subband can easily be used for the analysis of an arbitrary frequency subband. First, the spectral density of the required subband is determined as part of the spectrum of a very high order AR model that is estimated for the whole frequency range. Only the spectrum of the subband is transformed into an equivalent subband autocovariance function. Then, a reduced-statistics ARMA estimator uses the subband autocovariance to estimate a long AR model to compute dedicated subband ARMA and MA models. The reduced-statistics adaptations of order-selection criteria are required for proper selection of the statistical significance of details in a subband. A long AR model of the data is taken as a reference for the fit, instead of the observations.

Makhoul (1976) used the principle of frequency-selective or subband modelling. He started with a raw periodogram. He transformed the frequency band between ω_1 and ω_2 to the interval $0 - \pi$ and mirrored it around π for the complete interval $0 - 2\pi$. Use of the inverse Fourier transform computes an autocovariance-like function that belongs to this truncated spectrum. That autocovariance-like function in turn can be transformed into an AR model for the frequency subband by treating it as a true covariance function and by estimating the parameters with the Yule-Walker relations (5.24).

Instead of using a subband of the periodogram as the starting point, a better possibility has been developed by Broersen and de Waele (2003b). Their approach is completely within the area of time series analysis. It uses the full range spectrum calculated with a very long AR model $C_L(z)$, estimated from the data, as the basis for the frequency subband of interest. The AR model $C_L(z)$ is given the fixed order $L = N/2$. In practice, the value L is often restricted to the maximum value 2000 if N is greater than 4000, but that is not necessary; AR models have been computed with L as high as 100,000. The restriction is for computational purposes, and it is justified because higher order models are seldom selected for practical data. The long AR spectrum of a subband between ω_1 and ω_2 is transformed to the interval $0 - \pi$ by the mapping

$$\omega' = \pi(\omega - \omega_1)/(\omega_2 - \omega_1) \tag{10.17}$$

Afterward, this ω' is mirrored around π and the subband spectrum is sampled. The subband spectrum is a continuous function. It can be sampled with an arbitrary high frequency. That frequency is chosen by realising that the sampled subband spectrum should contain at least L discrete equidistant frequency points between zero and π to make certain that no significant detail is lost in the computations. Further, the subband spectrum is inversely Fourier transformed into an autocovariance-like function. That autocovariance-like function in turn is used in the Yule-Walker relations to calculate a new long AR model $D_L(z)$ for the subband between ω_1 and ω_2, with $0 < \omega' < \pi$. The long AR subband model $D_L(z)$ is the input information for the frequency-selective time series analysis. That can be done automatically with the reduced-statistics estimator, ARMAsel-rs.

Estimated models $D_L(z)$ should represent a **predefined** subband of the true spectrum. It is important that the subband is chosen *a priori*, independent of the data at hand. Under those circumstances, reliable statistical inferences are allowed. If the interesting subband is chosen after a preliminary full range spectral estimation, there is a probability that the chosen subband might contain details that only seem interesting in the single realisation of the process that is evaluated. The statistical reliability of order selection can be improved then by taking 4 or 5 for the penalty factor. In this way, some protection against subjective interference can be provided.

Using only the frequency band between ω_1 and ω_2 from the true spectral density, the subband transform can be used to determine a *true time series model* only for the frequency subband selected.

Simulations have been carried out for an example with two close peaks at frequencies $0.27f_0$ and $0.28f_0$ and white background noise. The peaks are generated as separate AR(2) processes, both with a pole at radius 0.99. The true power in the first peak is 1, the power of the second peak is 0.4, and the power of the additive white noise is 0.01. Figure 10.8 shows the true spectrum and the spectrum of a long

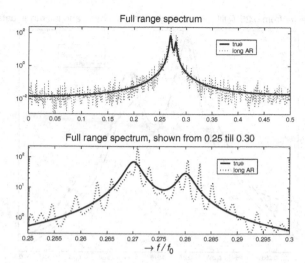

Figure 10.8. True spectrum and AR(500) spectrum of 2000 observations of a process with two peaks at $0.27f_0$ and $0.28f_0$ and a white noise background

AR(500) model, obtained from $N = 2000$ observations. Use of a still higher order AR model would give more details in the spectrum of the long AR, but no significant differences in the final subband spectrum. The order of the long AR model is not critical. The vertical scaling of the spectra is rather arbitrary, such that the surface under the spectrum is normalized to the number of calculated frequencies.

A visual inspection of the long AR spectrum might suggest a two-peak spectrum around $f = 0.275f_0$. However, the same visual inspection might suggest a valley around the frequencies $f = 0.257f_0$ and $f = 0.267f_0$ and a third peak around $f = 0.291f_0$. Hence, visual inspection is not reliable, and it would be a dubious starting point for defining the statistical significance of peaks. Using the same order, 500, that had been used for the complete spectrum and also for the long AR representation of only the subband guarantees that no information is lost in the transition to the subband, even if all estimated spectral details were concentrated in the subband.

Figure 10.9 shows the outcome of order selection. The true process has two peaks; the standard selected ARMAsel spectrum from the 2000 observations shows a single peak that is statistically significant. The range of the subband is 0.1 of the full range, starting at $0.25f_0$. Broersen and de Waele (2003b) showed that the effective number of observations to be used in subband order selection is therefore 200, the same fraction as the subband in the frequency domain. In other words, the degrees of freedom in the frequency range selected determine the effective number of observations. The model selected for the chosen subband of the frequency range has two peaks, whereas over the whole frequency range, the selected model fails to detect the double peak in this realisation.

Investigating a smooth subband of the full range spectrum without peaks gives flat frequency-selective spectra if the order is selected for the best subband model. Peaks are detected in a subband only if they are present. For a few observations, peaks cannot be detected at all. For some more observations, subband analysis

Spectra from 0.25*f_0 till 0.3*f_0, selected from subband and full range data

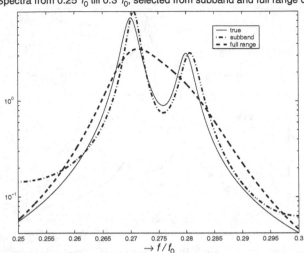

Figure 10.9. True spectrum and the ARMAsel spectrum, selected from the data of Figure 10.8 and the ARMAsel-rs spectrum, selected from the long AR(500) subband model. The selected full range model has a single peak, whereas the selected subband model displays two peaks close to the peaks in the true spectrum. The accuracy of the location of the peaks is better than the height of the peaks in the subband model selected.

shows details that are not significant in a model for the whole frequency range. If N becomes greater and greater, true twin peaks also become statistically significant in spectral models selected for the full freequency range. However, even then, the subband model will often be more accurate for the precise spectral shape.

10.4 Missing Data

The use of time series modeling is also successful if data are missing. Broersen *et al.* (2004, 2004b) give a survey of existing methods for missing data problems and present simulations showing that AR modeling applied to missing data outperforms other known methods in the literature. Broersen and Bos (2004) also use MA and ARMA models as candidates.

In real-life experiments, sensor failure or outliers lead to missing data problems. Weather conditions can disturb the equidistant sampling scheme in hydrologic, meteorologic, satellite, and astronomical observations. The pattern of missing data is important for the analysis. It may be random from sensor failure or outliers or it may have large gaps due to weather conditions.

The treatment of missing data has two different principles:

* reconstruct the missing data
* use only the remaining data.

The first group of methods is based on estimation algorithms that have been developed for contiguous, equidistant data. Two methods that extract equidistant information from missing data will be distinguished in this category. One is some sort of static interpolation between the remaining observations, linear, nearest-neighbour, sample and hold, or splines. Those methods have been studied extensively. The performance depends heavily on the characteristics of the data. It may be accurate for slowly varying low-frequency signals, but no interpolation method gives good results for all types of data. The other method reconstructs the missing data using a model for the covariance structure. This method is derived from the EM algorithm of Little and Rubin (1987) for missing data. The E step finds the conditional expectation of the missing data, given the observed data together with the currently estimated model; the M step computes maximum likelihood estimates for the parameters of the model from all data, observed and reconstructed. Reconstruction methods may be simple and accurate but only for small missing fractions, say, less than 10% missing.

The method of Lomb-Scargle (Lomb, 1976 and Scargle, 1982) uses only the available measured data. It computes Fourier coefficients as the least-squares fit of sinusoids and cosines to the available observations. The Lomb-Scargle spectrum is accurate in detecting spectral peaks. Bos *et al.* (2002) give an example that shows that this method is rather poor in describing slopes in the spectrum.

Rosen and Porat (1989) tried to find some autocovariance estimate for incomplete data and to use that for further analysis. This technique does not always guarantee that the sample covariance estimate is positive-semidefinite. Moreover, sample autocovariances are known to be inefficient estimators for the covariance structure and generally will not produce accurate spectral estimates, not even for equidistant data.

A third idea selects consecutive segments in the observed data and uses a special segment variant of the Burg algorithm of de Waele and Broersen (2000) to compute the parameters of an AR model. It can be successful if only a few parameters have to be computed or if large gaps alternate with longer consecutive segments. Its use is limited to rather small *randomly* missing fractions because only uninterrupted data segments can be used.

The final method fits a time series model directly to the available observations with maximum likelihood estimation. An exact maximum likelihood approach with Kalman filtering has been given by Jones (1980). Broersen *et al.* (2004, 2004b) describe a finite interval approximation. The numerical results of those two approaches are almost identical. The computational effort of the exact approach depends on the total observation length of missing and available data; the computation time for the approximation depends mainly on the available number of data. The computing time is the same if about 10% of the data remains and 90% is missing. If less is missing, the method of Jones (1980) is faster. The finite interval approach can be very much faster if less than 10% remains.

The original Burg algorithm for AR parameter estimation recursively estimates reflection coefficients k_p from equidistant data. The Burg algorithm for segments has been modified by de Waele and Broersen (2000) for simultaneous AR estimation from S segments. First, the functions $f_0^s(n)$ and $b_0^s(n)$ are made equal to the S segments $x^s(n)$, $s = 1,..., S$. The filtering operations in (6.6) and (6.7) are applied to all segments longer than the filter length. A segment can contribute as long as it is longer than the current filter order. The segments with different length consist of the consecutive observations, and the missing data separate the segments. Isolated observations are discarded. The first reflection coefficient k_1 can now be estimated with $p = 1$:

$$k_p = \frac{-2\sum f_{p-1}^s(n)b_{p-1}^s(n-p)}{\sum f_{p-1}^s(n)^2 + \sum b_{p-1}^s(n-p)^2} \tag{10.18}$$

All reflection coefficients have the properties of (6.12) and are guaranteed to be less one in magnitude. New functions, $f_0^s(n)$ and $b_0^s(n)$, called forward and backward residuals, can be computed with (6.8) for all segments that are long enough for stage p:

$$f_p^s(n) = f_{p-1}^s(n) + k_p b_{p-1}^s(n-p)$$
$$b_p^s(n) = b_{p-1}^s(n) + k_p f_{p-1}^s(n+p) \tag{10.19}$$

This means that only segments that already contributed at stage $p - 1$ can appear in the residuals (10.19). These are used to estimate a new reflection coefficient k_{p+1} with (10.18). Only segments longer than $p + 1$ points contribute to the estimate k_{p+1} of order $p+1$. Therefore, the number of active segments decreases with the AR order. The total number of contributing residuals should be at least 10 for sufficient statistical reliability. Parameters of AR models of all intermediate orders can be calculated by applying the Levinson-Durbin formulas (5.24) to the series of reflection coefficients obtained with (10.18). This method can give good results if the missing fraction is small. Otherwise, too few segments are long enough to allow for model orders higher than AR(1) or AR(2). In those cases, it becomes interesting to develop an algorithm that also uses the interrupted data.

The general theory of Gaussian variables gives the possibility of constructing ARMA(p,q) models from randomly available observations. Suppose that the scalar random variable U with mean μ_U and variance σ_U^2 has a joint multivariate normal distribution with a k-dimensional random variable Y with mean $\mu_Y \in R^k$ and covariance matrix $R_{YY} \in R^{k \times k}$. The cross covariance vector is denoted $R_{UY} \in R^{1 \times k}$. Then, the conditional density of U for a given vector Y is the normal distribution

$$f_{U|Y}(u \mid y) = \left(\frac{1}{2\pi\sigma_{U|Y}^2}\right)^{1/2} \exp\left\{\left[-\frac{(u - \mu_{U|Y})^2}{2\sigma_{U|Y}^2}\right]\right\} \tag{10.20}$$

where the conditional mean and variance are given by Anderson and Moore (1979):

$$\mu_{U|Y} = \mu_U + R_{UY}R_{YY}^{-1}(y - \mu_Y)$$

$$\sigma_{U|Y}^2 = \sigma_U^2 - R_{UY}R_{YY}^{-1}R_{YU}$$

$$s_{U|Y}^2 = \frac{\sigma_{U|Y}^2}{\sigma_U^2} \tag{10.21}$$

This conditional density can be used to find the ARMA parameters in a missing data problem. The mean in (10.21) can be used to predict any of the available observations U, given the $N-1$ available other observations that are together in Y.

The joint density $f_X(x)$ for arbitrary distributions can be written as a product of the pdf of the first observation with conditional density functions (2.26). According to this recursive formulation, it is sufficient to use only previous observations for the prediction in (10.21). Consider a realisation of a random vector of N observations X, consisting of $x_{t(1)},...,x_{t(N)}$, where $t(i)$ is a multiple kT of the sampling time T. They are the N available samples of a stationary stochastic process with mean value zero and variance σ_X^2. The conditional density of (10.20) will be used to predict the next available observation $x_{t(i+1)}$ from all available previous observations $x_{t(i)},..., x_{t(1)}$, which are in general not contiguous if data are missing. The probability density for the first observation $x_{t(1)}$, without the possibility of predicting, is given by

$$f_{X_{t(1)}}(x_{t(1)}) = N(x_{t(1)}, 0, \sigma_X^2) = \left(\frac{1}{2\pi\sigma_X^2}\right)^{1/2} \exp\left(-\frac{x_{t(1)}^2}{2\sigma_X^2}\right) \tag{10.22}$$

where the normal distribution of a random variable with mean μ and variance σ^2 is given by (2.18). The conditional density of $x_{t(i+1)}$ for given $x_{t(1)}, x_{t(2)},...,x_{t(i)}$ follows from (10.20) and (10.21) by substituting $x_{t(i+1)}$ for u and all available previous observations $(x_{t(1)}, x_{t(2)},...,x_{t(i)})$ for y. All elements of the covariance matrices required in (10.20) are determined symbolically by (4.62) for an ARMA(p,q) process. The product probability density function of the complete vector X can now be written as

$$f_X(x) = \left(2\pi\sigma_X^2\right)^{-N/2} \prod_{j=1}^N \left(s_{t(j)|t(j-1),...,t(1)}^2\right)^{-1/2}$$

$$\times \exp\left[-\frac{1}{2}\sum_{j=1}^N \frac{\left(x_{t(j)} - \hat{x}_{t(j)|t(j-1),...,t(1)}\right)^2}{\sigma_X^2 s_{t(j)|t(j-1),...,t(1)}^2}\right] \tag{10.23}$$

The values for the predictions $\hat{x}_{t(j)|t(j-1),...,t(1)}$ and the conditional variances $s_{t(j)|t(j-1),...,t(1)}^2$ are defined implicitly by the equations in (10.21), with $s_{t(1)}^2 = 1$.

The ARMA(p,q) parameter vector $\hat{\theta}$ with elements $\hat{a}_1,\cdots, \hat{a}_p$ and $\hat{b}_1,\cdots, \hat{b}_q$ is estimated in the case of missing data by minimising

$$L(X;\hat{\theta},\hat{\sigma}_x^2) = -2\log f_X(x) - N\log 2\pi \qquad (10.24)$$

with respect to $\hat{\theta}$ and to $\hat{\sigma}_x^2$, which becomes

$$L(X;\hat{\theta},\hat{\sigma}_x^2) = N\log\hat{\sigma}_x^2 + \sum_{j=1}^{N}\log s_{t(j)|t(j-1),\dots,t(1)}^2$$

$$+ \sum_{j=1}^{N}\frac{\left(x_{t(j)} - \hat{x}_{t(j)|t(j-1),\dots,t(1)}\right)^2}{\hat{\sigma}_x^2\, s_{t(j)|t(j-1),\dots,t(1)}^2} \qquad (10.25)$$

By using the maximum likelihood estimate for $\hat{\sigma}_x^2$,

$$\hat{\sigma}_x^2 = \frac{1}{N}\sum_{j=1}^{N}\left(x_{t(j)} - \hat{x}_{t(j)|t(j-1),\dots,t(1)}\right)^2 / s_{t(j)|t(j-1),\dots,t(1)}^2 \qquad (10.26)$$

it follows that substitution in (10.25) yields a constant N for the last term, and the first two terms of (10.25) are minimised together.

The algorithm with (10.25) for the exact likelihood would still require too much computing time in practice, even for moderate N. The number of contributions in the predictor $\mu_{U|Y}$ in (10.20) would become $N-1$ for the last observation and a $N-1 \times N-1$ matrix R_{YY} has to be inverted in (10.21) to determine $\sigma_{U|Y}^2$.

Generally, the prediction accuracy improves most with the nearest previous observations. Observations further away have much less influence on the prediction accuracy if the correlation function dies out fast enough, and they contain less useful information for the parameters to be estimated. A compromise in the algorithm limits the maximum interval in which previous observations are used to predict with an ARMA(p,q) model to a finite time interval $2(p+q)T/\gamma$ in missing data. The number γ denotes the remaining fraction of the data, with $0 < \gamma < 1$. Only the previous K_i observations are used, where K_i is the largest integer such that (10.27) holds:

$$t(i) - t(i-K_i) \le 2(p+q)T/\gamma \qquad (10.27)$$

This algorithm is called **ARMAFIL, ARMA,** for a Finite Interval Likelihood. The name is ARFIL if it is used for AR models. The number K_i of observations within the interval (10.27) that are used to predict $x_{t(i)}$ varies with the index of the predicted observation; the average number is $2p + 2q$. Using a larger interval in (10.27) would lead to longer computation times, but it would not generally influence the accuracy of the estimated spectrum. However, taking a much smaller interval will reduce the accuracy considerably. Only if estimated poles are very close to the unit circle, say, closer than $1/N$, may some useful information be lost because the autocorrelation function is still significant outside the finite interval $2(p+q)T/\gamma$. Thus, losing this information might give a bad approximation for the likelihood. However, it does not necessarily mean that the maximum of the finite interval likelihood is found at really different parameter values. The average number of observations in the interval turns out to be sufficient to yield good spectral estimates in practical examples.

To improve numerical robustness, the ARMAFIL algorithm uses unconstrained optimisation of tan $(\pi/2*k_i)$ for increasing orders p and q for both the AR and the MA polynomials. This guarantees that the estimated k_i is always in the range $-1 < k_i < 1$. Hence, all models computed by nonlinear numerical optimisation routines are stationary and invertible. The usual Yule-Walker or Burg algorithms keep previous reflection coefficients as constants in computing one new reflection coefficient. In contrast with those consecutive algorithms, however, all k_i are optimised afresh and simultaneously in ARMAFIL for every model order p and/or q.

As possible starting values for the nonlinear optimisation of the ARMA(p,q) model, the reflection coefficients of the ARMA($p-1, q-1$) model have been considered with additional zeros as a start for the new k_p and k_q. Broersen et al. (2004, 2004b) showed that those recursive starting values are successful in AR estimation. However, this method failed completely for MA and ARMA estimation.

ARMA models have a second problem in the maximum likelihood method. If the true process has a pole and a zero both of which have a rather small radius, they are almost canceling. The likelihood will mostly converge to an almost canceling pair of a pole and a zero, but their location is rather arbitrary in the complex plane. This will hinder the convergence. This problem also exists for uninterrupted data, but it is still more annoying if many data are missing.

The best practical solution that has been found for the starting values of MA(q) or ARMA(p,q) models is inspired by the reduced-statistics estimator for uninterrupted data. That algorithm can estimate MA and ARMA models using only a limited number of estimated AR parameters, instead of using the data themselves. The MA(q) or ARMA(p,q) model estimated from the AR parameters has been used as the starting point for a nonlinear search for that model. The model quality of the starting point has also been determined to see if and how much the nonlinear estimation with iterative minimisation of the likelihood function can improve it. In most cases, the reduced-statistics solution for the MA and ARMA models obtained from a long AR model was better than the result from the minimisation of the likelihood; see Broersen and Bos (2004).

Having obtained many AR models and also a number of MA and ARMA models, it is essential to develop a selection criterion to choose the best among the candidates. Order selection for models estimated by likelihood minimisation can be performed with a generalized information criterion (GIC), defined as:

$$GIC(p+q,\alpha) = L(X;\underline{\hat{a}}_p,\underline{\hat{b}}_q,\hat{\sigma}_x^2) + \alpha(p+q) \tag{10.28}$$

The best value for the penalty α has been investigated in the missing data case; $\alpha = 2$ is the famous AIC. Broersen and Bos (2004b) used simulations to conclude that the best choices for AR selection are between 3 and 5, 3 for a few missing data, 5 for less than 25% remaining and 4 if around 50% is missing. For MA or ARMA selection, $\alpha = 3$ is a good choice. In a theoretical missing data derivation, Cavanaugh and Shumway (1998) also showed that the best penalty factor for order-selection criteria will depend on the missing fraction.

Some simulation results will be presented here. N/γ equidistant contiguous observations are generated with an ARMA(3,2) process with the AR and MA parameter vectors

$$[1, -2.2053, 1.7265, -0.5018] \text{ and } [1, 0.7153, 0.1282].$$

Plotted on a double logarithmic scale, this process resembles a turbulence spectrum with constant slopes of $f^{-5/3}$ and f^{-7} in the power spectral density.

If only a few percent is missing, say, less than 25%, the algorithms described above give very good results, comparable to the accuracy obtained when no data are missing. The real possibilities of time series models in missing data problems are shown in examples where many data are missing. The simulations use γ equal to 0.01 and 0.1. Only 1% remaining means that 99% of the data is lost at random instants. This is realised by randomly discarding 99% of the $100N$ contiguous simulated data. If the distribution of the remaining times is a Poisson distribution, $\gamma\%$ remaining means also that also γN consecutive pairs of observations are found in the average, and γN couples with a gap of 1, 2, ... and every other gap. Therefore, γN can be called the effective number of observations. A couple of AR parameters can be estimated if at least 5 or 10 effective observations are available.

First, simulations have been carried out with $N = 100$ and variable values of γ. For γ greater than 0.75, the results were about the same as if no data were missing.

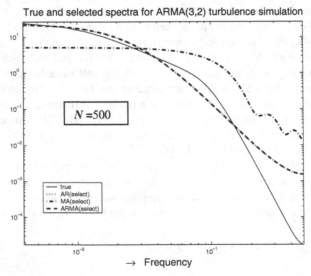

Figure 10.10. The true spectrum and the selected AR, MA, and ARMA spectra for a problem with 99% of the data missing. The MA and ARMA models are computed with ARMAsel-rs. The number of remaining observations is 500 of the ARMA(3,2) process with $\gamma = 0.01$. Selected models for the three types are AR(2), MA(3), and ARMA(2,1) which is equal to AR(2).

The smallest value of γ for which a reasonable spectral estimate has been found was $\gamma = 0.05$ for $N = 100$, with $\gamma N = 5$. The estimated spectra densities were almost identical with that in Figure 10.10, which have been estimated from the same *effective* number of observations. Taking $\gamma = 0.01$ leaves one effective observation for $N = 100$. That is not enough for useful parameter estimation. The ARMAsel program for missing data selected the white noise model in this case.

With γN equal to 5 for $N = 500$ in Figure 10.10, only two AR parameters could be estimated. MA models are not suitable for this type of spectra, and estimated MA models give a poor fit for all sample sizes. They also are inaccurate if no data are missing. AR and ARMA models give the same spectrum here, because the reduced-statistics algorithm can use only the AR(2) model as a basis. Then, the best ARMA(2,1) model fitting to an AR(2) model will have the MA parameter estimated as zero. For only five effective observations, the true spectral shape is already somewhat visible in the estimated and selected AR(2) model in Figure 10.10.

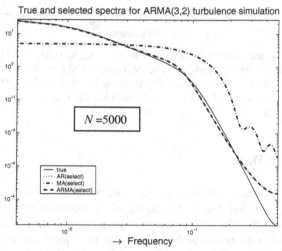

Figure 10.11. The true spectrum and the selected AR, MA, and ARMA spectra for a problem with 99% of the data missing. The MA and ARMA models are computed with ARMAsel-rs. The number of remaining observations is 5000 of the ARMA(3,2) process with $\gamma = 0.01$. Selected are AR(3), MA(5), and ARMA(3,2), which is equal to AR(3).

Figure 10.11 gives the result for $\gamma N = 50$ effective observations. The AR(3) model is selected here and is rather accurate, at least for 99% missing data.

Figure 10.12 has γN equal to 500 with 10% remaining. Selected are the AR(4) model and the MA(8) model for the AR and MA types, respectively. The most accurate ARMA(3,2) model is almost equal to AR(4). Those models are very close here, within the line width. Obviously, turbulence-like spectra can be represented by a few AR parameters and well-fitting AR and ARMA models can be estimated. The ARMAsel algorithm for missing data performs well for all processes that can be described by low-order AR models. That category includes many-real life data, where low-order AR models are very often at least a reasonable approximation.

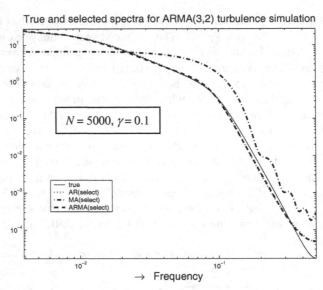

Figure 10.12. The true spectrum and the selected AR, MA, and ARMA spectra for a problem with 90% of the data missing. The MA and ARMA models are computed with ARMAsel-rs. The number of remaining observations is 5000 of the ARMA(3,2) process with $\gamma = 0.1$. Selected are AR(4), MA(8), and ARMA(3,2). The most accurate ARMA(3,2) model is almost equal to AR(4).

10.5 Irregular Data

Meteorological data or turbulence data obtained by laser-Doppler anemometry are often irregularly sampled due to the nature of the observation system. This may have the advantage that the highest frequency that can be estimated is higher than half the mean data rate, which is the upper limit for equidistant observations. The Lomb-Scargle estimator of Lomb (1976) and Scargle (1982) is a method that fits sinusoids to irregular data. It is equal to the periodogram if the data are equidistant. As for equidistant missing data problems, the Lomb-Scargle properties are not favourable for irregular data, unless for periodic signals with very little noise.

Martin (1999) applied autoregressive modeling to irregularly sampled data using a dedicated method. It was particularly good in extracting sinusoids from noise in short data sets. Söderström and Mossberg (2000) evaluated the performance of methods for identifying continuous-time autoregressive processes, which replace the differentiation operator by different approximations. Larsson and Söderström (2002) apply this idea to randomly sampled autoregressive data. They report promising results for low-order processes. Lahalle *et al.* (2004) estimate continuous- time ARMA models. Unfortunately, their method requires explicit use of a model for irregular sampling instants. The precise shape of that distribution is very important for the result, but it is almost impossible to establish it from practical data.

No generally satisfactory spectral estimator for irregular data has been defined yet. Continuous time series models can be estimated for irregular data, and they are the only possible candidates for obtaining the Cramér-Rao lower boundary, because the true process for irregular data is a continuous-time process. Jones (1981) has formulated the maximum likelihood estimator for irregular observations. However, Jones (1984) also found that the likelihood has several local maxima and the optimisation requires extremely good initial estimates. Broersen and Bos (2005) used the method of Jones to obtain maximum likelihood estimates for irregular data. If simulations started with the true process parameters as initial conditions, that was sometimes, but not always, good enough to converge to the global maximum of the likelihood. However, sometimes even those perfect and nonrealisable starting values were not capable of letting the likelihood converge to an acceptable model. So far, no practical maximum likelihood method for irregular data has solved all numerical problems, and certainly no satisfactory realisable initial conditions can be given. As an example, it has been verified in simulations that taking the estimated $AR(p-1)$ model together with an additional zero for order p as starting values for $AR(p)$ estimation does not always converge to acceptable $AR(p)$ models. The model with the maximum value of the likelihood might not in all cases be accurate and many good models have significantly lower numerical values of the likelihood. Martin (1999) suggests that the exact likelihood is sensitive to round-off errors. Broersen and Bos (2005) calculated the likelihood as a function of true model parameters, multiplied by a constant factor. Only the likelihood for a single pole was smooth. Two poles already gave a number of sharp peaks in the likelihood, and three or more poles gave a very rough surface of the likelihood. The scene is full of local minima, and the optimisation cannot find the global minimum, unless it starts very close to it.

Because the maximum likelihood principle does not lead to a practical estimator for continuous processes with irregular sampling instants, different estimators have to be developed from other principles. Approaches starting from continuous modeling have been mentioned. Broersen and Bos (2005) and Broersen (2005a) apply the best missing data algorithm to irregular data. It would be the ultimate goal to develop an estimator that can be applied to small and large data sets that requires no user interaction, that does not use any assumption about the distribution of the observation instants, and that estimates spectra with an accuracy close to the Cramér-Rao lower bound. An algorithm that can be used for arbitrary sample sizes will generally improve its variance performance with $1/N$. Hence, it will become better and better for increasing sample sizes. The missing data ARMAsel algorithm is an example that operates for any sample size. Hence, it might be a good starting point for an algorithm for irregular data. If the sample size and hence the effective number of observations are too small, it just selects the white noise model.

If an algorithm can be used only for very large sample sizes, it will be difficult to establish the accuracy as a function of sample size. The bias will often not become smaller for increasing N because the bias may be independent of the sample size and depend only on the slot width, the resampling frequency, or any other characteristic of the estimation method. Adrian and Yao (1987) showed that this applies to the sample and hold reconstruction of data. Britz and Antonia (1996) compared interpolation methods and slotted autocorrelation function estimation. They concluded that no method is able to compensate satisfactorily for a data rate that is too low. De Waele and Broersen (1999) showed that resampling after sample and hold reconstruction is much better than slotting, if the data rate is high. The bias of slotting remains important at low resampling rates. Benedict *et al.* (2000) give an extensive survey of many techniques for irregularly sampled data. Various slotted estimates of the autocovariance function are compared using postprocessing algorithms. No serious problems are encountered if the highest frequency of interest is less than 20% of the mean data rate. Then, the resampling bias is rather small and many methods give acceptable results. Problems arise when higher frequencies are of interest.

The challenge is to develop spectral estimation methods that can be used for the analysis of frequencies above the mean data rate. Slotting methods estimate an equidistant autocovariance function from irregularly sampled data. It is a lagged product estimator that gives a contribution for every two observations that have a distance within a certain slot width. For equidistant observations, the unbiased estimator (3.28) is a similar estimator. The problem with contiguous equidistant observations was that this specific estimator lacks the positive-definite property. The improved equidistant solution with the biased lagged product estimator (3.30) has no equivalent for irregular data. Slotting algorithms have been refined with local normalization and a variable window by Tummers and Passchier (1996, 2001). Local normalization reduces the variance of the estimated autocovariance function, and a variable window is necessary to get a positive spectral estimate for each frequency. They reported very good results for sample sizes of 200,000 for a simulation with an AR(2) process until $N = 1,000,000$ for a benchmark example with two spectral peaks. The examples show that the accuracy is excellent in that part of the frequency range where the power in the spectrum is concentrated and becomes lower in the weak parts of the spectrum. A large width of the variable spectral window is required there to obtain positive spectral estimates. Unfortunately, the algorithm may fail if it is applied to much smaller data sets. If an estimator for an autocorrelation function is not positive-definite, negative spectral estimates will be found for unfavourable spectral windows. It is possible to develop one single variable window algorithm for many different signals, all of which have constant and steep spectral slopes at higher frequencies. Hence, a single algorithm is suitable for large data sets of different types of decaying spectra. However, it will not be possible to detect peaks or other local details in the higher frequency range due to the width of the variable window there. Moreover, the method cannot be used for much smaller data sets. The choice of the variable window requires skill as well as (*a priori*) knowledge of the spectrum. Applying a wrong choice for the variable window will result in negative spectral estimates at some frequencies because the slotted autocorrelation estimate cannot be made positive-definite. Applying a different choice for the variable window results in a different spectral estimate from the same data. Theoretical considerations cannot give a unique preference for the variable window. Hence, it must be accepted that the experimenter has a strong influence on the estimated spectrum. The slotting method will always have a bias in the autocorrelation function because the arbitrary time intervals between observations are replaced by a multiple of the slot width.

Two types of slotted autocorrelation functions have been used as estimators. The first is the direct slotted estimate that is not positive-definite. Therefore, it does not have the properties of an autocorrelation function. Moreover, small spectral details and spectral slopes are not visible in that slotted autocorrelation. A proper measure like the cepstrum of (5.56) cannot be determined. The second is based on the Fourier transform of the direct slotted estimate. It uses a variable window over the spectrum to obtain a positive spectrum at all frequencies, and afterward the inverse Fourier transform of the spectrum is defined as the autocorrelation estimate. This has the advantage that the autocorrelation and the spectrum make a valid Fourier transform pair, which is a theoretical requirement. The disadvantage is that it is not possible to establish a variable window that is independent of the data and can be used for moderate sample sizes.

Van Maanen et al. (1999) introduced fuzzy slotting. This distributes a correlation product over two adjacent time slots and produces a smoother auto-covariance function with a smaller bias because the true irregular time differences are shifted less. They also report a reduction in the variance of the estimated autocovariance function. Therefore, fuzzy slotting improves the slotting performance for large data sets. Fuzzy slotting can be combined with the local normalization and the variable windowing technique. However, autocovariance functions estimated by the fuzzy slotting technique are still not guaranteed to be positive-definite. This improvement results in spectra that can become negative at a percentage of the frequencies where the power is weak. That will always happen if the algorithm is applied to smaller data sets. All methods that improve the performance of slotting autocovariance methods require very large data sets. Even then, examples can be given where an algorithm would fail. However, those counterexamples do not occur in turbulence spectra, meteorologic spectra, or spectra of other physical properties. For instance, mirroring a true spectrum in the frequency range replaces the spectrum at frequency f by the spectrum at $0.5f_0 - f$. This will still give a valid true spectrum that is positive everywhere, like the original spectrum was before mirroring. It is not realistic for any physical signal but feasible with simulated data. It would give strange results for a slotting algorithm with a variable window that was dedicated to the spectral shape before mirroring. Slotting algorithms with variable windows will not produce valid spectra for data of both the true spectral shape and the mirrored shape. This theoretical argument shows that all slotting algorithms will eventually fail for small data sets but may be successful if enough data are available. Applications have shown good results for very large data sets. Even then, a priori knowledge of spectral shape is necessary, and subjective experimental choices have to be made to produce successful spectral estimates.

A special method of van Maanen and Oldenziel (1998) uses a curvefit to the autocorrelation function. That method will be useful to extract information from the data if already very much is a priori known and if that knowledge is completely true for the given data. If a spectrum is known to be flat up to a certain frequency and further to have a slope of f^{-7} for higher frequencies, it is possible to develop a useful estimator for the cutoff frequency. But neither the flat part nor the slope result from the estimation procedure; only the cutoff frequency is estimated. That frequency has a meaning only if the assumptions are exactly true. The method of van Maanen and Oldenziel (1998) uses a flexible shape of the autocovariance function. That shape has been derived from a prototype turbulence spectrum. Although many different autocorrelation functions can be generated, many more autocorrelation functions do not belong to the class. If the class contains *exactly* the true shape, estimation of a few parameters can give very accurate spectral estimates. If it is merely an approximation of the true shape of the autocorrelation, it is not possible to give a theoretical analysis of the accuracy of the method. If noise is present in the measured data, the method can estimate only a turbulence model for the noise. The choice of this parametric model is a strong limitation, but it can be a good alternative if all methods fail that give the data more influence on the result. However, it should be realised that the class of candidate autocorrelation functions is not complete in any mathematical sense.

Some theoretical problems exist for the curve-fit method applied to an estimated autocorrelation function. It uses a weighted, squared sum of the difference between estimated and model autocorrelations. A much better autocorrelation measure would be based on the cepstrum, defined in (5.56). Absolute measures in the spectrum or in the autocorrelation give hardly any difference in Figure 3.4, whereas logarithmic differences are very important. If the spectrum at some frequency is less than 1% of the highest spectral value, errors in the estimate of the spectrum at that frequency have almost no influence on the sum of squared differences. This means that all details in the estimated curve-fit spectrum that are a factor of 100 or more smaller than the low-frequency level have no influence on the spectral estimate or on the estimated parameters. The result of the curvefit is completely independent of the true low-power parts of the spectrum of the data. The spectral estimate obtained with the curve-fit method for those low-power frequencies is completely determined by the estimation method and by the class of autocorrelation candidates and is completely independent of the data. Furthermore, the curve-fit method requires weighting factors that depend on the correlation coefficient for a proper answer. The weight factors of van Maanen and Oldenziel (1998) have a value between 1 and 200,000, and they have a strong but unknown influence on the resulting spectrum. Unfortunately, they cannot be derived from a theoretical concept.

Resampling techniques reconstruct a signal at equal time intervals. After resampling, the equidistant data can be analysed using the periodogram or time series models. Spectral estimates at higher frequencies will be severely biased. Adrian and Yao (1987) described sample and hold reconstruction as low-pass filtering followed by adding noise. These effects can in theory be eliminated using the refined sample and hold estimator of Nobach et al. (1998). The same principle has been used by Simon and Fitzpatrick (2004). The method explicitly uses the Poisson distribution of observation instants to eliminate bias. If that distribution is not exactly true, the method fails. In practice, all spectral details that are smaller than the bias are lost. They are not visible in the rough estimated spectrum and cannot be reconstructed by the refined estimator. The accuracy of the reconstruction method is amazingly good if it is applied to the theoretically distorted spectrum. However, the quality of the same method is rather disappointing if it is applied to measured spectra, unless the sample size is very large. Nearest-neighbour resampling has similar bias and noise characteristics (de

Waele and Broersen, 2000), but this reconstruction method gives many more problems in undoing the bias than sample and hold. The resampled spectra are strongly biased for frequencies higher than about 20% of the mean data rate. The noise and filtering effects of equidistant resampling set limits to the achievable accuracy of resampling methods. This precludes the accurate estimation of spectra at higher frequencies where the inevitable resampling noise exceeds all small details and hides spectral slopes. The bias of resampling for a given data rate is independent of the sample size (Adrian and Yao, 1987). Details below the level of the bias can never be estimated by those methods, no matter how many data are available. Moreover, any deviation from the assumed distribution of the irregular observation instants gives errors in the reconstruction. In turbulent flow, this certainly happens in bubbly liquid flow.

Bos *et al.* (2002) introduced a new idea for irregular data with time series analysis. Their estimator can be perceived as searching for uninterrupted sequences of data that are almost equidistant. The selected sequences of different lengths can be analysed with a slotted irregular version of the Burg (1967) algorithm for segments. The slotted nearest-neighbour Burg method uses an equidistantly resampled signal with many empty places where no original observation fell inside a slot. It has been demonstrated that the bias of *slotted* resampling is very much smaller than the bias of resampling without slotting. The reason is that a single original irregular observation can never appear at multiple resampled time instants. A disadvantage of this slotted resampling method is that still very large data sets are required to obtain some uninterrupted sequences of sufficient length for the irregular Burg algorithm. It operates well for sample sizes where slotted estimation of the autocorrelation, variable windows and refined reconstruction can also be successful.

There is an obvious transition from uninterrupted sequences of data that are almost equidistant to interrupted sequences of data that are almost equidistant. That in turn resembles an equidistant missing data problem. It turned out that the non-linear maximum likelihood missing data algorithm of Jones (1980), treated in Section 10.4, could also give a better solution for irregular data, if much less data are available. Whereas the slotted Burg method required about 200,000 irregular observations, the quasi-maximum likelihood method sometimes already converges to an accurate spectral estimate using less than 2000 observations.

The variance of the spectra of useful methods in spectral estimation becomes smaller if more data are available. Often, the bias is independent of sample size, as in sample and hold resampling. The variance is generally inversely proportional to sample size. Therefore, most existing methods may finally converge to the biased spectral result if enough data are available. The variations due to the estimation variance will become negligible; only the bias error remains. Only then, refined estimators (Nobach *et al.*, 1998; Simon and Fitzpatrick, 2004) can be applied successfully. Therefore, bias reduction algorithms require very large data sets. The purpose here is to develop a spectral estimator that can be used in small and in large data sets. It will be derived from the ARMAsel algorithm for missing data. Therefore, an irregular sampling scheme of the observations will be transformed into a regular time scheme on a fixed time grid with missing data. A resampling scheme will be developed where the resampling time and the slot width are not directly connected. The resampling time determines the highest frequency in the spectral density, and the slot width determines the bias that is due to moving the irregular times to a regular grid within the slot. The smaller the slot width, the higher the missing fraction because every empty slot represents a missing observation in the regular resampled signal.

10.5.1. Multishift, Slotted, Nearest-neighbour Resampling

Analysis of resampling methods shows that a common and important problem is the multiple use of a single irregular observation for more resampled data points. This immediately creates a bias term in the estimated covariance function because the autocovariance $R(0)$ leaks to estimated nonzero autocovariance lags. The analysis of Adrian and Yao (1987) shows that both the autocovariance and the spectrum suffer from bias in sample and hold resampling. De Waele and Broersen (2000b) evaluate nearest-neighbour (NN) resampling with the same problems. Bias is caused by

- the shift of irregular time intervals to a fixed grid
- multiple use of the same irregular observation.

Multiple use causes a correlation in the resampled signal, and it gives a coloured spectral estimate, even if the true irregular process were white noise. Multiple use will be eliminated in *slotted* NN resampling.

The signal $x(t)$ is measured at N irregular time instants $t_1, \ldots t_N$. The average distance T_0 between samples is given by $T_0 = (t_N - t_1) / (N - 1) = 1/f_0$, where f_0 denotes the mean data rate. The signal is resampled on a grid at kN equidistant time instants at grid distance $T_r = T_0 / k$ (for simplicity in notation, k or $1/k$ is limited to integral numbers). The resampled signal exists only for $t = nT_r$ with n integers. Therefore, the resampled signal is equidistant with or without missing data. The spectrum can be calculated up to frequency $k f_0 / 2$. The usual nearest-neighbour resampling substitutes the closest irregular observation $x(t_i)$ at all grid points $t = nT_r$ with

$$|t_{i-1} - nT_r| > |t_i - nT_r| \tag{10.29}$$

$$|t_{i+1} - nT_r| > |t_i - nT_r|$$

The uninterrupted resampled signal contains $k (N - 1) + 1$ equidistant observations. If k is greater than 1, that means that many of the original N irregular observations will be used for more resampled observations.

Slotted nearest-neighbour resampling accepts only a resampled observation at $t = nT_r$ if there is an irregular observation $x(t_i)$ with t_i within the time slot w

$$nT_r - 0.5w < t_i \leq nT_r + 0.5w \tag{10.30}$$

If there is more than one irregular observation within a slot, the one closest to nT_r is selected for resampling; if there is no observation within the slot, the resampled signal at nT_r is left empty. For small T_r and w equal to T_r, the number N_0 of nonempty resampled grid points nT_r becomes close to N because almost every irregular time point falls into another time slot. For larger values of T_r, hence with $k < 1$, more irregular observations may fall within one slot, and only the one closest to the grid point survives in the slotted NN resampled signal.

Taking $w = T_r / M$, with integer M in (10.30) gives disjunct intervals where some irregular times t_i are not within any slot of (10.30). Many observations can be lost completely in that resampling operation. Therefore, multi-shift, slotted, NN resampling is introduced, where M different equidistant missing data signals are extracted from one irregular data set:

$$nT_r+mw-0.5w < t_i \leq nT_r+mw+0.5w, \quad m=0, 1,..., M\text{-}1 \qquad (10.31)$$

Now, all slots of width w are connected in time, and all irregular observations fall within a slot. The number of possible grid points is $(N-1)*M*T_0 / T_r + 1$. Hence, the remaining fraction γ is given approximately by $1/Mk$. Experience with missing data problems of Broersen $et\ al.$ (2004b) and also in the previous section shows that time series models can be easily estimated for $\gamma > 0.1$. It may become somewhat difficult if γ is less than 0.01, unless the remaining number of observations is very large. That has been demonstrated in Figures 10.10–10.12. This limits the useful range of resampling time and slot width for a given number of observations.

10.5.2 ARMAsel for Irregular Data

Inputs for the algorithm are the M equidistant missing data sequences or segments obtained with the multishift, slotted, nearest-neighbour algorithm of (10.31). All segments are derived from the original observations in the same irregularly sampled time interval. In principle, the data in the different segments are correlated and not independent. However, the most influential parts of each segment are found at those dense time intervals where only few data are missing. Generally, the dense time intervals for the various segments are at different locations. Hence, the assumption that the segments are more or less independent is justified. The "likelihood" is computed separately for each segment that is treated as a missing data signal. The outcomes for the M segments are added afterward in the minimisation procedure. Therefore, not all contributions to the "true likelihood" are taken into account because some near observations are in different segments. Using all M segments, each with about N/M observations, gives much better accuracy than using only one segment. However, the method with resampled data will never be an exact maximum likelihood algorithm, not even approximately, because the time axis has been changed and not all near observations are used. Jones (1980) computed the likelihood of a missing data problem exactly by relating observations to all previous observations that are present in one single segment. In the irregular case, the data are distributed over different segments if M is greater than one.

All elements for an automatic ARMAsel algorithm for irregular data can be copied from the algorithm that has been developed for missing data by Broersen et

al. (2004, 2004b). Only the creation of equidistant segments with (10.31) had to be added by Broersen and Bos (2005).

- The "likelihood" for AR models is computed with the method of Jones (1980) or with ARfil (Broersen *et al.*, 2004b), depending on whether γ is greater or smaller than 0.15, respectively.
- The tangent of $\pi/2$ times the AR reflection coefficients is used in the minimisation to guarantee estimated reflection coefficients with absolute values less than one.
- The starting values for the AR(p+1) model are the estimated reflection coefficients of the AR(p) model with an additional zero for order $p + 1$.
- AR(p) order selection uses GIC(p) of (10.33) as a criterion to select the AR(K) model

  with $\alpha = 3$ for less than 25% missing,

  $\alpha = 5$ for less than 25% remaining,

  $\alpha = 4$ otherwise; see Broersen and Bos (2004b).
- The maximisation of the "likelihood" of MA and ARMA models gives problems with MA starting values and with order selection. Those models are much better estimated from the parameters of an intermediate AR model by a reduced-statistics method; see Broersen and Bos (2004a).
- The order of that intermediate AR model is chosen as the highest AR order with a spectrum close to the spectrum of the selected AR(K) model. The difference between the estimated spectra is measured with the model error (ME) of (5.40). The ME contribution should be less than two for each order higher than K.
- Order selection for MA and ARMA is based on GIC of (10.33) with penalty 3, the "likelihood", plus three times the number of estimated parameters. The same criterion is used to determine the preferred model type for irregular data; see Broersen and Bos (2004a).
- The quantity γN can be considered an effective number of observations. The remaining fraction γ is determined by the choice of the resampling period and the slot width.

10.5.3 Performance of ARMAsel for Irregular Data

Simulations with a known (aliased) spectrum are a first step in testing new algorithms. Test data were generated using the following procedure. First $128N$ equidistant data points were generated using a high-order AR process. Then, $127N$ data points were discarded randomly. Each data point had a probability of $127/128$ of being discarded. The process was the same as used for missing data with an additional peak at $0.75f_0$. This peak is at a frequency above half the mean data rate and would have been invisible or aliased if the observations were equidistant. The resulting data can be considered completely irregular, and the time intervals between the observations were roughly Poisson distributed. The global shape of the logarithmic spectrum has two constant slopes of $f^{-5/3}$ and f^{-7} and is representative of a possible turbulence spectrum. The peak is added to test whether details on the

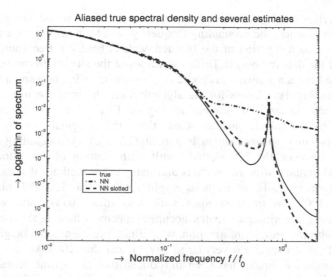

Figure 10.13. The true spectrum, the selected ARMAsel spectrum for data resampled with the nearest-neighbour scheme, and the ARMAsel missing data spectrum obtained by the multi-shift, slotted, nearest-neighbour resampling method. The sample size was 1000 irregular observations with $T_r = 0.25T_0$ and w = $0.5T_r$, with $T_0 = 1/f_0$. The NN resampled spectrum shows no detail at the location of the peak.

steep slope can also be detected. Leaving the peak away in the generated data discloses whether the peak has influence on the spectrum that is estimated by the various methods.

Figure 10.13 shows the result obtained with the ARMAsel-irreg algorithm for $N = 1000$. The algorithm is automatically estimating, selecting the model order, and selecting the model type with the algorithm of Section 10.5.2. The result, denoted NN slotted, is quite accurate. It shows the first $f^{-5/3}$ slope, the transition frequency to the second slope, and the peak at the true frequency. The only user interference is the choice of the resampling frequency and of the slot width. It is compared with the ARMAsel spectrum obtained by using nearest-neighbour resampling at all times nT_r. The nearest-neighbour spectrum is close to its biased expectation that follows from the theory of NN bias. Above $0.2f_0$, the estimated NN spectrum is greatly determined by the bias and hardly by the true spectrum at those frequencies. The peak at $0.75f_0$ is not detectable in the ARMAsel spectrum of Figure 10.13. The bias of sample and hold and of NN swallows the peak. Hence, refined reconstruction methods like that of Nobach et al. (1998) cannot reconstruct this peak for $N = 1000$. The bias is not dependent on the sample size, and the small narrow peak above the bias level will be very difficult to detect, even if 10,000 or more observations are used. If the true peak is still somewhat smaller, below the level of the biased spectrum, it will never be possible to reconstruct it with NN resampling, no matter how many data are available.

The resampling frequency determines the highest frequency for the estimated spectrum, and the slot width determines the bias. High resampling frequency and a small slot width give a resampled signal with a small remaining fraction. That may create numerical problems because large missing fractions are not computationally attractive. In practice, the resampling frequency is determined by the true process characteristics as a function of the frequency. It is hardly a free choice, but the character of the data imposes it. Different values of the slot width can be tried as a compromise between a small bias and a small remaining fraction and a larger bias with less missing data. Inputs for the algorithm are the irregular instants and the signal at those instants. It is clear in Figure 10.13 that the slotted, nearest-neighbour has a much smaller bias than the original nearest-neighbour interpolation where an observation is substituted at every resampling time. It is difficult to compare the NN slotted result with slotting of the autocorrelation function and variable window methods and many other methods that can be tuned by the experimenter. Those methods might need at least 100,000 observations instead of 1000. In small samples with less than 10,000 observations the ARMAsel-irreg algorithm estimates accurate spectra where other methods fail. However, with the present information, no definite preference can be given for an algorithm if the sample size is very large, say, greater than 100,000.

A new robust estimator has been introduced that fits a time series model to multishift, slotted, nearest-neighbour resampled segments obtained from irregularly sampled data. The new irregular ARMAsel algorithm combines a spectrum that is guaranteed to be positive with accurate results at higher frequencies. In simulations with few data, the results are much better than those that can be obtained from the same data by other known existing techniques. The order and type of the best time series model for the data are selected automatically without user interaction.

Multishift, slotted, nearest-neighbour resampling with ARMAsel-irreg can estimate accurate spectra if low-order AR models can reasonably approximate the true spectral shape. Many processes have this type of spectra, including turbulent flow data. For a large dynamic range, a small slot width will reduce the expectation of the bias. That requires very large data sets to obtain accurate estimates. The choice of the resampling frequency and the slot width is a compromise.

10.6 Exercises

10.1   Given is an AR(1) process. Derive
$$N \operatorname{var}\left[\hat{\rho}_{LP}(2)\right] = \frac{1+a^2-5a^4+3a^6}{1-a^2} .$$

10.2   Given is an AR(1) process. Derive
$$\frac{\operatorname{cov}\left[\hat{\rho}_{LP}(1),\hat{\rho}_{LP}(2)\right]}{\sqrt{\operatorname{var}\left[\hat{\rho}_{LP}(1)\right]\operatorname{var}\left[\hat{\rho}_{LP}(2)\right]}} = \frac{2a(1-a^2)}{\sqrt{1+a^2-5a^4+3a^6}}$$

10.3   Generate five observations of a MA(1) process with six normally distributed random numbers.

10.4   Generate 10 observations of a stationary AR(2) process with 10 normally distributed random numbers.

Bibliography

Adrian, R.J. and C. S. Yao (1987). Power spectra of fluid velocities measured by laser Doppler velocimetry. *Exp. Fluids*, vol. 5, pp. 17–28.

Akaike, H. (1969). Power spectrum estimation through autoregressive model fitting. *Ann. Inst. Stat. Math.*, vol. 21, pp. 407–419.

Akaike, H. (1970). Statistical predictor identification. *Ann. Inst. Stat. Math.*, vol. 22, pp. 203–217.

Akaike, H. (1970a). A fundamental relation between predictor identification and power spectrum estimation. *Ann. Inst. Stat. Math.*, vol. 22, pp. 219–223.

Akaike, H. (1974). A new look at the statistical model identification. *IEEE Trans. Autom. Control*, vol. AC–19, pp. 716–723.

Akaike, H. (1978). A Bayesian analysis of the minimum AIC procedure. *Ann. Inst. Stat. Math.*, vol. 30A, pp. 9–14.

Anderson, B.D.O. and J.B. Moore (1979). *Optimal Filtering*, Prentice–Hall, Englewood Cliffs, NJ.

Arato, M. (1961). On the sufficient statistics for stationary gaussian random processes. *Theory Probab. Its App.*, vol. 6, pp. 199–201.

Bauer, D. (2001). Order estimation for subspace methods. *Automatica*, vol. 37, pp. 1561–1573.

Bauer, D. and S. de Waele (2003). A finite sample comparison of automatic model selection methods. *Preprints 13th IFAC SYSID Symposium*, Rotterdam, pp. 1790–1795.

Benedict, L.H., H. Nobach, and C. Tropea (2000). Estimation of turbulent velocity spectra from laser Doppler data. *Meas. Sci. Technol.*, vol. 11, pp. 1089–1104.

Blackman, R.B. and J.W. Tukey (1959). *The Measurement of Power Spectra from the Point of View of Communication Engineering*. Dover, New York.

Bos, R., S. de Waele, and P.M.T. Broersen (2002). Autoregressive spectral estimation by application of the Burg algorithm to irregularly sampled data. *IEEE Trans. Instrum. Meas.*, vol. 51, no. 6, pp. 1289–1294.

Box, G.E.P. and G.M. Jenkins (1976). *Time Series Analysis: Forecasting and Control*. Holden–Day, San Fransisco.

Britz, D. and R.A. Antonia (1996). A comparison of methods of computing power spectra of LDA signals. *Meas. Sci. Technol.*, vol. 7, pp. 1042–1053.

Brockwell, P.J. and R.A. Davis (1987). *Time Series: Theory and Methods*. Springer Verlag, New York.

Broersen, P.M.T. (1985). Selecting the order of autoregressive models from small samples. *IEEE Trans. Acoust. Speech Signal Process.*, vol. ASSP-33, pp. 874–879.

Broersen, P.M.T. (1986). Subsets of reflection coefficients. *Proc. ICASSP Conf.*, Tokyo, pp. 1373–1376.

Broersen, P.M.T. (1990a). The prediction error of autoregressive small sample models. *IEEE Trans. Acoust. Speech Signal Process.*, vol. ASSP-38, pp. 858–860.

Broersen, P.M.T. (1990b). Selecting subsets of autoregressive parameters. *Signal Process.*, vol. 20, pp. 293–301.

Broersen, P.M.T. (1995). A comparison of transfer function estimators. *IEEE Trans. Instrum. Meas.*, vol. 44, pp. 657–661.

Broersen, P.M.T. (1998a). The quality of models for ARMA processes. *IEEE Trans. Signal Process.*, vol. 46, pp. 1749–1752.

Broersen, P.M.T. (1998b). Estimation of the accuracy of mean and variance of correlated data. *IEEE Trans. Instrum. Meas.*, vol. 47, pp. 1085–1091.

Broersen, P.M.T. (2000). Facts and fiction in spectral analysis. *IEEE Trans. Instrum. Meas.*, vol. 49, pp. 766–772.

Broersen, P.M.T. (2000a). Finite sample criteria for autoregressive order selection. *IEEE Trans. Signal Process.*, vol. 48, pp. 3550–3558.

Broersen, P.M.T. (2000b). Autoregressive model orders for Durbin's MA and ARMA estimators. *IEEE Trans. Signal Process.*, vol. 48, pp. 2454–2457.

Broersen, P.M.T. (2001). The performance of spectral quality measures. *IEEE Trans. Instrum. Meas.*, vol. 50, pp. 813–818.

Broersen, P.M.T. (2002). Automatic spectral analysis with time series models. *IEEE Trans. Instrum. Meas.*, vol. 51, no. 2, pp. 211–216.

Broersen, P.M.T. (2004). Mean square error of the empirical transfer function estimator for stochastic input signals. *Automatica*, vol. 40, pp. 95–100.

Broersen, P.M.T. (2005a). Time series analysis for irregularly sampled data. *Proc. IFAC World Conf.*, Prague, Czech Republic, paper 1696.

Broersen, P.M.T. (2005b). The uncertainty of measured autocorrelation functions. *Proc. AMUEM 2005, Int. Workshop Adv. Methods Uncertainty Estimation Meas.*, Niagara Falls, Canada, pp. 90–95.

Broersen, P.M.T. (2005c). ARMAsel as a language for random data. *Proc. IEEE/IMTC Conf.*, Ottawa, Canada, pp. 1531–1536.

Broersen, P.M.T. and R. Bos (2004a). Estimation of time series spectra with randomly missing data. *Proc. IEEE/IMTC Conf.*, Como, Italy, pp. 1718–1723.

Broersen, P.M.T. and R. Bos (2004b). Autoregressive order selection in missing data problems. *Proc. Eusipco Conf.*, Vienna, Austria, pp. 2159–2162.

Broersen, P.M.T. and R. Bos (2005). Estimating time series models from irregularly sampled data. *Proc. IEEE/IMTC Conf.*, Ottawa, Canada, pp. 1723–1728.

Broersen, P.M.T. and S. de Waele (2000a). Detection of methacholine with time series models of lung sounds. *IEEE Trans. Instrum. Meas.*, vol. 49, pp. 517–523.

Broersen, P.M.T. and S. de Waele (2000b). Some benefits of aliasing in time series analysis. *Proc. Eusipco Conf.*, Tampere, Finland, 4pp.

Broersen, P.M.T. and S. de Waele (2002). Selection of order and type of time series models estimated from reduced statistics. *Proc. IEEE/IMTC Conf.*, Anchorage, AK, pp. 1309–1314.

Broersen, P.M.T. and S. de Waele (2003a). Generating data with prescribed power spectral density. *IEEE Trans. Instrum. Meas.*, vol. 52, pp. 1061–1067.

Broersen, P.M.T. and S. de Waele (2003b). Time series analysis in a frequency subband. *IEEE Trans. Instrum. Meas.*, vol. 52, pp. 1054–1060.

Broersen, P.M.T. and S. de Waele (2004). Finite sample properties of ARMA order selection. *IEEE Trans. Instrum. Meas.*, vol. 53, pp. 645–651.

Broersen, P.M.T. and S. de Waele (2005). Automatic identification of time series models from long autoregressive models. *IEEE Trans. Instrum. Meas.*, vol. 54, pp. 1862–1868.

Broersen, P.M.T., S. de Waele and R. Bos (2004). Application of autoregressive spectral analysis to missing data problems. *IEEE Trans. Instrum. Meas.*, vol. 53, pp. 981–986.

Broersen, P.M.T., S. de Waele, and R. Bos (2004b). Autoregressive spectral analysis when data are missing. *Automatica*, vol. 40, pp. 1495–1504.

Broersen, P.M.T. and A.H. Weerts (2005). Automatic error correction of rainfall–runoff models in flood forecasting systems. *Proc. IEEE/IMTC Conf.*, Ottawa, Canada, pp. 963–968.

Broersen, P.M.T. and H.E. Wensink (1993). On finite sample theory for autoregressive model order selection. *IEEE Trans. Signal Process.*, vol. 41, pp. 194–204.

Broersen, P.M.T. and H.E. Wensink (1996). On the penalty factor for autoregressive order selection in finite samples. *IEEE Trans. Signal Process.*, vol. 44, pp. 748–752.

Broersen, P.M.T. and H.E. Wensink (1998). Autoregressive model order selection by a finite sample estimator for the Kullback–Leibler discrepancy. *IEEE Trans. Signal Process.*, vol. 46, pp. 2058–2061.

Burg, J.P. (1967). Maximum entropy spectral analysis. *Proc. 37th Meet. Soc. Exploration Geophysicists*, Oklahoma City, pp. 1–6.

Burnham, K.P. and D.R. Anderson (1998). *Model Selection and Inference. A Practical Information–Theoretic Approach.* Springer–Verlag, New York.

Byrnes, I.B., P. Enqvist, and A. Lindquist (2001). Cepstal coefficients, covariance lags and pole–zero models for finite data strings. *IEEE Trans. Signal Process.*, vol. 49, pp. 677–693.

Cavanaugh, J.E. (1999). A large–sample model selection criterion based on Kullback's symmetric divergence. *Stat. and Probaby Lett.*, vol. 42, pp. 333–343.

Cavanaugh, J.E. (2004). Criteria for linear model selection based on Kullback's symmetric divergence. *Autr. N. Z. J. Stat.*, vol. 46, pp. 257–274.

Cavanaugh, J.E. and R.H. Shumway (1998). An Akaike information criterion for model selection in the presence of incomplete data. *J. Stat. Plann. Inference*, vol. 67, pp. 45–65.

Choi, B.S. (1992). *ARMA Model Identification.* Springer–Verlag, New York.

Cooley, J.W. and J.W. Tukey (1965). An algorithm for the machine calculation of complex Fourier series. *Math. Computation*, vol. 19, pp. 297–301.

Crutzen, P.J. and V. Ramanathan (2000). The ascent of atmospheric sciences. *Science*, vol. 290, pp. 299–304.

Davidson, J.E.H. (1981). Problems with the estimation of moving average processes. *J. Econometrics*, vol. 16, pp. 295–310.

Durbin, J. (1959). Efficient estimation of parameters in moving average models. *Biometrika*, vol. 46, pp. 306–316.

Durbin, J. (1960). The fitting of time series models. *Revue Inst. Int. Stat.*, vol. 28, pp. 233–243.

Erkelens, J.S. (1996). *Autoregressive Modelling for Speech Coding: Estimation, Interpolation and Quantisation.* PhD Thesis, Delft University Press, Delft.

Erkelens, J.S. and P.M.T. Broersen (1995). Equivalent distortion measures for quantisation of LPC model. *Electron. Lett.*, vol. 31, pp. 1410–1412.

Erkelens, J.S. and P.M.T. Broersen (1997). Bias propagation in the autocorrelation method of linear prediction. *IEEE Trans. Speech Audio*, vol. 5, pp. 116–119.

Findley, D.F. (1984). On some ambiguities associated with the fitting of ARMA models to time series. *J. Time Ser. Anal.*, vol. 5, pp. 213–225.

Friedlander B. and B. Porat (1984). A general lower bound for parametric spectrum estimation. *IEEE Trans. Acoust. Speech Signal Process.*, vol. ASSP-32, pp. 728–733.

Friedman, B.H., I.C. Christie, S.L. Sargent, and J.B. Weaver (2004). Self–reported sensitivity to continuous non–invasive blood pressure monitoring via the radial artery. *J. Psychosomatic Res.*, vol. 57, pp. 119–121.

Godolphin, E.J. and J.G. de Gooijer (1982). On the maximum likelihood estimation of the parameters of a gaussian moving average process. *Biometrika*, vol. 69, pp. 443–451.

Graupe, D., D.J. Krause and J.B. Moore (1975). Identification of autoregressive moving-average parameters of time series. *IEEE Trans. Autom. Control*, vol. AC-20, pp. 104–107.

Gray, A.H. and J.D. Markel (1976). Distance measures for speech processing. *IEEE Trans. Acoust. Speech Signal Process.*, vol. ASSP-246, pp. 380–391.

Hamilton, J.D. (1994). *Time Series Analysis*. Princeton University Press, Princeton, NJ.

Hannan, E.J. and M. Deistler (1988). *The Statistical Theory of Linear Systems*. Wiley, New York.

Hannan, E.J. and B.G. Quinn (1979). The determination of the order of an autoregression. *J. R. Stat.. Soc.*, Series B-41, pp. 90–195.

Harris, F.J. (1978). On the use of windows for harmonic analysis with the discrete Fourier transform. *Proc. IEEE*, vol. 66, pp. 51–83.

Hernandez, G. (1999). Time series, periodograms, and significance. *J. Geophys. Res.*, vol. 104, no A5, pp. 10,355–10,368.

Hocking, R.R. (1976). The analysis and selection of variables in linear regression. *Biometrics*, vol. 32, pp. 1–49.

Hurvich, C.M. and C.L. Tsai (1989). Regression and time series model selection in small samples. *Biometrika*, vol. 76, pp. 297–307.

Hurvich, C.M., R. Shumway, and C.L. Tsai (1990). Improved estimators of Kullback–Leibler information for autoregressive model selection in small samples. *Biometrika*, vol. 77, pp. 709–719.

Jones, R.H. (1976). Autoregression order selection. *Geophysics*, vol. 41, pp. 771–773.

Jones, R.H. (1980). Maximum likelihood fitting of ARMA models to time series with missing observations. *Technometrics*, vol. 22, pp. 389–395.

Jones, R.H. (1981). Fitting a continuous time autoregression to discrete data. *Applied Time Series Analysis II,* (Ed., D.F. Findley), pp. 651–682, Academic Press, London.

Jones, R.H. (1984). Fitting multivariate models to unequally spaced data. *Time Series Analysis of Irregularly Spaced Data,* (Ed., E. Parzen), pp. 158–188, Springer-Verlag, New York.

Kay, S.M. (1988). *Modern Spectral Estimation*. Prentice–Hall, Englewood Cliffs, NJ.

Kay, S.M. and J. Makhoul (1983). On the statistics of the estimated reflection coefficients of an autoregressive process. *IEEE Trans. Signal Process.*, vol. 31, pp. 1447–1455.

Kay, S. M. and S.L. Marple (1981). Spectrum analysis – a modern perspective. *Proc. IEEE*, vol. 69, pp. 1380–1419.

Khintchine, A. (1934). Korrelationstheorie der stationären stochastischen Prozessen. *Mathematische Annalen*, vol. 109, pp. 604–615.

Kim, K.H., S.W. Bang,and S.R. Kim (2004). Emotion recognition system using short–term monitoring of physiological signals. *Med. Biol. Eng. Computing*, vol. 42, pp. 419–427.

Klees, R., P. Ditmar, and P.M.T. Broersen (2003). How to handle colored noise in large least–squares problems. *J. Geodesy*, vol. 76, pp. 629–640.

Kullback, S. (1959). *Information theory and Statistics*. Wiley, London.

Lahalle, E., G. Fleury, and A. Rivoira (2004). Continuous ARMA spectral estimation from irregularly sampled observations. *Proc. IEEE/IMTC Conf.*, Como, Italy, pp. 923–927.

Larsson, K.L. and T. Söderström (2002). Identification of continuous–time AR processes from unevenly sampled data. *Automatica*, vol. 38, pp. 709–718.

Legius, H.J.W.M. (1997). *Propagation of Pulsations and Waves in Two–Phase Pipe Systems*. PhD Thesis, Delft University of Technology.

Liley, D.T., P.J. Cadusch, M. Gray, and P.J. Nathan (2003). Drug–induced modification of the system properties associated with spontaneous human electroencephalographic activity. *Phys. Rev.*, vol. E 68, art 051906 pp. 1–15.

Little, R.J.A. and D.B. Rubin (1987). *Statistical Analysis with Missing Data*. Wiley, New York.

Ljung, L. (1987). *System Identification. Theory for the User*. Prentice–Hall, Englewood Cliffs, NJ.

Lomb, N.R. (1976). Least–squares frequency analysis of unequally spaced data. *Astrophys. Space Sci.*, vol. 39, pp. 447–462.

Maanen, H.R.E. van and A. Oldenziel (1998). Estimation of turbulence power spectra from randomly sampled data by curve–fit to the autocorrelation function applied to laser–Doppler anemometry. *Meas. Sci. Technol.*, vol. 9, pp. 458–467.

Maanen, H.R.E. van, H. Nobach and L.H. Benedict (1999). Improved estimator for the slotted autocorrelation function of randomly sampled LDA data. *Meas. Sci. Technol.*, vol. 10, pp. L4–L7.

Makhoul, J. (1976). Linear prediction: A tutorial review. *Proc. IEEE*, vol. 63, pp. 561–580.

Mallows C.L. (1973). Some comments on Cp. *Technometrics*, vol. 15, pp. 661–675.

Mann, H.B. and A. Wald (1943). On the statistical treatment of linear stochastic difference equations. *Econometrica*, vol. 11, pp. 173–220.

Markel, J.D. and A.H. Gray (1976). *Linear Prediction of Speech*. Springer–Verlag, Berlin.

Marple, S.L. (1987). *Digital Spectral Analysis with Applications*. Prentice–Hall, Englewood Cliffs, NJ.

Martin, R.J. (1999). Autoregression and irregular sampling: spectral estimation. *Signal Process.*, vol. 77, pp. 139–157.

Martin, R.J. (2000). A metric for ARMA processes. *IEEE Trans. Signal Process.*, vol. 48, pp. 1164–1170.

Melton, B. S. and P.R. Karr (1957). Polarity coincidence scheme for revealing signal coherence. *Geophysics*, vol. 22, pp. 553–564.

Mentz, R.P. (1977). Estimation of first–order moving average model through the finite autoregressive approximation. *J. Econometrics*, vol. 6, pp. 225–36.

Miller, A.J. (1990). *Subset Selection in Regression*. Chapman and Hall, London.

Mood, A.M., F.A. Graybill and D.C. Boes (1974). *Introduction to the Theory of Statistics*. McGraw–Hill, Tokyo.

Ninness, B. (2003). The asymptotic CRLB for the spectrum of ARMA processes. *IEEE Signal Process. Lett.*, vol. 11, pp. 293–296.

Ninness, B. (2004). On the CRLB for combined model and model–order estimation of stationary stochastic processes. *IEEE Trans. Signal Process.*, vol. 51, pp. 1520–1531.

Nobach, H., E. Müller, and C. Tropea (1998). Efficient estimation of power spectral density from laser–Doppler anemometer data. *Exp. Fluids*, vol. 24, pp. 499–509.

Osborn, D.R. (1976). Maximum likelihood estimation of moving average processes. *Ann. Econ. Soc. Meas.*, vol. 5, pp. 75–87.

Papoulis, A. (1965). *Probability, Random Variables and Stochastic Processes*. McGraw–Hill, New York.

Pavageau, M., C. Rey and J.–C. Eliser–Cortes (2004). Potential benefit from the application of autoregressive spectral estimators in the analysis of homogeneous and isotropic turbulence. *Exp. Fluids*, vol. 36, pp. 847–859.

Parzen, E. (1974). Some recent advances in time series modelling. *IEEE Trans. Autom. Control*, vol. AC–19, pp. 723–730.

Percival, D.B. and A.T. Walden (1993). *Spectral Analysis for Physical Applications: Multitaper and Conventional Univariate Techniques*. Cambridge University Press, Cambridge.

Petit J.R. *et al.* (1999). Climate and atmospheric history of the past 420,000 years from the Vostok ice core, Antarctica. *Nature*, vol. 399, pp. 429–436.

Pfurtscheller, G. (2004) Brain–computer interface — state of the art and future prospects. *Proc. EUSIPCO Conf.*, Vienna, Austria, pp. 509–510.

Porat, B. (1994). *Digital Processing of Random Signals*. Prentice–Hall, Englewood Cliffs, NJ.

Pollock, D.S.G. (1999). *A Handbook of Time–Series Analysis, Signal Processing and Dynamics*. Academic Press, San Diego.

Priestley, M.B. (1981). *Spectral Analysis and Time Series*. Academic Press, London.

Rissanen, J. (1978). Modelling by shortest data description. *Automatica*, vol. 41, pp. 465–471.

Rissanen, J. (1986). A predictive least–squares principle. *IMA J. Math. Control Inf.*, vol. 3, pp. 211–222.

Rosen, Y. and B. Porat (1989). Optimal ARMA parameter estimation based on the sample covariances for data with missing observations. *IEEE Trans. Inf. Theory*, vol. 35, pp. 342–349.

Scargle, J.D. (1982). Statistical aspects of spectral analysis of unevenly spaced data. Studies in astronomical time series analysis II. *Astrophys. J.*, no. 263, pp. 835–853.

Schuster, A. (1898). On the investigation of hidden periodicities with application to a supposed twenty–six day period of meteorological phenomena. *Terrestrial Magnetism*, vol. 3, pp. 13–41.

Searle, S.R. (1982). *Matrix Algebra Useful for Statistics*. Wiley, NewYork.

Shibata, R. (1976). Selection of the order of an autoregressive model by Akaike's information criterion. *Biometrika*, vol. 63, pp.117–126.

Shibata, R. (1984). Approximate efficiency of a selection procedure for the number of regression variables. *Biometrika*, vol. 71, pp. 43–49.

Shumway, R.H. and D.S. Stoffer (2000*). Time Series Analysis and its Applications*. Springer, New York.

Simon, L. and J. Fitzpatrick (2004). An improved sample–and–hold reconstruction procedure for estimation of power spectra from LDA data. *Exp. Fluids*, vol. 37, pp. 272–280.

Slutsky, E. (1937). The summation of random causes as the source of cyclical processes. *Econometrica*, vol. 5, pp. 105–146.

Sobera, M.P., C.R. Kleijn, H.E.A. van den Akker, and P. Brasser (2003). Convective heat and mass transfer to a circular cylinder sheathed by a porous layer. *AIChE J.*, vol. 49, pp. 3018–3028.

Sobera, M.P., C.R. Kleijn, P. Brasser, and H.E.A. van den Akker (2004). Hydrodynamics of the flow around a circular cylinder sheathed by a porous layer. *Proc. 2004 ASME Heat Transfer/Fluids Eng. Conf.*, Charlotte, NC, pp. 1–7.

Söderström, T. and M. Mossberg (2000). Performance evaluation of methods for identifying continuous–time autoregressive processes. *Automatica*, vol. 36, pp. 53–59.

Solo, V. (2001). Asymptotics for complexity regularized transfer function estimation with orthonormal bases. *Proc. IEEE/CDC Conf. Decision Control*, pp. 4766–4769.

Stoica, P. and R. Moses (1997). *Introduction to Spectral Analysis*. Prentice Hall, Upper Saddle River, NJ.

Tjøstheim, D. and J. Paulsen (1983). Bias of some commonly–used time series estimates. *Biometrika*, vol. 7, pp. 389–399.

Tomczak, M. and E.H. Djermoune (2002). A subband ARMA modeling approach to high–resolution NMR spectroscopy. *J. Magn. Resonance*, vol. 158, pp. 86–98.

Tummers, M.J. and D.M. Passchier (1996). Spectral estimation using a variable window and the slotting technique with local normalization. *Meas. Sci. Technol.*, vol. 7, pp. 1541–1546.

Tummers, M.J. and D.M. Passchier (2001). Spectral analysis of biased LDA data. *Meas. Sci. Technol.*, vol. 12, pp. 1641–1650.

Ulrych, T. J. and T.N. Bishop (1975). Maximum entropy spectral analysis and autoregressive decomposition. *Rev. Geophys. Space Phys.*, vol. 13, pp. 183–200.

Viswanathan, R. and J. Makhoul (1975). Quantization properties of transmission parameters in linear predictive systems. *IEEE Trans. Acouts., Speech Signal Process.*, vol. 23, pp. 309–321.

Waele, S. de (2003). *Automatic Inference from Finite TimeOobservations of Stationary Stochastic Signals*. PhD Thesis, Delft University of Technology.

Waele, S. de and P.M.T. Broersen (1999). Reliable LDA–spectra by resampling and ARMA modeling. *IEEE Trans. Instrum. Meas.*, vol. 48, pp. 1117–1121.

Waele, S. de and P.M.T. Broersen (2000a). The Burg algorithm for segments. *IEEE Trans. Signal Process.*, vol. 48, pp. 2876–2880.

Waele, S. de and P.M.T. Broersen (2000b). Error measures for resampled irregular data. *IEEE Trans. Instrum. Meas.*, vol. 49, pp. 216–222.

Waele, S. de and P.M.T. Broersen (2001). Multirate autoregressive modelling. *Selected Top. Signals Syst. Control*, pp. 75–80.

Waele, S. de and P.M.T. Broersen (2002). Finite sample effects in vector autoregressive modeling. *IEEE Trans. Instrum. Meas.,* vol. 51, pp. 917–922.

Waele, S. de and P.M.T. Broersen (2003). Order selection for vector autoregressive models. *IEEE Trans. Signal Process.*, vol. 51, pp. 427–433.

Wensink, H.E. (1996). *Autoregressive Model Inference in Finite Samples*. PhD Thesis, Delft University of Technology.

Wiener, N. (1930). Generalised harmonic analysis. *Acta Mathematica*, vol. 35, pp. 117–258.

Wilson,G. (1969). Factorization of the covariance generating function of a pure moving average process. *SIAM J. Numer. Anal.*, vol. 6, pp. 1–7.

Whittle, P. (1953). Estimation and information in stationary time series. *Arkiv for Matematik*, vol. 2, pp. 423–434.

Wolff, S.S., J.B. Thomas ,and T.R. Williams (1962). The polarity-coincidence correlator: a nonparametric detection device. *IEEE Trans. Inf. Theory*, vol. 8, pp. 5–9.

Yang, D. *et al.* (2001). USPIO-enhanced dynamic MRI: Evaluation of normal and transplanted rat kidneys. *Magn. Resonance Med.*, vol. 46, pp. 1152–1163.

Yule, G.U. (1927). On a method of investigating periodicities in disturbed series with special reference to Wolfer's sunspot numbers. *Philos. Trans. R. Soc., Series A*, vol. 226, pp. 267–298.

Zacks, S. (1971). *The Theory of Statistical Inference*. Wiley , New York.

Index

absolute measure 106
accuracy measures 99
additive noise 56, 169, 223
AIC criterion 168, 179, 198
AIC_C 180
aliasing 6, 238, 265, 288
AR process 63
    AR(1) 64
    AR(2) 69
    AR(p) 72
    asymptotic theory 128
    Burg 126
    estimation methods 124
    finite-sample 130
    least-squares 125
    ML estimation 121
    order selection 167
    relations 96
    representations 95
    subset selection 208
    Yule-Walker 124
ARMA process 74
    ARMA(p,q) 74
    estimation methods 140
    filter 74
    first-stage, long AR 142
    first-stage, long MA 143
    first-stage, long COV 143
    first-stage, long Rinv 144
    least-squares 141
    ML estimation 123
    order selection 209
    reduced-statistics 141
    residuals 225
    second-stage 144
ARMASA 250
ARMASA toolbox 250

ARMAsel 250
arma2cor 250
arma2pred 250
arma2psd 250
asymptotic order selection 176
asymptotic theory AR 128
autocorrelation 31
    accuracy of time series 253
    estimation of lagged product 44
    estimation of time series 160
    partial 90, 92
    sampling properties 44, 45, 252
autocovariance 32
    aliased 50
    estimation of lagged product 43
    estimation of time series 161
    generating function 77
    matrix 33
    positive-definite 41, 278, 279
    sampling properties 44, 45, 251
autoregressive process 63
autoregressive moving average process 74

backward difference operator 61
backward residuals 61, 125, 224
bias 24
    of autocorrelation 44
    of periodogram 52, 54
    propagation 108
    triangular 36, 109
BIC 180
Bivariate 12, 15, 30
Burg's method 126
Burg's method for segments 270

central moment 12
cepstrum 107